High-Pressure Physics

Scottish Graduate Series

High-Pressure Physics

Edited by
John Loveday
The University of Edinburgh, UK

CRC Press
Taylor & Francis Group
Boca Raton   London   New York

CRC Press is an imprint of the
Taylor & Francis Group, an **informa** business

A TAYLOR & FRANCIS BOOK

CRC Press
Taylor & Francis Group
6000 Broken Sound Parkway NW, Suite 300
Boca Raton, FL 33487-2742

First issued in paperback 2016

ISBN 13: 978-1-138-19910-1 (pbk)
ISBN 13: 978-1-4398-1428-4 (hbk)

Library of Congress Cataloging-in-Publication Data

Loveday, John, 1963-
   High-pressure physics / editor, John Loveday.
     p. cm. -- (Scottish graduate series)
   Includes bibliographical references and index.
   ISBN 978-1-4398-1428-4 (hardcover : alk. paper)
   1. High pressure (Science) 2. High pressure chemistry I. Title.

QC281.L68 2012
541'.363--dc23
                                                    2011036556

Visit the Taylor & Francis Web site at
http://www.taylorandfrancis.com

and the CRC Press Web site at
http://www.crcpress.com

SUSSP Schools

/continued

SUSSP Schools (continued)

/continued

SUSSP Schools (continued)

Lecturers

Stefan Klotz	Institut de Minéralogie et de Physique des Milieux Condensés, Paris, France
Konstantin Kamenev	The University of Edinburgh, United Kingdom
Samuel T. Weir	Lawrence Livermore National Laboratory, United States
Wilfried B. Holzapfel	Universität Paderborn, Germany
John B. Parise	Stony Brook University, United States
Mario Santoro	LENS, European Laboratory for Non-linear Spectroscopy, Florence, Italy
Federico A. Gorelli	LENS, European Laboratory for Non-linear Spectroscopy, Florence, Italy
Alexander F. Goncharov	Geophysical Laboratory, Carnegie Institution of Washington, United States
F. Malte Grosche	The University of Cambridge, United Kingdom
Chrystéle Sanloup	Université Pierre et Marie Curie, Paris, France, and The University of Edinburgh, United Kingdom
Paul F. McMillan	University College London, United Kingdom
Martin C. Wilding	Aberystwyth University, United Kingdom
Russell J. Hemley	Geophysical Laboratory, Carnegie Institution of Washington, United States

Organising Committee

Malcolm McMahon	University of Edinburgh	*Director*
Konstantin Kamenev	University of Edinburgh	*Secretary*
Ingo Loa	University of Edinburgh	*Treasurer*
John Loveday	University of Edinburgh	*Editor*
Leanne O'Donnell	University of Edinburgh	*Administration*

Preface

High-pressure science has undergone a revolution in the last 15 years. The development of intense new x-ray and neutron sources, improved detectors, new instrumentation, greatly increased computation power and advanced computational algorithms have enabled researchers to determine the behaviour of matter at static pressures in excess of 400 GPa. And shock-wave techniques have allowed access to the experimental P-T range beyond 1 TPa and 10,000 K. Although there have been a number of recent Summer Schools on high-pressure topics, these have focused on specific topics such as crystallography, mineralogy or general high-pressure research. There have been no Summer Schools specifically in high-pressure physics since the highly successful Enrico Fermi Conference on *"High Pressure Phenomena"* in Varenna in 2001. It was thus becoming very timely to hold another international physics-specific school. A bid was made to SUSSP to hold such a school in the summer of 2008, and the 63rd Scottish Universities Summer School in Physics on *High Pressure Physics* ran from 26th May to 6th June 2008 at the Sabhal Mhor Ostaig Gaelic College on The Island of Skye.

There were 55 registered students and postdocs at the school, who attended 42 lectures, 6 hands-on workshops and 10 evening research seminars spread over 8 days. Lectures covered experimental and computational physics, and both static and dynamic compression techniques, and were given by leading researchers from the UK, Europe and the US. Attendees were also able to enjoy two weeks of weather, which even the locals classed as miraculous (sunburn was common). However, the warm weather also brought out the midges in their millions, encouraging interactions as participants compared their individual tallies of insect bites.

These proceedings contain papers summarising the lectures given during the School, and cover a wide range of topics and techniques, as outlined above. We would like to thank all the contributors for their hard work in preparing their papers, and our CSEC colleagues for their help in editing them.

Malcolm McMahon (Director)
Konstantin Kamenev (Secretary)
Ingo Loa (Treasurer)
John Loveday (Editor)

Contents

Chapter 1
High-Pressure Devices

Stefan Klotz

IMPMC, CNRS-UMR 7590, Université Pierre et Marie Curie, 4 Place Jussieu, Paris, France, 75252

1 Introduction

This article is a review of high-pressure devices used in solid state research up to the Mbar-range. For obvious reasons this article cannot go in very much depth, and I have to refer to literature sources for more details. Being a contribution to a "School", the article is rather intended to be an introduction for newcomers and focus onto the basic principles and terminology of high-pressure devices. As such it might be useful as a primer to more exhaustive and specialized literature [1, 2, 3, 4, 5, 6, 7]. The article is organized as following: The first part (sections 2–5) will cover various large-volume devices, starting with the most common high-pressure apparatus, the cylindrical pressure vessel, followed by a discussion of Belt- and Drickamer-type devices, Bridgman- and multianvil-cells. Section 6 is entirely devoted to the diamond anvil cell and section 7 will give an overview of pressure transmitting media.

2 The cylinder: the most common high-pressure device

It is useful and instructive to discuss in more detail the cylinder (more precisely the "thick-walled" cylinder) since this is not only the most common high-pressure apparatus, but also a mechanical problem which can be solved analytically in form of the Lamé equations [8]. Given an open or closed cylinder with inner and outer radii r_i and r_o respectively, which is subjected to a hydrostatic pressure p inside and zero pressure outside, the stress distribution inside the cylinder wall is given by [1, 4, 9]:

$$\sigma_r = \frac{p}{K^2 - 1}[1 - \frac{r_o^2}{r^2}] \tag{1}$$

$$\sigma_\theta = \frac{p}{K^2 - 1}[1 + \frac{r_o^2}{r^2}] \tag{2}$$

where $K = r_o/r_i$. As for the longitudinal stress, $\sigma_z = 0$ for an open cylinder, $\sigma_z = p/(K^2 - 1)$ for a closed cylinder, and $\sigma_z = 2p\nu/(K^2 - 1)$ for a longitudinally-constrained cylinder, where ν is Poisson's ratio ($\rightarrow$ Glossary). The quantity σ_θ is known as the *hoop stress*. It is, by definition, positive, which corresponds to traction, contrary to the radial stress, σ_r, which is negative and which indicates compression. It is seen that both the hoop and radial stresses are maximal at the bore of the cylinder ($r = r_i$) with values $\sigma_\theta = p(K^2 + 1)/(K^2 - 1)$ and obviously $\sigma_r = -p$, and that they fall off as $\sim 1/r^2$, see Fig. 1. This immediately illustrates the fundamental problem of a simple cylinder: the maximal stresses occur at the bore and an increase of the wall thickness (K) does not significantly improve its strength. For this reason, cylinders with $K \geq 3$ are rarely used since the considerable increase in external diameter and weight does not justify the marginal increase in strength. Given the fact that the largest stresses occur for σ_θ (Fig. 1), one might take this quantity as an indicator for the strength of a cylinder, i.e. assume that plastic deformation will occur if σ_θ reaches the yield stress σ^* ($\rightarrow$ Glossary) of the cylinder material. This is sometimes called the Rankine plasticity criterion. The maximal pressure is then derived from equ. 2:

$$p_{max} = \sigma^* \frac{(K^2 - 1)}{(K^2 + 1)} \tag{3}$$

The elastic limit σ^* (yield stress) of a high-tensile steel or alloy being at most 1.8 GPa, equation 3 indicates that the maximal pressure a simple ("monobloc") cylinder with K=3 can sustain is ~ 1.5 GPa. A more sophisticated yield criterion (von Mises criterion [10]) predicts an onset of plastic deformation at a pressure of [1]:

$$p_{max} = \frac{\sigma^*}{\sqrt{3}} \frac{(K^2 - 1)}{K^2} \tag{4}$$

For a cylinder with $K = 3$, equation 4 predicts a limit 36% lower than equation 3, and this is strongly supported by experiment. Equation 4 gives the maximum pressure at which a monobloc cylinder can be used in a fully elastic regime. This does not necessarily mean that it will burst at this pressure. This is because a metal might support stresses well beyond the yield stress if the material can be sufficiently overstrained, i.e. if it is sufficiently ductile ($\rightarrow$ Glossary). In this case the cylinder will be deformed after the experiment, its lifetime will be severely reduced, and it might burst under load without prior warning.

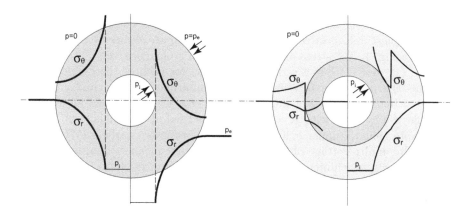

Figure 1. *Left drawing: Stress distribution in a simple (monobloc) cylinder. The left-half shows the situation for a cylinder subjected to an internal pressure p_i and no external pressure ($p_e = 0$), the right half with both internal and external pressures p_i and p_e. Right drawing: Stress distribution in a fretted (compound) cylinder. The left-half shows the situation in the unloaded state ($p_i - p_e = 0$), the right-half under internal load.*

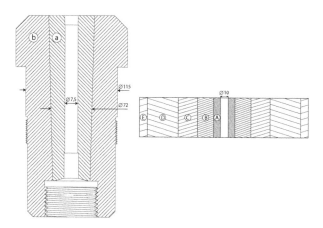

Figure 2. *Examples of fretted and multi-fretted cylinders. Left: Cylinder for use in a hydraulic intensifier generating pressures up to 1.2 GPa (Basset, France). (a) liner (HRC 52), (b) jacket (HRC 42). Right: Multi-fretted cylinder with $K = 22$ for pressures in the 3–5 GPa range. (A): tungsten carbide liner. (B)-(D) steel binding rings with hardness 60, 50, and 42 HRC, respectively; (E): Safety ring (steel). After Bradley [6].*

A way out of this limitation is known as *frettage*, a special case being *auto-frettage*. In the first case, the pressure vessel consists of several cylinders, in the simplest and most common case two, which are shrink-fitted into each other, see Fig. 1. The stress distribution is then no longer given by the simple formula (1) and (2) but a series of equations for each of the cylinders. As a result, the maximum pressure of such a multiple-fretted cylinder can be increased by a factor ~ 2 at most, i.e. ~ 5 GPa. Figure 2 shows an extreme example of a cylinder with $K = 22$ which was used in experiments up to ~ 5 GPa. Note that the inner cylinder is made of tungsten carbide (WC liner). Practically, multi-frettage of long cylinders is difficult and expensive since it involves precise machining to respect the tight tolerances. Auto-frettage is an inexpensive alternative. In this case a simple cylinder is strained to beyond the elastic limit of its inner part. After this procedure the inner part (typically 1/3 of the wall thickness) is under radial compression from the outer part, since the former was plastically deformed. The cylinder is then machined to its final dimension before use. Autofrettage requires the use of sufficiently ductile ($\rightarrow$ Glossary) steels, i.e. steels which can be substantially deformed without breaking. The relevant practical formula for frettages and autofrettages can be found in refs. [4, 9].

To conclude, the maximum pressure a cylinder can withstand, even with frettage, is practically limited to $\sim$3 GPa. Its main advantages are the large sample volume and ease of use. Cylinders are hence used in high-pressure sciences where a few GPa is the relevant pressure range, for example in material research on organic conductors and superconductors, soft condensed matter, and bio- and food sciences. Historically, this pressure limit prevented its use for diamond synthesis, which was the main stimulus for the development of high-pressure devices of the "belt" type, which will now be discussed.

3   Belt type apparatuses

For piston-cylinder devices routinely reaching pressures to 4–5 GPa, it is the piston which becomes the main limitation, even if it is made of tungsten carbide. It fails systematically at the entrance to the cylinder where it is no longer radially supported. This observation lead to the design of "belt-type" devices, originally developed by T. Hall and subsequently modified by numerous groups (Fig. 3). Its characteristic feature is a tapered piston made of tungsten carbide, and a short, multi-frettaged cylinder (girdle), in most cases with a tungsten carbide liner. The space between the piston and the girdle is filled with an adequate gasket material which extrudes under load as the pistons advances. Standard materials for gasket assemblies are pyrophyllite, MgO, and teflon, which limits the extrusion. The standard maximum pressure of operation is in the 5–10 GPa range, but Bundy reported measurements up to 20 GPa [11]. The belt-type apparatus is nowadays mainly used for synthesis under high P/T conditions, and, as such, the working horse for large-scale

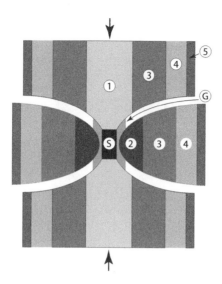

Figure 3. *Schematic view of a belt-type apparatus. S: sample, G: gasket, (1)-(2) tungsten carbide piston and liner, (3)-(5) steel binding rings.*

diamond production since it provides large sample volumes at considerably high-pressures. But this device has also been used for a number of physical properties measurements, such as specific heat [12], resistivity [13] (even at low temperatures down to 77 K [14]), magnetic permeability [16], differential thermal conductivity [17], and DTA [15]. An interesting recent application, using a miniature belt-type device, are rheological measurements under high pressure and high temperatures applying in-situ neutron diffraction [19].

4 Opposed anvil devices: Bridgman, Drickamer and profiled anvils

4.1 Drickamer cells

Drickamer anvils are a further variant/evolution of the belt geometry, see Fig. 4. Pressure cells of this kind (Drickamer cells) are characterized by tapered pistons (cone half angle $\approx 40°$) and a girdle (inset) with a cylindrical bore of ≈ 3 mm diameter in which the central part of the pistons (WC) slightly penetrate. In the Drickamer I cell, the tip of the WC piston is flat (as in Fig. 4), whereas in Drickamer II devices the tip is conical with a small central truncation where the sample is located. The conical half angle is typically $84°$, and the space between the front faces of the two anvils is completely filled with gasket material. The pressure generated [4] in this part of the cell supports the piston and acts effectively as a first stage for the compression

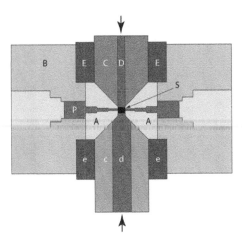

Figure 4. *Schematic view of a Drickamer cell. (C) & (D) movable and (c) & (d) stationary piston; (B) body, (E) and (e) piston guides, (A) insert (girdle), (S) sample. P is an optical window/plug, usually made of NaCl. Redrawn after ref. [20].*

in the second stage, the sample chamber on the flat tip of the anvils. An interesting aspect in such Drickamer cells is that they can provide optical access to the sample through a conical opening in the girdle, see Fig. 4. This opening is plugged with NaCl, and optical measurements up to 5 GPa are reported. Since the advent of the diamond anvil cell and laser radiation, the Drickamer cell has become obsolete. However, it is, as far as I am aware, still the only large-volume cell which provides optical access, and, as such, might still be useful for certain applications which rely on large sample volumes.

4.2 Bridgman anvils

Bridgman introduced a type of device which is based on two opposed anvils as shown in Fig. 5(a), very similar to Drickamer-II anvils, but without a girdle. The sample is compressed between the flat tips of the anvils which are made from a hard material, tungsten carbide or sintered diamond, supported radially by a steel binding ring. The gasket can be pyrophyllite, catlinite, or a metal, depending on the type of measurement. This kind of anvil assembly exploits the method of "massive support" [1] provided by the anvil material around the sample chamber. The performance hence depends on the conical half angle α. For an angle of $\alpha=85°$, the strengthening factor is 4, i.e. for anvils made of WC which has a compressional strength of 5 GPa, pressures up to 20 GPa can be generated. For $\alpha=70°$, this factor decreases to 2. Bridgman anvils work only for thin gaskets/samples, similar to diamond anvils as will be discussed later, and Bridgman used them essentially for resistivity measurements. For α close to 90° (i.e. almost flat anvils), the thickness of the gasket/sample assembly has to be small, since otherwise the deformation will rapidly increase its area and thereby reduce the pressure efficiency. Bridgman anvils are still used in the 0–25 GPa range, for low-temperature transport [21, 22], magnetic [21] and specific heat measurements, for example. In this pressure range they are an attractive alternative, due to the low costs and the considerable sample volume, they can provide compared to DACs.

4.3   Profiled opposed anvils

The use of profiled opposed anvils is an attempt to increase the sample volume. Anvils with deepenings were used already by Hall [18] who claims that even Bridgman experimented with such anvils. Systematic use started at the beginning of the 1960's in the former Soviet Union [2] for material synthesis, and a design with the name "Conac" is still widely applied. Such anvils can have considerable size: Conac-28 anvils have a sample chamber volume of 200 mm^3 to reach 10 GPa, whereas Conac-40 anvils can accommodate sample of up to 1 cm^3 in volume to be compressed to 4 GPa. The outer diameter of the WC dies is $\approx$ 90 mm. Small anvils of this geometry have recently gained considerable importance for high-pressure high-temperature in situ x-ray and neutron diffraction. Similar to Bridgman anvils, they are made of steel-enforced tungsten carbide or sintered diamond dies. The gasket material for x-ray scattering is boron-epoxy due its transparency for x-rays, and pyrophyllite for neutron scattering. Pressures of up to 17 GPa are reported for x-ray scattering using conoidal anvils as shown in Fig. 5(b) [23]. For high P/T neutron scattering, the limit so far is 7 GPa, mainly due the requirement of relatively large sample volumes (50 mm^3). A disadvantages of the anvil design in Fig. 5(b) is that the gap between the anvils, and hence the accessible window for x-ray and neutron beams, decreases dramatically under load. This also complicates electrical feed-throughs or thermocouples which have to pass through the gaskets.

A particular type of profiled opposed anvils are toroidal anvils as shown in Fig. 5(c)–(d), which were first proposed by Khvostantsev et al. [24] and then extensively used in various East European laboratories. Although the mechanical behaviour of such anvils under load has been theorized to a certain extent [2], the role of the toroid remains somehow obscure. Apart from material synthesis, this type of anvils has been used for a number of physical property measurements, among others resistivity, ultrasonics, and specific heat. Originally, pyrophyllite was used as the gasket material, but more recently, metallic gaskets became the material of choice for the extensive use of toroidal anvils for neutron scattering [25]. Contrary to the behaviour of oxides, metallic gaskets deform without fracturing and the change of dimensions under load is perfectly reproducible. This behaviour is crucial for attenuation corrections to obtain reliable intensities in diffraction experiments. Toroidal anvils can have more than one toroidal groove. Figure 5(d) shows double-

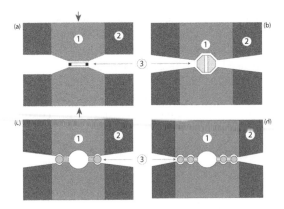

Figure 5. *Schematic view of Bridgman-type opposed-anvil cells. (a) standard Bridgman anvils, (b) conoidal anvils, (c) simple toroidal anvils, (d) double toroidal anvils; (1) anvil, (2) steel binding ring, (3) gasket and sample.*

toroidal anvils which are now routinely used for neutron scattering to 30 GPa, using sintered diamond as anvil material [26]. Experience reveals a numbers of considerable practical advantages of toroidal anvils of the type shown in Fig. 5(c,d). The gasket assembly consists of two or more parts ("rings") which are all of cylindrical shape and which can be easily and quickly machined on any lathe. In practice, the thickness of the various rings is chosen in a such a way that they are compressed successively starting from the inner one. Contrary to conoidal anvils, the gap between toroidal anvils remains appreciable even at very high loads, typically several tenth of millimeters. And, finally, the pressure efficiency of toroidal anvils can be considerably higher compared to Bridgman anvils of the same size. This depends strongly, however on details of the profile geometry, in particular the position of the groove with respect to the flat and conical part of the anvil faces.

5   Multi-anvil devices

These are pressure devices with more than two anvils and in almost all cases "large-volume" apparatuses for samples of 1 mm^3 or more. These cells require considerable forces of typically 500–2000 tn and the presses required for such loads can be impressive. Multi-anvil cells can be classified according to the geometry of the pressure chamber, i.e. tetrahedral, cubic, and octahedral devices, see Fig. 6. The forces act normal to the faces of the respective polyhedron. Such devices play a major role in research in the Earth sciences, which is high-pressure - high temperature mineral physics and chemistry, as well as in material synthesis.

5.1 Tetrahedral pressure cells

These are cells as shown in Fig. 6(a) with the forces on the anvils usually provided by four hydraulic rams. This type of press was developed by T. Hall to circumvent the patent rights owned by General Electric for diamond syntheses using belt devices, which were invented also by Hall. The well-known problem of tetrahedral presses are the balanced movements of the four rams, and pressure generation was limited to approximately 10 GPa. A driving mechanism which uses a single ram and sliding wedges to transfer the axial load onto the four anvils has been proposed by Lloyd and Hutton [28]. The general observation is that tetrahedral presses have become less and less popular compared to cubic and octahedral presses.

5.2 Cubic pressure cells

These compress simultaneously the six faces of a cube which forms the solid pressure transmitting medium for the sample and contains all other elements such as furnace, electrical contacts, thermal insulation etc. The force is usually provided by a hydraulic press and transferred to the six faces via sliding mechanisms of various types. Cubic systems are widely used for in-situ x-ray diffraction, mostly on synchrotron sources, but also using laboratory sources. The anvils being usually of tungsten carbide, the only available window for x-rays is the gap between the anvils which is filled with gasket material (pyrophyllite, MgO, boron epoxy). Presses for such x-ray applications have become known under the names MAX80 and MAX90 ("Multi-anvil press for X-ray diffraction designed in the 1980s/1990s") in Japan or SAM85 ("Six-anvil machine designed in 1985") in the US. Cubic pressure cells have also been used for low-temperature measurements of resistivity and magnetic susceptibility down to 2 K [29]. For this purpose only the anvil-module is cooled, and thermally insulated from the hydraulic system which remains at ambient temperature. Pressures are determined by the pressure-induced shift of a superconducting material (Pb, Sn). The helium consumption is reported to be about 20 litres to cool the system from 77 to 4 K.

5.3 Octahedral pressure cells

In principle it would be possible to construct an octahedral cell in the same way as tetrahedral and cubic cells, i.e. using eight pistons each driven by a ram. Such a device would be rather complex. Instead, octahedral cells are built as a two-stage device. The first stage is a multianvil cell of cubic type with six anvils as discussed above, but providing relatively large cube dimensions of up to 10 cm edge length to fit in the second stage. The second stage is an assembly of eight small cubes, each of them truncated at one corner to form an octahedral cavity in the center (Fig. 6(c)). This cavity contains the solid pressure transmitting medium of octahedral shape with the sample

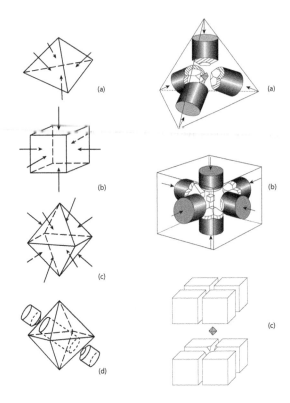

Figure 6. *Gasket-sample assembly (left) and corresponding anvil configuration in tetrahedral (a), cubic (b) and octahedral (c) multianvil cells. (d) shows the setup in the 6-8-2 cell with the two squeezers inside the octahedron. Guide and thrust mechanisms are not shown.*

and the various other elements inside. For this reason such pressure cells are also named 6-8 devices; they were designed by Kawai and Endo in Japan [30]. Kawai-type devices are nowadays among the most common pressure cells for high P/T synthesis. If the length (in mm) of the truncated cube edge is TEL and the length of the sample octahedron is OEL, frequently used combinations OEL/TEL are 10/4 and 7/2 for maximal pressures of 10 and 25 GPa, approximately, if WC anvils are used. The cube assembly may be oriented either with the cubic [100] or the [111] direction vertically, depending on the thrust mechanism which acts on it. In the "Walker-module" [31], a commercially available system for high P/T synthesis, the [111] direction is vertical. The same applies for the Stony Brook "T-Cup" device [32] used for synchrotron x-ray diffraction. Octahedral devices are frequently used for in situ synchrotron radiation experiments up to very high-pressures, despite the fact that the two stages provide very limited access to the sample. If the cubes are made from sintered diamond , pressures well beyond 30 GPa can be generated, under the condition that the thrust mechanism is improved. A common problem of all multi-anvil cells is that in the standard setup using a single ram and guiding blocks, the precise movement of the cubes is difficult to control. Due to the deformation of the guide blocks, the upper and lower cubes approach faster than the other 4 anvils, which leads to a loss in pressure performance. Ito et al. [33] have recently developed a 6-8 Kawai multianvil press controlled by six independent rams of 700 tonnes capacity, equipped with a servo mechanism which allows to control the position of the cubes to a precision of 2 μm. Using such a load frame, pressures close to 90 GPa could be generated on samples of several 0.1 mm^3 at 300 K.

A drawback of all multianvil presses is the limited view on the sample due to the opacity of the WC or sintered diamond anvils which prevents angle-dispersive diffraction. A way out of this problem is the use of sintered cubic boron nitride cubes which are transparent to x-rays. A miniature 6-8 device using cBN anvils for in situ x-ray diffraction, and a geometry with the [111] axis vertical, is the "T-cup" module developed at Stony Brook [32]. Pressures of up to 20 GPa can be generated with this device. Since it requires "only" forces of a few hundred tonnes, it can be combined with a V7 Paris-Edinburgh press to give an extremely compact and portable multi-anvil device [34]. In combination with oscillating radial collimators, such a setup enables angular dispersive diffraction on synchrotron sources with relatively clean diffraction patterns which can be analyzed by Rietveld methods.

The idea of high-pressure cells with several stages (multi-staging) has been pushed one step further in the 6-8-2 multi-anvil cells to reach the highest pressures attained so far in large volume devices [35]. In this setup (Fig. 6(d)), the octahedron of a 6-8 press contains an opposed anvil cell (squeezer) which compresses the sample. The pressure performance depends strongly on the squeezer material. Comparative measurements using single-crystal diamond, sintered diamond and nano-polycristalline diamond (NPD) show that pressures of up to 80 GPa can be achieved with NPD-squeezers. Although the sample volume of ~ 0.02 mm^3 (excluding furnace and other elements) appears to be small, it is still 2-3 orders of magnitude larger that what is used in standard DAC experiments. Pressures of almost 1 Mbar can be maintained even under heating to 1200 K using this material, but then decrease drastically upon further heating [35].

6   The diamond anvil cell

The diamond anvil cell (DAC) has become undoubtedly the most successful and versatile high-pressure device. It was invented in 1958, but the widespread use started only from the early 1970s on. By that time, the ruby fluorescence method was introduced which allowed a rapid and convenient determination of pressures. Also, the use of metallic gaskets allowed measurements to be carried out under hydrostatic conditions using adequate high-pressure media (methanol-ethanol mixtures). The fact that diamond is transparent in the visible was historically an enormous stimulus for the development of the DAC as a tool for high-pressure research. In fact, diamond is opaque only in the 5–5000 eV range, i.e. from the ultraviolet to the soft x-ray range. In the far-infrared, transparency is limited by diffraction from the gasket hole. An overview of 50 years of research using DACs has recently been given by W Bassett [36] It is impossible to cover all different designs and applications which have emerged during the last five decades, and the reader is referred to more exhaustive literature [1, 2, 3, 36], including other contributions in this book. I will here give only the basic principles, and focus on more recent developments. The DAC is conceptually an extremely simple device. Any DAC, whatever design and for whatever application, contains three essential elements: (1) the anvil/gasket assembly, (2) the backing plates which might include an alignment mechanism, (3) a thrust-generating mechanism.

6.1   Anvils

Figure 7 shows the anvil/gasket setup which illustrates simultaneously the principle of a DAC: Two diamonds with a flat tip (culet) of typically 300 μm diameter squeeze on a metallic gasket which has a bore of typically 100 μm diameter. The cavity formed by the bore is filled with the sample, immersed in a fluid which acts as pressure-transmitting medium. A small piece of ruby is added to measure the pressure. As the load onto the anvils increases, the gasket deforms plastically and the volume of the cavity becomes smaller, thereby increasing the pressure. Traditionally, anvils of either 8 or 16 facets are used, but since the advent of modern machining techniques, also the conical shapes used in the "Boehler-Almax" anvils [37] (Fig. 7) are possible. The advantage of such a geometry will be discussed further below. For anvils with a flat culet of diameter d, the maximum pressure P_{max} which can be "safely" obtained is given by $P_{max} = 10/d$ (d in mm and P_{max} in GPa), i.e. for culet of 0.5 mm a pressure of 20 GPa. This relation is correct for gems of approximately 60 mg (0.3 carat) and a table size of 3 mm, and between about 5 and 50 GPa. This concerns probably 90% of all experiments carried out today. The majority of diamonds used in DACs are natural gem stones which contain impurities. These gems can be classified in type I and II diamonds. Type I contain small amount of nitrogen, which causes an intense absorption between 1100 and 1500 cm^{-1}. Those of type II contain no nitrogen and show only an intrinsic absorption band around 2000 cm^{-1}. The luminescence of diamond anvils can be a serious problem for Raman measurements and depends on the gem. For this reason it should be specified on the purchase order. Recently, impurity-free synthetic diamonds of sufficient size have become available. To give a rough idea, the current price, a type I anvils of 0.3 carat is about 1200 Euro, for type II anvils 1500 Euro, and for a synthetic anvils 2000–2500 Euro.

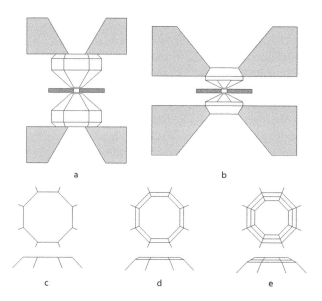

Figure 7. *Anvil geometry and backing seats in DACs: (a) conventional design; (b) Boehler-Almax design; (c) standard culet cut with 8 facets; (d) single-beveled culet cut with 8 facets; (d) double-beveled culet cut with 8 facets.*

6.2 Backing seats

Backing seat are necessary to transfer the load, typically 0–5 kN, from the metallic load frame (next section) onto the anvils. Without the backing plates the anvils would indent the metal contact surface. Backing plates are therefore made of hard materials with a compressive strength well beyond those of metals. Traditionally tungsten carbide (WC) is used, which has a compressive strength of 60 tn/cm^2. Tungsten carbide is however opaque to x-rays, and is also magnetic due to the cobalt binder. For diffraction studies which need large angular openings, low-Z elements are preferable, such as cubic BN, boron carbide B_4C, or sintered diamond containing low-Z binders. For small loads, hard beryllium has been used. However, any of these materials absorb to a certain extent, and the ideal geometry would be a seat with conical openings as large as possible. This is the major advantage of the Boehler-Almax design [37], since it allows an x-ray accessible opening of 70 to 90 degrees. The particular seating also allows smaller gems to be used, for the same culet size, hence similar sample volume. These are the main reasons why this type of anvil-seat assembly has become more and more popular in the community dealing with x-ray and light scattering.

It was realized by the inventors of the DAC that care has to be taken in aligning the culets of the anvils. This becomes particularly important for experiments beyond 10 GPa. Therefore, most of the DACs contain a more or less sophisticated aligning mechanism which allows at least one anvil to be (i) horizontally translated by a few tenths of a mm, and (ii) oriented by a few degrees. This mechanism acts in most cases on the backing seats. In the easiest case, such as in the Merrill-Bassett type-cells, (i) is achieved by small screws which move one of the two backing plates, and (ii) is achieved simply by aligning the anvils initially and trying to keep the alignment by tightening the load bolts coherently. More sophisticated mechanisms use hemispherical and hemi-cylindrical backing seats (Fig. 8). These need to be polished to fit precisely into their metallic counterpart on the load frame. For very small DACs, such as used for measurements under high magnetic field, the aligning mechanism needs to be simple and compact, and a wedge-type system is appropriate. Here, the faces of the two backing plates are not exactly parallel but inclined to each other by a degree or less. In turning the plates there are two positions where the culets of the two diamonds are parallel (but not necessarily strictly perpendicular to the thrust axis).

6.3   Thrust mechanisms

The required thrust on a DAC is a few kN ("a few hundred kilograms"). This can be achieved by 3-4 bolts such as used in Merrill-Bassett cells. But for most applications, in particular low- and high temperature measurements, a more sophisticated mechanism is needed (Fig. 8). The conceptually simplest design is the "nutcracker" mechanism applied by Mao and Bell, and still used in many high-temperature experiments. Piermarini incorporated a hydraulic ram into the load frame in one of the first DACs [36]. However this cannot be used at very low temperatures, due to the sealing problems under such conditions. The "Syassen-Holzapfel" cell applied a "toggle-latch" (German: Kniehebel; French: genouillère) mechanism. In this method, the thrust is applied by turning a single pin oriented 90° to the thrust axis. This allows scattering measurements to be carried out at low temperatures, with the cell in a cryostat, the thrust axis horizontally (as required in most optical and x-ray setups), and the thrust being manipulated by a rod introduced into the cryostat from above. A very popular thrust mechanism applies a metallic membrane to generate the load on the anvils which is controlled by the gas pressure inside the membrane. Such devices are very successful for low-temperature measurements. It requires, however, a gas handling system which might be more expensive than the cell itself, and which poses safety problems for air and rail travel. A compact computer controlled gas-handling system for membrane cells has recently become commercially available (Sanchez Technologies, France).

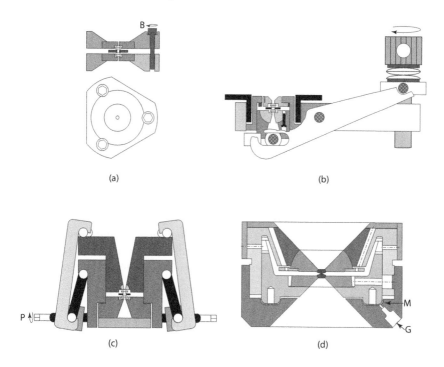

(a)

(b)

(c)

(d)

Figure 8. *Thrust mechanism for DACs: (a) bolt mechanism such as in Merrill-Bassett type cells [38]. (b) nutcracker mechanism such as applied in the NBS (NIST) design (see for example Ref. [39]) or the Mao-Bell cell [40], (c) latch mechanism such as applied in the Syassen-Holzapfel DAC [41], (d) membrane mechanism such applied in the Paris desings by Letoullec [42] and Chervin [43]. B: bolt, P: pin, G: gas inlet, M: membrane.*

7 Other gem anvil cells: sapphire, moissanite and zirconia cells

There are a few transparent hard materials which can replace diamond for certain experiments, in a geometry which is essentially the same as for DACs. Sapphire is one of them. Its Knoop hardness ($\rightarrow$ Glossary) is approximately 1200 (Mohs hardness: 9) and artificial gem stones are available in large sizes, with low levels of impurities and imperfections, at considerably less cost than diamond. Sapphire anvil have therefore been used for measurements where relatively large sample volumes are needed, i.e. $0.1 - 1$ mm^3 for pressures up to several GPa. Neutron scattering is a typical application [44, 45, 46]. Sapphire spheres of 10 mm diameter are available at a cost of approximately 100 Euro, and can be used as anvils after machining two flat culets. The pressure limit of carefully polished and aligned sapphire anvils under non-

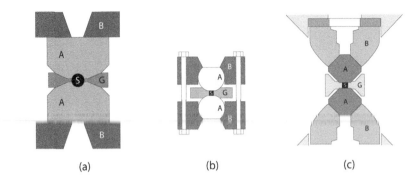

(a)   (b)   (c)

Figure 9. *Examples of sapphire cells. (a) cylindrical anvils with conical/rounded tips such as in ref. [44], (c) spherical with flattened front and back, such as in refs. [45, 51, 52, 50], (b) conoidal such as in refs. [46]. S: sample, G: gasket, B: backing seats. The diameter of the anvils vary between 5 and 15 mm.*

hydrostatic conditions is approximately 15 GPa. It is interesting to note that for large samples beyond 0.1 mm^3 volume, sapphire anvils perform better than diamond anvils [44]. This is due to the observation that the strength of diamond anvils decreases extremely rapidly with the size, probably due to flaws and impurities.

Moissanite is a gemstone of α-SiC (hexagonal lattice). Its Knoop hardness is 3000 (Mohs hardness: 9.25) and it provides a clear window for optical studies between 0.4 and 5.5 μm. Large single crystals in cm size are commercially available, but are much more expensive than sapphire and tend to have flaws. Pressure generation of 52 GPa are reported using small DAC-size anvils under nonhydrostatic conditions, and 10 GPa using methanol-ethanol [48].

Cubic zirconia is ZrO_2 stabilized with $\approx 10\%$ Y_2O_3 and is an inexpensive diamond simulant. It has a Knoop hardness of 1370, and its optical absorption is essentially flat between 2000 and 20000 cm^{-1}. Therefore, similar to sapphire, it is interesting for optical measurements in the 1500 - 3000 cm^{-1} range where diamond shows an intrinsics absorption band, see above. Another interesting property of cubic zirconia is its thermal stability to beyond 2000 °C. Zirconia anvils of 4 mm diameter have been reported to generate 13 GPa under nonhydrostatic conditions and 1 GPa using methanol-ethanol [49].

8   Pressure transmitting media

Bridgman stated that "...the most important immediate problem of technique in this field is to find methods of producing stress systems which are truly hydrostatic" [7]. This is still true, since the achievable pressures have increased and the measuring techniques have become more sensitive. There are numerous reports of "new" or "anomalous" effects which later turned out to be a result of non-hydrostatic conditions, see refs. [54, 55] for a few examples. Requirements for pressure transmitting media (PTM) are: (i) zero shear stress, (ii) chemical inertness, (iii) easy to load, (iv) low cost. Undoubtedly the most common PTM is the 4 : 1 methanol-ethanol mixture, which at 300 K vitrifies at 10.5 GPa. The addition of water in the 16:3:1 methanol-ethanol-water mixture does not seem to increase this pressure range, but even modest heating definitely does [58]. Other commonly used fluids are: silicone oils (denoting a whole range of fluids) including DAPHNE [56] (used for low-temperature clamp measurements), Fluorinert (a perfluoro-carbon liquid), and elemental gases such as argon, nitrogen, neon, and helium. The hydrostatic pressure limits of these have recently been systematically revisited [57]. Interestingly, nitrogen, neon and helium produce no detectable shear stresses even in their solid phases, well beyond their solidification pressure. Helium shows first signs of pressure gradients at approximately 20 GPa (300 K). These gases need to be loaded cryogenically, or at 300 K under pressures of 0.1-0.2 GPa using a gas-loader. Solid rare gases appear to be the best choice for low temperature experiments, in particular when the pressure is changed at low temperatures as it is often done in DAC experiments. Gas-loading techniques for DACs are now well established and the equipment can even be obtained commercially. For large-volume experiments, such as done in Paris-Edinburgh presses, the use of the 4 : 1 methanol-ethanol mixture has become standard. But in the case of low-temperature measurements, a change in pressure in the liquid (i.e. at high temperatures) is strongly recommended. An alternative and convenient (but not perfect) option is the use of lead as a pressure transmitting medium, in particular for single-crystal measurements. Lead and indium have very low shear strengths which vary only little with pressure and temperature [1]. Lead has the additional advantage that it is almost transparent to neutrons and that its equation of state is well known and convenient for experiments in the 0–10 GPa range. Single-crystal data on reasonably robust samples indicate tolerably non-hydrostatic conditions even when the pressure is increased at low temperatures [59].

9 Glossary

- **Young modulus:** One of the two properties (the other is Poisson's ratio) which characterizes the elasticity of an isotropic material. It is defined as the ratio between the uniaxial stress (force divided by cross section) over the uniaxial strain (relative elongation), determined in a traction measurement. Its unit is hence Pa. A typical value for steel is 200 GPa. A stress of 1 GPa on a steel rod will hence produce an elongation of $1/200 = 0.5\%$.

- **Poisson's ratio:** When an object is stretched in one direction it usually contracts in the direction perpendicular to it. The ratio of relative expansion over relative compression is called Poisson's ratio ν. Steels have typical values of $\nu=0.3$.

- **Yield stress** (German: Elastizitätsgrenze, Streckgrenze; French: limite élastique): The stress (force/cross section) applied to a material at which it starts to deform plastically. It is determined in standardized traction tests using cylindrical test specimens. Usually the yield stress is defined as the stress which produces a 0.2% permanent elongation on the specimen. The yield stress is the most important characteristics of a construction material.

- **Tensile stress** (German: Bruchgrenze; French: limite de tenue): The traction stress applied to a test specimen which causes complete rupture. It is hence also called rupture stress.

- **Ductile-brittle:** A material is said to be ductile if it can be substantially deformed (10-20% in length) before rupture. The contrary is called brittle.

- **Toughness** (German: Bruchzähigkeit; French: tenacité): Toughness (or rupture toughness) measures the energy a material can absorb during plastic deformation before it breaks. It is hence proportional to the area under the stress-strain curve determined in traction tests. Toughness is related to the ability of a material to withstand the propagation of cracks. Toughness can be expressed by the stress concentration factor K_{Ic}. Low K_{Ic} values (0-20) indicate brittle material, high values (100-200) tough materials.

- **Hardness:** Hardness characterizes the resistance of a material against penetration of an undeformable object. It can be converted to the tensile strength, which provides an easy and non-destructive way to determine the strength of metallic objects. Three widely used definitions of hardness are Brinell-, Vickers- and Rockwell hardness denoted HB, HV, and HC (Scale C: HRC), respectively. Values are determined by applying a defined load onto an indentor and then measure the distance it penetrates. High tensile steels have HRCs of 50-60. Knoop hardness (HK, unit: Pa) is determined by a micro indentation test and used to measuring the hardness of minerals and ceramics. HK values are typically between 100 and 1000 (in units of kg/mm^2), with HK=8000 for diamond and HK=1200 for sapphire. A historical and still widely used measure for hardness of minerals is the Mohs scale which spans from 1 to 10. The hardness of diamond is defined to be 10, that of talc 1, and that of 8 other minerals are given intermediate values. The hardness of a mineral is then determined by a scratch test using these gauge materials.

10 Acknowledgements

I am grateful to A. Jeanne-Michaud (IMPMC) for preparing the figures for this article and F. Datchi for proof reading the manuscript.

References

[1] Sherman W.F. and Stadtmuller A.A. *Experimental Techniques in High-Pressure Research*, John Wiley & Sons LTD, 1987.

[2] Eremets M. *High Pressure Experimental Methods*, Oxford University Press, Oxford, 1996.

[3] Isaacs, N.S., Holzapfel W.B., *High Pressure Physicochemistry: A Practical Approach*, Oxford University Press, Oxford,1997.

[4] Spain I.L., Paauwe J. (edts.), *High Pressure Technology*, vol. 1, Marcel Dekker Inc., New York/Basel, 1977.

[5] Ito E. *Theory and Practice: Multianvil Cells and High Pressure Experimental Methods*, in *Treatise of Geophysics*, G. Schubert ed., Elsevier, Amsterdam, Vol. 2, 2007, pp. 197-230.

[6] Bradley C.C., *High Pressure Methods in Solid State Research*, Paul W. and Warshauer D.M. eds., McGraw-Hill, New York, 1969.

[7] Bridgman P.W., in: *Solids under Pressure*, Plenum Press, New York, 1963.

[8] Lamé G. and Clapeyron B.P.E., Memoire sur l'équilibre intérieure des corps solides homogènes, *Memoires présentés par divers savants à l'Academie des Sciences* 18, 733, 1868.

[9] Langlois P., Frettage et autofrettage, in: *Materiaux et joints d'étanchéité pour les hautes pression*, Publications de l'Université de Saint Etienne, Saint-Etienne, 2004 (ed. S. Mottin).

[10] Von Mises R., Mechanik der festen Körper im plastisch deformablen Zustand, *Göttinger Nachrichten, Math.-phys. Kl.*, 582, 1913.

[11] Bundy F.P., Direct conversion of graphite to diamond in static pressure apparatus, *J. Chem. Phys.*, 38, 631, 1963.

[12] Loriers-Susse C., Bastide J.P., Bäckström G., Specific heat measured at high pressure by a pulse method, *Rev. Sci. Instr.* 44, 1344, 1973.

[13] Bundy F.P., Pressure-temperature phase diagram of iron to 200 kbar, 900°C, *J. Appl. Phys.* 36, 616, 1965.

[14] Leger J.M., Bastide J.P., Variation de la température de Curie de Cr_3Te_4 sous très haute pression, *phys. stat. sol. (a)*, 29, 107, 1975.

[15] Leger J.M., Transformations de phases dans les solides ferromagnétiques sous très haute pression, *Thesis*, Université Paris, 1970.

[16] Leger J.M., Lorier-Susse C., Vodar B., Pressure effect on the Curie temperature of transition metals and alloys, *Phys. Rev. B*, 6, 4250-4261, 1972.

[17] Claussen W.F., Detection of the α-γ iron phase transformation by differential thermal conductivity analysis, *Rev. Sci. Instrum.* 31, 878, 1960.

[18] Hall H.T., Some high-pressure, high-temperature apparatus designs considerations: equipment for use at 100 000 atmospheres, *Rev. Sci. Instr.*, 29, 267, 1958.

[19] Dobson D.P., Mecklenburgh J., Alfe D., Wood, I.G., Daymond M.R., A new belt-type apparatus for neutron-based rheological measurements at gigapascal pressures, *High Press. Res.*, 25, 107, 2005.

[20] Fitch R.A., Slykhouse T.E., Drickamer H.G., An apparatus for optical studies under very high pressures, *J. Opt. Soc. Am.* 47, 1015, 1957.

[21] Fasol G., Schilling J.S., New hydrostatic pressure cell to 90 kilobars for precise electrical and magnetic measurements at low temperatures, *Rev. Sc. Instr.*, 49, 1722, 1978.

[22] Jaccard D., Holmes A.T., Behr G., Inada Y., Onuki Y., Superconductivity of ε-iron: complete resisitive transition, *Phys. Letters A* 299, 282-286, 2002.

[23] Morard G., Mezouar M., Rey N., Poloni R., Merlen A., Le Floch S., Toulemonde P., Pascarelli S., San-Miguel A., Sanloup C., Fiquet G., Optimization of Paris-Edinburgh press cell assemblies for in situ monochromatic X-ray diffraction and X-ray absorption, *High Press. Res.* 27, 223-233, 2007.

[24] Khvostantsev L.G., Vereshchagin L.F., Novikov, N., Device of toroid type for high pressure generation, *High Temp. High Press.* 9, 637, 1977.

[25] Besson J.-M., Nelmes R.J., Hamel G., Loveday J.S., G. Weill, Hull S., Neutron powder diffraction above 10 GPa, *Physica B* 180 & 181, 1735, 1992.

[26] Klotz S., Besson J.-M., Hamel G., Nelmes R.J., Loveday J.S., Marshall W.G., Wilson R.M., Neutron powder diffraction at pressures beyond 25 GPa, *Appl. Phys. Lett.* 66, 1735, 1995.

[27] Hall H.T., Anvil guide for multiple-anvil high pressure apparatus, *Rev. Sci. Instr.*, 33, 1278, 1962.

[28] Lloyd E.C., Hutton U.O., Johnson D.P., Compact multi-anvil wedge-type high pressure apparatus, *J. Res. Nat. Bur. Stand.* 63C, 59, 1959.

[29] Mori N., Takahashi H., Takeshita N., Low-temperature and high pressure apparatus developed at ISSP, University of Tokyo, *High Press. Res.*, 24, 225, 2004.

[30] Kawai N., Endo S., The generation of ultrahigh hydrostatic pressures by a split sphere apparatus, *Rev. Sc. Instr.*, 41, 1178, 1970.

[31] Walker D., Carpenter M.A., Hitch C.M., Some simplifications to multianvil devices for high pressure experiments, *Am. Miner.* 75, 1020, 1990.

[32] Vaughan M.T., Weidner D.J., Wang Y., Chen J., Koleda C.C., Getting I.C., T-CUP: A new high-pressure apparatus for x-ray studies, *Rev. High Pres. Sci. Technol.* 7, 1520, 1998.

[33] Ito E., Katsura T., Yamazaki D., Yoneda A., Tado M., Ochi T., Nishibara E., Nakamura A., A new 6-axis apparatus to squeeze the Kawai-cell of sintered diamond cubes, *Phys. Earth and Planetary Interiors*, 174, 264, 2009.

[34] Le Godec Y., Hamel G., Martinez-Garcia D., Hammouda T., Solozhenko V.L., Klotz S., Compact multianvil device for in situ studies at high pressure and temperatures, *High Press. Res.* 25, 243, 2005.

[35] Kunimoto T., Irifune T., Sumiya H., Pressure generation in a 6-8-2 type multi-anvil system: a performance test for third-stage anvils with various diamonds, *High Press. Res.* 28, 237-244, 2008.

[36] Bassett W., Diamond anvil cell, 50th birthday, *High Press. Res.*, 29, 163-186, 2009.

[37] Boehler R., De Hantsetters K., New anvil design in diamond-cells, *High Press. Res.* 24, 391-396, 2004.

[38] Merrill L., Bassett W.A., Miniature diamond anvil pressure cell for single crystal x-ray diffraction studies, *Rev. Sc. Instr.*, 45, 290, 1974.

[39] Piermarini G.J., Block S., Ultrahigh pressure diamond-anvil cell and several semiconductor phase transition pressures in relation to the fixed point pressure scale, *Rev. Sc. Instr.*, 46, 973, 1975.

[40] Mao H.-K., Bell P.M., Design of a diamond windowed high-pressure cell for hydrostatic pressures in the range 1 bar to 0.5 Mbar, *Carnegie Institute Washington Year Book*, 74, 402, 1974/75.

[41] Huber G., Syassen K., Holzapfel W.B., Pressure dependence of f-levels in europium pentaphosphate up to 400 kbar, *Phys. Rev. B*, 15, 5123, 1977.

[42] Letoullec R., Pinceaux J.P., Loubeyre P., The membrane diamond anvil cell: A new device for generating continuous pressure and temperature variations, *High Press. Res.*, 24, 193, 2004.

[43] Chervin J.-C., Canny B., Besson J.-M., Pruzan Ph., A diamond anvil cell for IR microspectroscopy, *Rev. Scientific Instrum.* 66, 2595, 1995.

[44] Goncharenko I.N., Neutron diffraction experiments in diamond and sapphire anvils, *High Press. Res.*, 24, 193, 2004.

[45] Kuhs W.F., Ahsbahs H., Londono D., Finney J.L., In-situ growth and neutron diffraction four-circle diffractrometry under high pressure, *Physica B*, 156-157, 684, 1989.

[46] Kuhs W.F., Bauer F.C., Hausmann R., Ahsbahs H., Dorwarth R., Hölzer K., Single crystal diffraction with x-rays and neutrons: high quality at high presure, *High Press. Res.*, 14, 341, 1996.

[47] Xu J., Huang E., Graphite-diamond transition in gem anvil cells, *Rev. Sc. Instr.*, 65, 204, 1994.

[48] Xu J., Mao H.-K., Moissanite: A window for high pressure experiments, *Science* 290, 783, 2000.

[49] Russell T.P., Piermarini G.J., A high pressure optical cell utilizing single crystal cubic zirconia anvil windows, *Rev. Sc. Instr.*, 68, 1835, 1997.

[50] Klotz S., Hamel G., Frelat J., A new type of compact large-capacity press for neutron and x-ray scattering, *High Press. Res.*, 24, 219, 2004.

[51] Daniels W.B., Lipp M., Strachan D., Winter D., Yu Z.-H., Simple sapphire ball cell for high pressure studies in the ultraviolet, in *Recent Trends in High Pressure Research - Proceedings of the XIII AIRAPT Int. Conf. on High Pressure Science and Technology*, A.K. Singh ed., Oxford & IBH Publishing Co., New Delhi, 2007, pp. 809-811.

[52] Takano K. J., Wakatsuki M., An optical high pressure cell with spherical sapphire anvils, *Rev. Sc. Instr.*, 62, 1576, 1991.

[53] Patterson D.E., Margrave J.L., The use of gem-cut cubic zirconia in the diamond anvil cell, *J. Phys. Chem.*, 94, 1094, 1990.

[54] Takemura K., New structural aspects of elements appearing under hydrostatic pressure, *Proceedings of the Joint 20th AIRAPT and 43th EHPRG Conf.*, Karlsruhe 2005, ISBN 3-923704-49-6, 2005.

[55] Takemura K., Hydrostatic experiments up to ultrahig pressures, *J. Phys. Soc. Jpn.*, 76, Suppl. A 202, 2007.

[56] Murata K, Yokogawa K, Yoshino A, Klotz S, Munsch P, Irizawa A, Nishiyama N, Iizuka K, Nanba T, Okada T, Shiraga Y, Aoyama S., Pressure transmitting medium Daphne 7474 solidifying at 3.7 GPa at room temperature, *Rev. Scientific Instrum.*, 79, 085101, 2008.

[57] Klotz S., Chervin J.C., Munsch P., Le Marchand G., Hydrostatic limits of 11 pressure transmitting media, *J. Phys. D: Appl. Phys.*, 42, 075314, 2009.

[58] Klotz S., Paumier L., Munsch P., Le Marchand G., The effect of temperature on the hydrostatic limit of 4:1 methanol-ethanol under pressure, *High Press. Res.*, 29, 649, 2009.

[59] Rotaru G.-M., Padmanabhan B., Gvasaliya S.N., Roessli B., Strässle Th., Cowley R.A., Lushnikov S.G., Klotz S., Study of diffuse scattering under hydrostatic pressure in $PbMg_{1/3}Nb_{2/3}O_3$, *J. Phys. Conf. Series.*, 251, 012011, 2010.

Chapter 2
Instrumentation Development for High-Pressure Research

Konstantin Kamenev

School of Engineering and Centre for Science at Extreme Conditions, The University of Edinburgh, United Kingdom

1   Introduction

Historically high-pressure research has been an area heavily dependent on the availability of the experimental equipment. In fact, many of the discoveries in high-pressure science followed promptly breakthroughs in instrumentation development which provided researchers with higher pressure limits or larger sample volumes. However, the opposite is also true and the biggest developments in instrumentation were driven by anticipated discoveries in science. Indeed a lot of the development of high-pressure equipment in the past two hundred years have been driven by the quest to create synthetic diamonds [1]. Another good example of the interconnection between high-pressure cell development and scientific research is the continuing search for metallic hydrogen. The first prediction of hydrogen becoming a metal under pressure has been published in 1935 [2]. Since then there was a number of estimates of the pressure at which hydrogen would become metallic but every time experimentalists managed to achieve these pressures and could not find the metallic behaviour [3], theorists would refine their calculations to predict a new higher pressure of metallization [4]. And so the quest for building pressure cells capable of achieving yet higher pressures continues.

The aim of this paper is not to review the variety of the existing pressure cells. Instead its goal is to provide the outline of engineering approach to design work. It introduces generic tools such as computer aided design (CAD) and finite element analysis (FEA) and their application to high-pressure cell development. It reviews the relevant material's properties and provides the information on some conventional materials commonly used in construction of high-pressure equipment. And finally it looks into the manufacturing stage of high-pressure cell development, discusses the issue of tolerances and surveys the relevant types of machining techniques.

The paper is aimed at non-engineering readers with background in Physics, Chemistry or Geosciences, who are involved in high-pressure work and who require either to modify the existing pressure cell or to design a completely new piece of high-pressure equipment for their research. The need for this might arise for a number of reasons such as, for example, particular sample requirements. For example, studying gaseous samples at high pressure has its specific requirements as compressibility of gases is much higher than that of liquids and solids. This means that either the initial sample volume needs to be large or the gas needs to be precompressed or liquefied and loaded at cryogenic temperatures. Often the sample volume needs to be adjusted to suit the sensitivity of the measurement as dictated by the sample properties. For example, the sample can be weakly/strongly scattering or have weak/strong response to magnetic field, etc. For the samples with weak response to the measurement technique the volume needs to be increased in order to be able to sense the sample through the high-pressure cell and to gather the measurement statistics in reasonable time. If the sample has a strong response its volume can often be decreased in order to achieve a larger pressure, and this in return affects the design of the pressure cell. Another common reason for developing a new pressure cell is the need to fit it around the existing sample environment such as cryostats, magnets, heaters, spectrometers, neutron or synchrotron stations.

These are just a few examples as to why the need in the development of new equipment might arise. A limited availability of commercial pressure cells means that it is likely that anyone working in this field will at some point engage in designing pressure cells or auxiliary equipment in order to remain at the cutting edge of high-pressure research. This paper aims to help to get the reader started in this process and to provide some useful references.

2   Design flow

There are several books dedicated to the mechanical design process of which the following two can be recommended [5, 6]. Figure 1 summarises a typical design flow diagram for developing a piece of mechanical equipment - in this case a pressure cell. The purpose of building a new (or modifying the existing) high-pressure apparatus is obviously clear from the beginning of the design process, but there are a number of factors to be considered along the way such as the type of the pressure cell, other sample environment, the materials to be used in the construction, producing technical drawings, testing, safety issues, etc.

The following chapters will address each of the stages of the design process as applied to high pressure equipment in more detail.

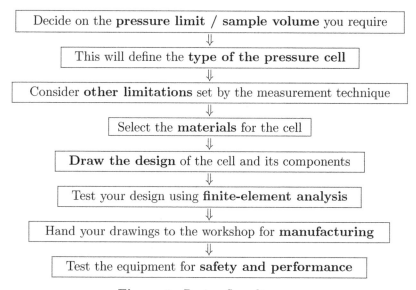

Figure 1. *Design flow diagram.*

3 Pressure generation and the types of pressure cells

This chapter provides a quick summary of the generic techniques for pressure generation (more information on this can be found in a number of review books and papers [7, 8, 9, 10]). Pressure is the force divided by the area to which this force is applied and this definition sums up the underlying approach to classifying high-pressure equipment. Thus the design brief in its basic approach can be reduced to answering the two questions - (i) how the force is generated, and (ii) how the surface area is defined. The most common methods for generating static force or load are summarised below:

- Use of *opposed anvils* is by far the most common way of generating the load. In this method the force is transmitted through two or more anvils onto the sample volume. In cases when a large load is required the force can be generated by the means of a hydraulic press. In the case of a medium load the force can be generated by a lever arm. When a low load is required a screw mechanism or a gas membrane can be used.

- *Compressors* can be used to generate pressure in the experiments where gas is either a sample under study or a pressure transmitting medium.

- In *hydrothermal* pressure cells the force is generated by the means of changing the temperature of the liquid sample or liquid pressure transmitting medium while its volume is kept constant.

In the opposed anvils setting the sample cannot be completely enclosed by the anvils and requires some support to stop it from escaping from between the anvils. This support can be provided in two ways outlined below:

- *Enclosing the sample and the anvils.* This method lies at the core of the piston-cylinder design which has been introduced in 19th century [11] and has been revolutionised by the development of an unsupported area seal [12]. In the piston-cylinder design one or two pistons slide inside a monoblock cylinder to compress the pressure medium. The pressure cells based on this design are limited to about 4-5 GPa limit

- *Enclosing the sample but not the anvils.* Unlike in the piston-cylinder design, in which the load is applied directly to the pressure medium, in this method the load is applied onto a gasket surrounding the sample. Thus, pressure generation here depends on the deformation of the gasket, which leads to the change in the sample volume, while in the case of the piston-cylinder design the cylinder ideally should not deform at all in order to provide a solid support for the pressure medium and the sample. This method is used in a variety of the pressure cells such as diamond anvil and indenter cells, large Bridgman anvil cells, multianvil cells, belt apparatus, *etc.* The highest pressures of an order of several megabar can be achieved in diamond anvil cells.

4   Materials properties

Before we consider the key mechanical properties of materials we introduce the notion of stress. By definition, *stress* is a measure of the average amount of force exerted per unit area and this is why so often it gets confused with *pressure*. The distinction between these two notions is that pressure is a surface phenomenon, *e.g.* it is the force exerted by the anvil divided by the surface area of the contact interface between the anvil and the object. Stress is a volume phenomenon showing how the pressure propagates through the bulk of the material. Stress analysis plays a major role in mechanical design and it will be discussed in more detail in Section 7.

The key mechanical properties and their relevance to high-pressure equipment design are outlined below.

Strength is the maximum stress that a part can withstand without failure. Strength is the first requirement to be considered when it comes to selecting the material for use in high-pressure equipment. There are three distinct types of strengths to be considered depending on the application in which the material is to be used. These are compressive, tensile and shear strengths. The compressive strength is important in parts such as pistons, tensile strength is of relevance to cylinders, and the shear strength should be considered in applications where the shear force is created. The strength values are derived from stress-strain measurements conducted in the industry-standard way. As

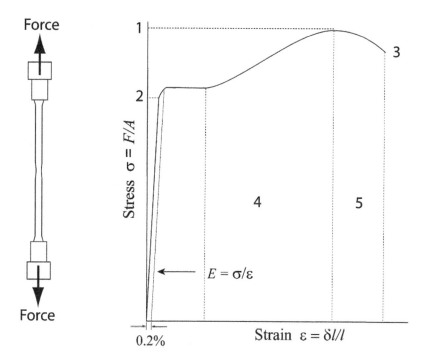

Figure 2. *Tensile strength testing and a stress-strain diagram.*

an example we will consider a tensile strength test and a stress-strain diagram typical for steels and alloys. Compressive strength analysis is treated in a similar way with the only difference that the material is subjected to a compressive force.

The tensile strength test is conducted on a dumbbell-shaped rod of material (Figure 2). The axial force is applied to the ends of the rod and the stress $\sigma = F/A$ is plotted as a function of strain $\epsilon = \delta l/l$ which is the relative elongation of the rod.

Up to point 2 on the diagram the rod experiences elastic deformation, i.e. if the load is released the rod will return to its original shape and dimensions. The initial slope of the straight line reflects on the response of the material to the stress in the elastic deformation regime and is called *Young's or elastic modulus, E*. The maximum stress that the material can withstand and still remain in the elastic deformation regime is called the *yield strength*. Yield strength is defined at the point at which the stress-strain curve deviates by a strain of 0.2% from the linear-elastic line (Figure 2). It is the stress at which dislocations first move large distances, and in metals it is the same in tension and compression.

Beyond point 2 the material enters the plastic deformation mode in which the deformation is irrecoverable. The material then undergoes a transition into the strain hardening region (4, Fig. 2) and the necking region (5, Fig. 2), and finally the failure occurs at the rupture point (3, Fig. 2). The maximum stress that can be achieved in the material provides the figure for the *ultimate strength* (1, Fig. 2).

For safety reasons elastic deformation regime is the one in which the pressure equipment is meant to operate. Therefore, the yield strength of the material is the most important parameter in considering the material's suitability for use in high-pressure application. However, in some applications the strength of the part can be increased beyond the yield strength, though obviously not beyond the ultimate strength. This can be done by the means of work-hardening (or autofrettage) during which a plastic deformation is created locally. One common application of this method is enhancing the strength of the cylinder in a piston-cylinder cell design. The bore of the cylinder is loaded with a pressure creating the stress in the walls of the cylinder larger than the yield strength. This creates the plastic deformation which propagates from the bore outwards but does not reach its outer wall. After the pressure is released, a boundary which separated plastically and elastically deformed parts of the cylinder remains and the elastic tensile stresses remain in the outer part of the cylinder and compressive stresses in the inner part.

The autofrettage can be done in two ways. One of them is to use the hydrostatic medium inside the pressure cell to create the over-pressure. This method should be used with extreme caution as there is a chance that due to the large load the plastic deformation can propagate through the whole cylinder and cause a failure of the cylinder with potentially catastrophic consequences. This failure can be further facilitated by internal faults and defects in the cylinder. The alternative method for autofrettage is to push an oversized object made from a harder material through the bore of the cell, e.g. tungsten carbide spheres used in ball-bearings which are commercially available in a variety of sizes. As the ball passes through the bore of the cylinder it induces local plastic deformation. This method is safer than the hydrostatic loading with over-pressure described above.

There's yet another way to create the plastic-elastic deformation boundary mentioned above. Two cylinders with interference fit between the inner diameter of one and the outer diameter of the other can be combined into a single cylinder. This can be done either by thermal shrink-fitting or by forcing the smaller cylinder into the outer cylinder. The stress created on the interface of the two cylinders will act to counteract the stress created by the hydrostatic pressure generated at the bore. For more details on the theory of the cylinder stress analysis see [10, 13].

Although strength is the key mechanical property that should be considered in high-pressure design work, there are also other parameters which are of equal significance.

- *Toughness* – the resistance to fracture of a material when stressed. For high-pressure applications high toughness is desirable as it reduces the likelihood of a sudden failure of the part.

- *Hardness* – resistance to elements such as indentation or scratch. High hardness means low deformation under load which is certainly an advantage in practically all high-pressure applications. However, it is worth remembering that hardness and toughness are often related in an adverse way as harder materials tend to lack toughness and be more brittle. This means that in terms of strain the rupture point is situated near the yield point of the material (Fig. 2) and when the yield point is reached the failure happens almost immediately.

- *Stiffness* – force-to-deflection ratio. This is the resistance of an elastic body to deflection or deformation. Stiffness is related to hardness but while the hardness is the property of the material, stiffness is the property of the part made of it and will vary with shape.

- *Fatigue resistance* – the number of stress-cycles before failure. This is self-explanatory and it is just worth mentioning here the importance of keeping a record of use of high-pressure equipment. The log will help to keep track of the number of times the apparatus has been exposed to stress.

5   Materials selection

The key mechanical engineering chart used for selection of materials for specific applications is the *Young's modulus-strength diagram* shown in Fig. 3.

The materials which are of interest for high-pressure applications are situated in the right top corner of the map. These materials, combining a high Young's modulus with high strength, can be grouped into three classes - alloys, ceramics and composites.

- *Engineering alloys* combine ductility and formability with good electrical and thermal conductivity. Their high tensile and compressive strengths are of the same order of magnitude which means that they can be used in tension (e.g. in cylinders) as well as in compression (e.g. in pistons). There are three types of engineering alloys commonly used in high-pressure equipment: maraging steels, superalloys and copper alloys.

  Maragning steels are iron-based alloys which are known for possessing superior strength and toughness without losing malleability and shapeability. They have high resistance to corrosion and crack propagation. However, they contain iron and cannot be used in applications where a weak magnetic background is expected from the pressure cell. Some common examples of maraging steels are Böhler W720 (Fe-Mo-Ni-Co) steel

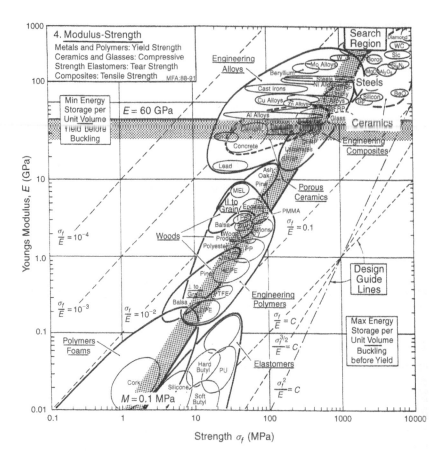

Figure 3. *Young's Modulus–Strength Diagram (from [14]).*

with the ultimate tensile strength (UTS) of 2260 N/mm^2 and Aubert & Duval 819AW (Fe-Mo-Ni-Cr) with the UTS of 1900 N/mm^2.[1]

Superalloys are Ni-, Co- or Co/Fe-based alloys with face-centered cubic crystal structure. They have excellent mechanical strength and creep resistance at high temperatures, good surface stability, as well as corrosion and oxidation resistance. An example of a superalloy is CrNiAl (Cr 39-41%, Al 3-4%, Ni balance) with the UTS of 2300 N/mm^2.

Copper alloys combine strength, toughness and low coefficient of friction. They are excellent for use in high-pressure equipment used with cryogenics, as their strength is enhanced at low-temperatures. Examples: phosphor bronze (P 0.03%, Zn+Sn+Fe 14%, Cu balance) with the UTS of 800 N/mm^2 and beryllium copper BERYLCO-25 (Be 1.8-2.0%,

---

[1]Note, 10^3 N/mm^2 = 1 GPa.

Co+Ni+Fe 0.6%, Cu balance) with the UTS of 1500 N/mm^2. Beryllium copper alloys are perfect for use in pressure cells for magnetic applications as they have low magnetic susceptibility. Also they are not prone hydrogen embrittlement and can be used for studies of hydrogen reach materials or with hydrogen gas as pressure transmitting medium.

- *Ceramic materials* have compressive strength higher than that of metals but their tensile strength is lower, which makes them suitable for use mainly in parts under compression such as pistons, anvils or backing plates. They are electrical and thermal insulators, resistant to high temperature, but can be brittle. Here are some examples of engineering ceramics with the values of their compressive strengths: zirconia (ZrO_2) - 2100 N/mm^2, alumina (Al_2O_3) - 2900 N/mm^2, tungsten carbide (WC) - 7000 N/mm^2, sintered diamond - 15000 N/mm^2.

- Unlike ceramics *composites* have high tensile strength but are weak in compression. Fig. 4 shows the comparison of beryllium copper alloy and ZrO_2 and WC ceramics with some modern composites in terms of their strength. It is clear that when it comes to the tensile strengths the composites are superior to the conventional materials.

  Engineering composites is a relatively new type of materials which has not yet had many applications in high-pressure research. The main reason for this is that the composites are not easy to shape into parts, they are difficult to machine or to thread. However, with the new developments in materials science and engineering these difficulties might well be overcome soon and we will see the composites revolution in high-pressure instrumentation.

For more information on mechanical properties of the materials and material selection the reader is referred to the following books [14, 15]. For novel materials the best source of up-to-date information is often the manufacturers' websites and online material property databases such as the MatWeb [16].

6 Technical drawings

Making technical drawings is the important part of any instrumentation development. The process of producing drawings starts at the stage of the design ideas and well before the overall design of the system is finalised. Technical drawings were used to be made by hand using a drawing board but with the advances in personal computing equipment *computer aided design* (CAD) packages became a unanimous tool for the job. The CAD packages most commonly used for mechanical engineering work are Solid Edge [17] and Solidworks [18]. The production of high-quality drawings is central to any design process, and even if the design of a particular piece of equipment is seen by

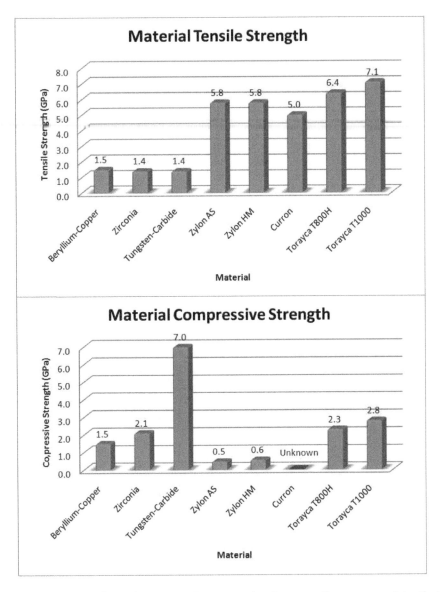

Figure 4. *Tensile and compressive strength of composites compared to that of alloys and ceramics.*

the developer as a one off exercise that can be done 'on the back of the enve-
lope', it is still worth while investing ones time into learning how to use the
CAD software. The software comes with tutorials that can help you learn it
in a short period of time. Although some packages can be costly the chances
are that the engineering department at your institution holds the site licence

which would give you access to the software. There are many advantages of using a CAD package some of which are listed below.

- Solid *three-dimensional parts* can be produced and manipulated aiding the visualisation process. The parts can be rotated and scaled on the screen with different projections and cross-sections exposed to the viewer.

- Separately drawn parts can be joined in an *assembly*. This is an important step in the design process which helps to check that various parts fit well together. The assembly sequence can be recreated and followed on screen to ensure that there will not be any undesirable interference between the parts when the same procedure is followed on manufactured parts. A movement of parts with respect to each other occurring during normal operation of the equipment can also be modelled and reviewed.

- A lot of *useful information* about parts and assemblies can be gathered from their drawings such as the weight or the centre of mass of the system. Each part can have the material properties assigned to it, which makes the calculation of the physical and mechanical properties possible.

- 3D parts and assemblies can be easily exported into 2D *technical drawings* showing them in various cross-sectional views with the relevant dimensions. Although CAD packages make it easy to place virtually any dimensions on the drawings, there is a certain convention that should be followed to ensure that the set of dimensions shown is correct and complete [19, 20, 21]. A professionally produced technical drawing takes into account the manufacturing processes involved into making each part and understanding of the required tolerances (see more on machining methods in Section 8).

- The 3D drawings of parts can be exported into finite element analysis (FEA) software for computer based testing of the system (see Section 7).

- In some instances the CAD drawings of a part can be converted directly into the code for computer numerically controlled (CNC) machining.

Apart from all of the abovementioned advantages of using CAD software there is another one - it is great for creating still images and animations for use in presentations and webpages. Figure 5 shows an example of the use of CAD in producing the drawings of the pressure cell.

7 Finite element analysis

Although mathematical solutions for stress analysis have been developed over a hundred years ago [22] they can only be applied to simple geometrical objects

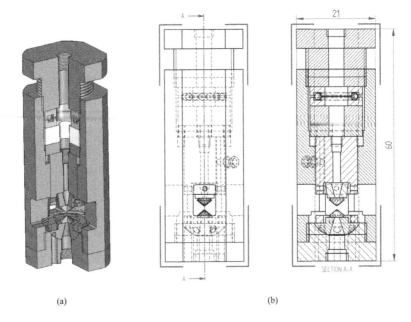

(a)   (b)

Figure 5. *Drawing of a DAC produced using SolidEdge CAD package: (a) a cross-sectional view of the pressure cell assembly, (b) technical drawing.*

such as long cylinders or plates [10, 13]. Stress calculations on arbitrary shaped objects can be done using finite element analysis (FEA).

FEA is based on the finite element method for finding approximate solutions of partial differential equations. The approximation is based on discretization, i.e. on replacing infinite dimensional problem with a finite dimensional one. FEA can be used to model stress distribution, deformation under load, temperature gradients, magnetic field distribution, fluid dynamics, *etc.* There are a number of FEA packages with the following being most frequently used in mechanical engineering: ANSYS [23], Abaqus [24], Femap [25] and Nastran [26]. The FEA software can be run on desktop personal computers although using it on workstations shorten the time required to complete the analysis. Just as in the case of the CAD software it is likely that the engineering department of your institution holds a licence for an FEA package.

Stress and deformation are the two types of analysis most relevant to designing high-pressure equipment. The overview of the key steps in FEA are outlined below. For more detailed information on the method the reader is referred to the textbooks [27, 28, 29].

- The part or the assembly to be analysed can be either *created* using the FEA software or *imported* from a CAD package. If it is an assembly, the *type of contact* between the constituent parts should be defined (e.g. bonded, frictionless, *etc*).

- Each part should have the basic *materials properties assigned* to it with the most important of them being the Young's modulus. The software should have the materials database with the relevant properties entered.

- The structure is then broken down into elements in the process called *meshing*. The size of the elements defines the accuracy with which the analysis will be performed. A coarse mesh with large elements is suitable for a quick analysis which will be lacking the detail. A finer mesh will produce more accurate results however at the cost of longer computing time. Using the symmetry of the system can help to increase the number of elements, reduce the size of the mesh and keep the computing time reasonable. For example, if the system has axial symmetry it would make sense to model it in a 2D axisymmetric fashion rather than to perform the analysis on its 3D model. In every situation the FEA software also recognises the areas where large gradients of properties can occur and adjusts the mesh size accordingly.

Figure 6 shows a model of a small DAC with a medium meshing. The model is symmetric with respect to the vertical axis. It is also symmetric with respect to the horizontal plane, i.e. there's an identical opposing anvil. So the simulation can effectively be performed on a quarter of the complete system.

- The next step is to *apply loads* and *boundary conditions*. The load (force or pressure) can be applied to the selected area, while the boundary conditions specify the supported surface(s).

- The FEA software solves the model for *displacements* resulting from the load application. This step is followed by the *postprocessing*, in which from the displacement of each element the reaction forces, stresses and deformations are found.

An example of an equivalent stress distribution in a diamond anvil and support is shown in Figure 7. In this model the pressure of 25 GPa is applied to the cullet of the diamond.

The stress distribution is shown as a colour map with the colour scale provided for guidance in the left-hand side of the figure. The exact figure for the stress at any point of the model can be found by clicking on it. The analysis of the stress distribution is performed by comparing the maximum stresses in various parts with the yield stresses of the corresponding materials. For example, the maximum stress produced in the anvil is 17 GPa which is well below the diamond's yield strength of 130-140 GPa [30]. The maximum stress in the diamond support is close to 2.1 GPa, which means that the material which can be used in this part will need to have its yield strength higher than this figure. A suitable material for this part can be maraging steel or a superalloy. The maximum stress in the external part is lower (0.95 GPa),

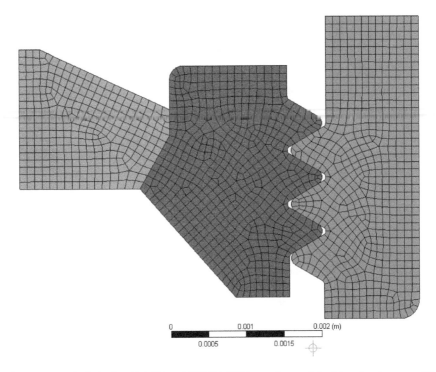

Figure 6. *Model of a DAC in ANSYS FEA package. Meshing is shown with the element size of an order of* 10^{-4} *m.*

which widens the range of suitable materials for this part to include copper alloys (as discussed in Section 5).

The FEA also solves the model for deformation. Figure 8 shows the total deformation of the system when 25 GPa is applied to the culet of the diamond, i.e. how parts of the DAC shift under pressure.

After the FEA analysis has been completed the technical drawings can be made and passed on to the workshop.

8 Machining and tolerances

For success of any design work it is important to know about the machining techniques and machines themselves which are to be used in the manufacturing of the parts. Here is a brief list of the common and specialised types of machining techniques commonly used for making parts of high-pressure equipment (for more information refer to [31, 32, 33]).

- *Turning* is the most common machining technique for making axisymmetric cylindrical parts. During turning the workpiece rotates with re-

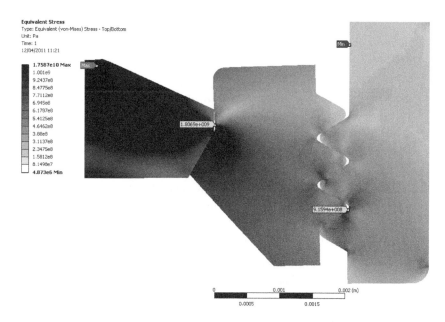

Figure 7. *FEA of the stress distribution in the DAC.*

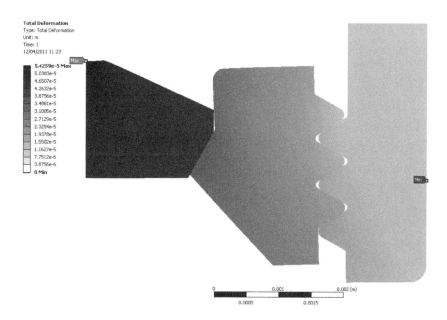

Figure 8. *Total deformation in the diamond anvil cell. The wire frame around each part shows the position of underformed parts before the pressure was applied.*

spect to the cutting tool.

- In *milling* the machining tool is a rotating cutter. It can be used to produce complex paths on the part.

- *Drilling* or *reaming* is commonly used for making straight or tapered holes. However, the length of the drill is commonly limited to 6- to 12-times its diameter. For making longer holes a lesser known technique called gun-drilling can be used. Gun drills have a groove along its length for taking the metal swarf away and prevents it from getting between the tool and the workpiece.

- In *grinding* the machining tool is a spinning abrasive wheel. Grinding is used to provide higher accuracy of machining and better surface finishes compared to that of turning and milling. Grinding can be plain surface as well as cylindrical (external and internal).

- If yet a better accuracy, surface finish and parallelism are required the technique called *honing* can be used. Honing is a variety of grinding in which a honing stone is supported by a tool called mandrel. Using the mandrel in the honing machine one can vary the pressure on the stone to do the machining in a more controlled fashion.

Table 1 summarises the tolerances and surface finishes that can be achieved by using various types of machining techniques.

Table 1. *Manufacturing tolerances and surface finishes.*

Operation (machine)	Tolerance, μm	Surface finish, μm
Turning (lathe)	± 20	3.2
Grinding (surface grinder)	± 1	0.8
Grinding (cylindrical grinder)	± 5	0.8
Milling (manual)	± 10	3.2
Milling (CNC)	± 5	3.2
Honing	± 1	0.025

Tolerances higher than $\pm 20\mu$m will not be required for majority of the parts. In fact the increase in the required tolerance beyond what the basic types of machining, i.e. turning and milling, can achieve will double or triple the time required to manufacture the part.

However there are situations in which precision machining will be crucial for successful operation of the pressure equipment. One such example is the machining of pistons and cylinders. In a large volume hydrostatic cell or in a DAC based on the piston-cylinder principle a close fit should be achieved between the piston and the cylinder. This allows the piston to slide inside the cylinder freely but without slackness which can lead to a leak of pressure medium in the hydrostatic cell or misalignment of the diamonds in the DAC. In engineering this fit is called a 'transition' fit. Another example is when the interference fit is required between two parts (as discussed in Section 4). These fits indeed require high accuracy of machining (down to a few microns), which should be noted on the technical drawings.

In order to achieve the transition or interference fit often several techniques will be combined in making a single part. For example, for making a piston the part will first need to be turned and then ground or honed until the transition fit with the cylinder and high surface finish to reduce friction are achieved.

9 Testing and safety certification

The final stage of the design development is experimental testing during which it can be established whether the equipment performs as expected and whether it is safe to use. Safety testing requirements to a large extent predetermined by the potential risks associated with using the equipment. To assess these risks it is important to estimate the amount of energy stored in the high-pressure medium and to analyse the potential scenarios of a sudden release of this energy in the result of a failure. For the same sample volume compressed to the same pressure the amount of energy stored in the liquid medium will be higher than that stored in a solid, but will be much lower than the energy of the compressed gas.

Once the risks have been assessed the relevant safety measures should be implemented to protect the user and the rest of the equipment from the consequences of a possible failure. Although DACs achieve extremely high pressures the sample volume is so small that they represent virtually no risk to users in case of a failure. Large volume pressure cells are much more of a concern and the safety precautions can be implemented by making sufficiently strong enclosures, shields and screens.

In terms of the official safety regulations pressure cells and auxiliary equipment for high-pressure research are a "grey area". There are documents on national, European and international levels that regulate the safety issues, however they are limited mainly to the cases of industrial hydraulic systems and gas pressure vessels [34]. Therefore, high-pressure labs develop their own procedures for safety certification of high-pressure equipment.

The most common way of ensuring the safety of the equipment is the overpressure test. The apparatus under test is used to create pressure which is 50% higher than the nominal working pressure for this equipment. The over-pressure is maintained for several minutes and then released. Where possible strain-gauges should be used to monitor the deformation of the parts during the test.

After the test the equipment is disassembled and examined for deformation and cracks. The test should be performed with all the necessary safety precautions and witnessed by at least by two competent persons. After the successful completion of the test the safety certificate can be issued and the equipment can be used by competent staff.

10 Acknowledgements

The author would like to thank the members of his group Artur Bocian, Gaétan Giriat and Somchai Tancharakorn for providing him with some of the graphics material used in this paper.

References

[1] Hazen, R.M., *The Diamond Makers*, Cambridge University Press, 1999.

[2] Wigner, E. and Huntington, H.B., On the Possibility of a Metallic Modification of Hydrogen , *J. Chem. Phys.*, 3, 764, 1935.

[3] Narayana, C., et al., Solid hydrogen at 342 GPa: no evidence for an alkali metal, *Nature*, 393, 46, 1998.

[4] Bonev, S.A. et al., A quantum fluid of metallic hydrogen suggested by first principles calculations, *Nature*, 431, 669, 2004.

[5] Kutz, M., *Mechanical Engineers' Handbook: Materials and Mechanical Design*, 3rd ed., John Wiley & Sons, 2005.

[6] Childs, P., *Mechanical Design*, 2nd ed., Butterworth-Heinemann, 2003.

[7] Eremets, M.I., *High Pressure Experimental Methods*, Oxford University Press, 1996.

[8] Sherman, W.F. and Stadtmuller, A.A., *Experimental Techniques in High-pressure Research*, John Wiley & Sons, 1987.

[9] Holzapfel, W.B. and Isaacs, N.S., *High Pressure Techniques in Chemistry and Physics: A Practical Approach*, Oxford University Press, 1997.

[10] Klotz, S., *High Pressure Devices*, in this Book of Proceedings.

[11] Amagat, E.H., Sur La Compressibliti et La Dilation des Gaz, *C. R. Acad. Sci., Paris*, 68, 1170, 1869.

[12] Bridgman, P.W., *The Physics of High Pressure*, Dover Publications, New York, 1970.

[13] Rees, D.W.A., *Mechanics of Solids and Structures*, Imperial College Press, 2000.

[14] Ashby, M.F., *Materials Selection in Mechanical Design*, 2nd edition, Butterworth-Heinemann Ltd, 2004.

[15] Higgins, R.A., *Materials for Engineers and Technicians*, 4th ed., Newnes, 2006.

[16] MatWeb Material Property Data, http://www.matweb.com/.

[17] Solid Edge, http://www.solidedge.com/.

[18] SolidWorks, http://www.solidworks.com/.

[19] Giesecke, F.E., *Technical Drawing*, 11th ed., Prentice Hall, 1999.

[20] Bielefeld, B. and Skiba, I., *Basics Technical Drawing (Basics)*, Birkhauser Verlag AG, 2006.

[21] Madsen, D.A., *Engineering Drawing and Design*, 4th ed., Delmar Cengage Learning, 2006.

[22] Lamé, G. and Clapeyron, B.P.E., Memoire sur l'équilibre intérieure des corps solides homogènes, *Memoires presenteés par divers savants à l'Academie des Sciences*, 18, 733, 1868.

[23] ANSYS (by ANSYS, Inc.), http://www.ansys.com/.

[24] Abaqus (by Simulia), http://www.simulia.com/.

[25] Femap (by Siemens PLM Software), http://www.nenastran.com/.

[26] Nastran (by MSC), http://www.mscsoftware.com/products/msc_nastran.cfm.

[27] MacDonald, B.J., *Practical Stress Analysis with Finite Elements*, Glasnevin Publishing, 2007.

[28] Logan, D.L., *A First Course in the Finite Element Method*, 4th ed., CL-Engineering, 2006.

[29] Moaveni, S., *Finite Element Analysis Theory and Application with ANSYS*, 3rd ed., Prentice Hall, 2007.

[30] Eremets, M.I. et al., The strength of diamond, *Appl. Phys. Lett.*, 87, 141902, 2005.

[31] Meyers, A.R. and Slattery, T.J., *Basic Machining Reference Handbook*, 2nd revised ed., Industrial Press, 2001.

[32] Walker, J.R., *Machining Fundamentals: From Basic to Advanced Techniques*, Goodheart-Willcox Publicing, 1993.

[33] Carvill, J., *Mechanical Engineers Data Handbook*, 3rd ed., Butterworth Heinemann, 2001).

[34] *PD 5500:2009 - Specification for unfired fusion welded pressure vessels*, British Standards, 2008.

Chapter 3
Electrical Transport
Experiments at High Pressure

Samuel T. Weir

Lawerence Livermore National Laboratory, United States

1 Introduction

High-pressure electrical measurements have a long history of use in the study of materials under ultra-high pressures. In recent years, electrical transport experiments have played a key role in the study of many interesting high pressure phenomena including pressure-induced superconductivity, insulator-to-metal transitions, and quantum critical behavior. High-pressure electrical transport experiments also play an important function in geophysics and the study of the Earth's interior.

Besides electrical conductivity measurements, electrical transport experiments also encompass techniques for the study of the optoelectronic and thermoelectric properties of materials under high pressures. In addition, electrical transport techniques, i.e., the ability to extend electrically conductive wires from outside instrumentation into the high pressure sample chamber have been used to perform other types of experiments as well, such as high-pressure magnetic susceptibility and de Haas – van Alphen Fermi surface experiments. Finally, electrical transport techniques have also been used for delivering significant amounts of electrical power to high pressure samples, for the purpose of performing high-pressure and –temperature experiments. Thus, not only do high-pressure electrical transport experiments provide much interesting and valuable data on the physical properties of materials extreme compression, but the underlying high-pressure electrical transport techniques can be used in a number of ways to develop additional diagnostic techniques and to advance high pressure capabilities.

2   Electrical measurement techniques with diamond anvil cells

Electrical transport experiments with diamond anvil cells (DAC's) are very challenging. Because of the very small sample sizes involved and because of the presence of a metal pressure gasket in most experiments, signal wires leading to the high-pressure sample must be well insulated to avoid short circuits. A large number of different approaches have been developed over the years to address this key problem. One solution is to replace the metal gasket with an insulating gasket. Mica-MgO composite gaskets, for example, have been successfully used to 40 GPa [1] The metal gasket can also be coated with an electrically insulating layer by either sputtering [2] , plasma-spraying [1], or by coating the gasket with mixture of alumina or cubic boron nitride mixed with epoxy [3] (Figure 1)

Several approaches have also been used to make the electrically conductive paths to the high-pressure sample. Fine metal wires (<25 μm diameter) or very flat metal foils of gold or platinum are often used. Recent years have also seen the development and application of a variety of advanced microfabrication techniques for placing electrically conductive paths on or within the diamond anvils themselves [4, 5, 6, 7] . These include the use of boron implantation, focused ion beam (FIB) equipment, and the combined use of microlithography and diamond chemical vapor deposition (CVD).

Electrical contacts between the sample and the conductive wires can be made in several ways. Simple pressed contacts are often sufficient to ensure good electrical contact between the wires and the sample, provided that the sample's surfaces are clean of oxide or contamination layers. In situations where secure contacts are required, small amounts of silver paste or silver epoxy can be used to bond the wire to the sample. These contacts may be applied directly to the sample, but contact resistances can often be reduced if thin-film metal contact pads (e.g., Au, Pt) are first evaporated or sputter deposited onto the sample. If the sample is a semiconductor, depositing metal contact films onto it is also beneficial since this will tend to reduce the tendency for the metal-to-semiconductor contacts to be rectifying (or "Shottky") contacts. Generally, non-rectifying or "ohmic" contacts are desirable for delivering electrical currents to semiconductors and getting voltage signals out of them. Finally, laser welding or spark welding have occasionally been used to make contacts to samples. The drawback with these methods is that special equipment is needed to deal with DAC-sized samples, and that special care must be taken to avoid overheating the sample.

The van der Pauw method [8] is frequently used for measuring the resistivity of arbitrarily shaped samples. There are several requirements associated with the van der Pauw method: the contacts to the sample must be located on the boundary of the sample, the contact areas should be small in comparison to the sample size, and the sample should be of some known uniform thick-

(a)   (b)

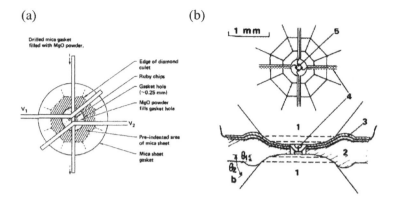

Figure 1. *Various techniques for electrically insulating electrical wires from the gasket. (a) Mica gasket with MgO insulator assembly of R. Reichlin [1]. Reprinted with permission from Rev. Sci. Instrum. 54, 1674, 1983. Copyright 1983, American Institute of Physics. (b) Metal gasket with an alumina insulating layer assembly of Gonzalez and Besson [2]. Reprinted with permission from Rev. Sci. Instrum. 57, 106, 1986. Copyright 1986, American Institute of Physics.*

ness. If these conditions are satisfied, the van der Pauw method is a powerful technique for determining the sample resistivity of samples of arbitrary shape (subject to the restriction that the thickness is uniform and known). If small electrical contacts are made at four points A, B, C, and D, on the periphery of the sample, a current I_{AB} can be applied from contact A to contact B while measuring the voltage drop V_D-V_C. If we define $R_{AB,CD} = (V_D$-$V_C)/I_{AB}$ and, analogously, $R_{BC,DA} = (V_A$-$V_D)/I_{BC}$ then the resistivity ρ is given by the equation [8]

$$\exp\left(-\frac{\pi d}{\rho}R_{AB,CD}\right) + \exp\left(-\frac{\pi d}{\rho}R_{BC,DA}\right) = 1 \qquad (1)$$

The resistivity ρ cannot be solved for as a closed–form expression from this equation, but the resistivity can be calculated by numerical methods for any given $R_{AB,CD}$ and $R_{BC,DA}$. The equation is simplified considerably if the points of electrical contact are symmetrically arranged around the circumference of a disc of uniform resistivity, then $R_{AB,CD} = R_{BC,DA}$ and eqn. 1 reduces to

$$\rho = \frac{\pi d}{\ln(2)}R_{AB,CD} \qquad (2)$$

In practice, however, the van der Pauw method, is usually difficult to apply to DAC resistivity experiments. The contact areas are often a significant fraction of the sample size, the samples are often completely irregular in shape, and the *in situ* thickness of a high pressure is usually very difficult to determine with much accuracy. However, a rough estimate of the absolute resistivity can usually be made based on the sample dimensions and the locations of the contact points by using 4-wire resistance measurements. The absolute resistivity of the sample can also be estimated by simulating the 3D current flow through the sample using the actual sample geometry and the positions of the probes on the sample [9].

3   Superconductivity under high pressure

The study of superconductivity under high pressures is motivated by both scientific and technological concerns. The dependence of the superconducting transition temperature on applied pressure can be used to test and confirm theories. The behavior of the transition temperature on applied pressure can also be used to focus efforts to develop new superconductors with higher transition temperatures. A superconductor which exhibits a relatively high transition temperature under high pressures, for example, is an obvious candidate for further study to see if higher transition temperatures can also be achieved by applying "chemical pressure" through selective impurity doping of its lattice.

4   Iron

The appearance of superconductivity in iron might thought to be unlikely since ferromagnetism and superconductivity are expected to be mutually exclusive according to conventional BCS theory. Strong magnetic fields tend to align the spins of each Cooper electron pair in the same direction and thus break up the Cooper pairs. Above 10 GPa, however, iron transforms to a non-magnetic structure, and conventional superconductivity becomes a possibility at low temperatures. Shimizu, et. al. [10], performed high-pressure electrical conductivity experiments on high purity iron samples by using an arrangement of gold and platinum foils and wires, with a thin layer of alumina covering the metal gasket for electrical insulation. NaCl, which is much weaker than alumina, was used as the pressure medium to cushion the iron sample. Spot welding was used to securely attach gold electrodes to the sample (Figure 2a). A sharp 10% drop in the resistance associated with the appearance of superconductivity is clearly seen at a temperature of approximately 1K at a pressure of 25 GPa. (Figure 2b). The fact that the resistance does not drop entirely to zero could be due to pressure inhomogeneities in the sample, or small contact resistances in the gold-to-iron spot welds.

The magnetic field dependence of the resistivity behavior confirms the existence of superconductivity in the iron sample. Figure 3 shows that the resistivity drop decreases with increasing magnetic field, until superconductivity is entirely quenched at a magnetic field of 1.8 Tesla. Finally, Figure 4

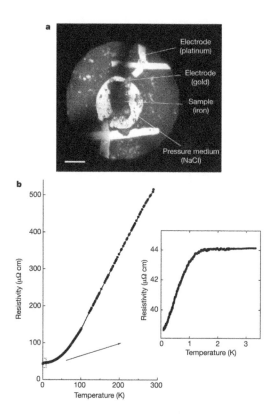

Figure 2. *(a) Assembly of the iron conductivity experiment of Shimizu, et. al. [10]. (b) The resistivity of iron versus temperature at 25 GPa, showing a drop in resistivity below 1 K attributed to superconductivity. Reprinted by permission from Macmillan Publishers Ltd: Nature 412, 316, 2001.*

shows the pressure dependence of the superconducting transition discovered between 15 and 30 GPa.

5 Oxygen

Based on high-pressure optical absorption edge and reflectivity experiments [11] , oxygen metallizes at approximately 95 GPa. The transition from insulating molecular oxygen to metallic molecular oxygen is accompanied by a structural phase transition. Shimizu, et. al. [12] , performed electrical conductivity experiments and discovered superconductivity in solid molecular oxygen in the pressure range from 98 GPa to at least 125 GPa. Figure 5 shows the preparation of the sample chamber, with four platinum foil electrodes contact-

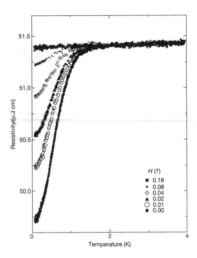

Figure 3. *The resistivity of an iron sample at 25 GPa under various magnetic field strengths [10]. Reprinted by permission from Macmillan Publishers Ltd: Nature 412, 316, 2001.*

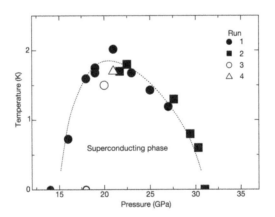

Figure 4. *A phase diagram indicating the superconducting region of iron found in the experiments of Shmizu, et. al. [10].*

ing the oxygen sample. Figure 6 shows the magnetic field dependence of the resistance transition, and confirms the existence of superconductivity with a critical field of about 0.2 Tesla at 120 GPa.

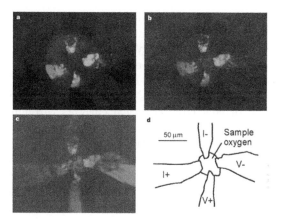

Figure 5. *Assembly for a high-pressure conductivity experiment on oxygen by Shimizu, et. al. [12]. Reprinted by permission from Macmillan Publishers Ltd: Nature 412, 316, 2001.*

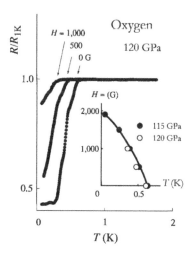

Figure 6. *Normalized resistance of an oxygen sample under various magnetic field strengths [12]. Reprinted by permission from Macmillan Publishers Ltd: Nature, 393, 767, 1998.*

6 Conductivity experiments at high pressure and very high temperatures

Relatively little work has been performed involving DAC electrical conductivity measurements at conditions of high pressures and very high temperatures

(>> 1000 °C), a temperature regime where externally heated DAC's start to give way to laser-heated DAC's and internal resistive heating techniques. Pioneering work on using internal resistive heating to heat tiny iron wires in a DAC while monitoring changes in the resistances of the wires was performed by Liu and Bassett [13] , Mao, et. al. [14] , and Boehler, et. al. [15] . In all these experiments, very fine iron wires with diameters ranging from 5 to 20 μm were heated by driving large electrical currents through them. Typical currents were in the 1-3 ampere range for a power dissipation in the range of a few watts and temperatures up to 2500 K.

Knittle and Jeanloz [16] performed electrical conductivity experiments on a laser heated FeO sample at pressures of about 70 GPa and temperatures of over 1000 °C, and confirmed the existence of a metallic phase of FeO under simultaneous conditions of high-pressure and –temperature which was first identified with shock-wave experiments. More recently, Li, et. al. [17], have performed conductivity measurements on high-pressure, laser-heated samples of $(Fe_{0.125}, Mg_{0.875})_2SiO_4$ using a specially prepared diamond anvil with an electrical thin-film circuit and a 3 μm thick alumina thermal insulating layer on it.

7   Single-crystal experiments

Electrical transport experiments on single crystal samples require special care and preparation due to the need to avoid excessive shearing stresses on the sample while at the same time maintaining secure electrical connections to the sample. Electrical connections may be made by using silver or gold paint or epoxy to bond thin wires to the sample. In order to provide good ohmic contacts to some samples, thin film pads of gold may need to be sputter deposited onto the sample before bonding wires to it. Either photolithographic masking or shadow masking can be used to prepare the sample for sputter deposition.

A good pressure medium is needed to minimize shearing stresses on the sample. Ideally, the medium should remain fluid to very high pressures and possess a very small strength even after it solidifies. Additionally, it should be chemically inert and not react with either the sample or the metal wires. Thermal stability and non-toxicity are also desirable properties. Methanol-ethanol mixtures, Fluorinert 3M, argon, and helium are commonly used media.

The inset of Figure 7a shows the setup of an electrical conductivity experiment by Cui, et. al. [18] on a single crystal sample of an organic semiconductor, tetramethyltetratelluronaphtalene (TMTTeN). Four thin gold wires (5 μm diameter) were attached to the sample with gold paint, and a fluorocarbon mixture of Fluorinert 3M, FC70, and FC72 was used as a pressure medium. Figure 7b shows the resistivity as a function of temperature for pressures up to 25.4 GPa, at which point the activation energy for conduction has

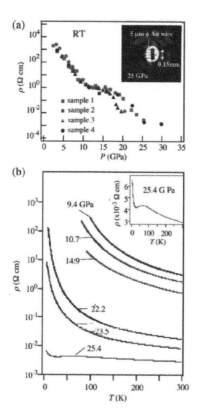

Figure 7. (a) Resistivity versus pressure of a single-crystal sample of tetramethyltetratelluronaphtalene (TMTTeN) to 30 GPa. The inset shows the single-crystal sample with attached wires at a pressure of 25 GPa. (b) Temperature dependence of the resistivity of TMTTeN at various pressures.Reprinted with permission from J. Am. Chem. Soc., 130, 3738, 2008. Copyright 2008 American Chemical Society.

been reduced to about 5 meV.

8   Hall effect and magnetoresistance

Hall effect and magnetoresistance experiments have been performed on single crystal and thin-film samples in diamond anvil cells. Hall effect experiments are useful for measuring the charge carrier density and also giving information on the sign of the charge carriers. The Hall coefficient is defined by

$$R_H = \frac{E_y}{(j_x B)} = \frac{V_H d}{(IB)} \tag{3}$$

Where E_y is the transverse electric field, j_x is the current density, B is the magnetic field, V_H is the Hall voltage, I is the current, and d is the sample thickness. For materials in which the electric current is carried by a single band, if n is the density of carriers and e is the electron charge then the Hall coefficient is $-(1/ne)$ or $+(1/ne)$, depending on whether the current is carried by electrons or holes, respectively. For semiconductors in which both electrons and holes may be present the Hall coefficient is

$$R_H = \frac{-e^2(n\mu_e^2 + p\mu_h^2)}{e(n\mu_e + p\mu_h)^2} \qquad (4)$$

where n and p are the electron and hole densities, and μ_e and μ_h are the electron and hole mobilities, respectively.

For Hall effect experiments, thin samples are desirable, since for a given magnetic field B the magnitude of the Hall voltage is directly proportional to the current density. Patel, et.al. [19] , studied the Hall coefficient and carrier mobility of a single crystal of GaAs under pressures up to 6 GPa by evaporating gold-germamium contacts onto a GaAs wafer, and then cleaving and polishing the wafer to obtain a sample 150 μm x 150 μm x 40 μm. Gold wires were then attached to the sample with silver epoxy, and glycerol was used as a pressure medium. The Hall coefficients and carrier mobility of GaAs were then measured to 6 GPa.

Hall effect and magnetoresistance experiments have also been performed by Boye, et.al. [20, 21] on nickel samples using diamond anvil cells. Their approach is unique in that they combined lithographic and electroplating techniques to produce free—standing $Ni_{0.985}O_{0.015}$ thin film samples having precise dimensions and geometry. The samples measured 50 μm x 50 μm x 15 μm and pressure contact was made to four 125 mm diameter Pt wires using a pyrophyllite piece. Hall coefficient and magnetoresistance experiments were performed to pressures up to 6 GPa using 10 Tesla magnetic fields. A decrease in the high field magnetoresistance was attributed to a reduction in electron–magnon scattering due to spin wave damping under high magnetic fields.

9   Photoconductivity

High pressure photoconductivity data is sometimes used to complement electrical resistivity data when studying materials which may be in the vicinity of an expected insulator-to-metal transition. Photoconductivity can be a convenient add-on experiment since the major difficulties associated with setting up a conductivity experiment have already been overcome, and only a small additional investment in time and equipment (e.g.,an illumination laser, optical chopper, lock-in amplifier) is needed to perform photoconductivity experiments. In a resistivity experiment on hydrogen to 210 GPa which showed no measurable conduction to 210 GPa, for example, Eremets, et. al. [22], looked

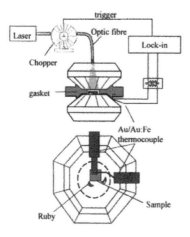

Figure 8. *The experimental setup for high-pressure specific heat experiments by Braithewaite, et. al., [24] showing the thermocouple assembly used for monitoring temperature variations in a sample as it is pulse heated with a laser. With kind permission of Springer Science and Business Media.*

for evidence of small hydrogen bandgap by illuminating the sample with a 647 nm laser (1.9 eV) while monitoring the conductance of the sample. No measurable photoconductivity was observed, indicating that the bandgap was still above 1.9 eV at 210 GPa.

Photoconductivity as well as resistivity experiments were also performed on hydrogen iodide (HI) under high pressure by van Straaten and Silvera [23]. Here photoconductivity measurements with a 10 mW green laser (5145 Å) were used to confirm the onset of HI metallization by band-overlap at 45 GPa.

10   Other uses of electrical transport techniques

Electrical transport techniques have also been used to develop a wide range of diagnostic tools for applications such as the measurement of specific heats, magnetic susceptibilities, and thermoelectric powers under high pressures. These techniques rely on the ability of electrical circuits within the sample chamber to detect changes in temperature (e.g., using thermocouple or resistive sensors), to detect induced voltages, or to transport electrical power.

Specific heat experiments with diamond anvil cells are difficult because of the highly nonadiabatic nature of the sample environment. The diamond anvils sandwiching the sample are outstanding thermal conductors, and the

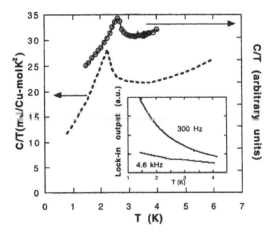

Figure 9. *Specific heat of a $Sr_{2.5}Ca_{11.5}Cu_{24}O_{41}$ sample as a function of temperature using the DAC specific heat technique of Braithewaite, et. al. [24]. With kind permission of Springer Science and Business Media.*

small size of the sample chamber means that any thermal insulating layer must be rather thin, perhaps not more than a few microns thick. Coupled with the small size of the sample, this means that the thermal relaxation time of the sample is very fast, perhaps in the neighborhood of a millisecond or less. Consequently, in order to detect changes in the specific heat of the sample, ac calorimetry methods are required in which thermal power to the sample is varied at a very rapid frequency of several kHz or more while the magnitude of the sample's temperature oscillations is monitored. D. Braithwaite, et. al. [24], developed a high pressure specific heat technique in which a pair of Au/Au-Fe thermocouples was introduced into the sample chamber to measure the sample temperature while it was heated by a series of laser pulses delivered via an optic fiber. A lock-in amplifier was used to detect the temperature oscillations at the frequency of the chopped laser. The experimental setup is shown in Figure 8. Figure 9 shows a peak in the specific heat of a $Sr_{2.5}Ca_{11.5}Cu_{24}O_{41}$ sample going through an antiferromagnetic phase transition.

A magnetic susceptibility technique developed by Alireza and Julian [25] involved placing a small wire signal coil inside the sample chamber (Figure 10). The signal coil was made of 12 μm diameter copper wire which was wound into a coil just 300 μm in diameter. This technique results in a very high sample filling factor with the size of the coil very well matched to the size of the sample. A magnetic excitation coil outside of the sample chamber was used to generate an alternating magnetic field, and magnetic induction then generates a voltage in the signal coil inside the sample chamber, with the magnitude of the induced voltage varying with the magnetic susceptibility of the sample. Because of the high sensitivity and low noise level of the sensing coil, it has also

(a)   (b)

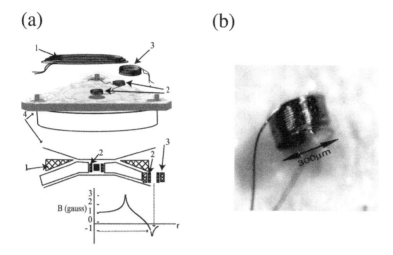

B (gauss)

Figure 10. *(a) High-pressure magnetic susceptibility assembly of Alireza and Julian [25], showing the placement of the excitation, compensation, and sensing coils. (b) A picture of a sensing coil made from 12 μm insulated copper wire. Reprinted with permission from Rev. Sci. Instrum., 74, 4728, 2003. Copyright 2003, American Institute of Physics.*

been successfully used to perform de Haas – van Alphen (dHvA) experiments at high pressure [26]. In these experiments, small oscillations in the magnetic susceptibility of a sample as a function of applied magnetic field are used to gain information on extremal cross-sectional areas of Fermi surfaces.

Thermoelectric power experiments have also been performed with diamond anvil cells by Polvani, et. al., [27] by introducing a pair of thermocouples into the sample chamber in order to measure the average temperature gradient along the length of a small sample (Figure 11). Good electrical as well as thermal contact between the thermocouple junctions and the sample is required, since the thermocouples are used to measure both the voltage gradient as well as the temperature gradient along the sample. The temperature gradient is established by heating one end of the long sample with an infrared laser.

The ability to place electrical wires leading to the sample chamber also offers the possibility of transporting large amounts of electrical power to heat samples to high temperatures [13,14 ,15]. These internal resistive heating methods are capable of reaching sample temperatures of thousands of degrees, but it has proved difficult to reach pressures of more than about 20 – 30 GPa because the electrical wires are susceptible to breaking or short-circuiting as the wires, insulating layers, and gasket deform under high pressure loading. Since it is difficult to introduce much thermal insulation between the internal resistive heating element and the diamond anvils, considerable amounts

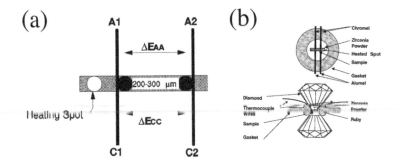

Figure 11. *(a) Thermoelectric power measurement assembly of Polvani, et. al. [27] showing the pair of thermocouple junctions used to measure the temperature gradient along the length of the sample. (b) Schematic diagram of the overall DAC assembly. Reprinted with permission from Rev. Sci. Instrum., 70, 3586, 1999. Copyright 1999, American Institute of Physics.*

of electrical power and electrical current must be delivered to the sample. Recently, Zha and Bassett [28] have performed internal resistive heating experiments examining the high pressure melting temperature of gold samples embedded in small rhenium wires. More recent work involving a modification of their original technique was successful in measuring the P-V-T equation-of-state of platinum to pressures of 80 GPa and temperatures up to 1900 K [29].

11   Future directions

High pressure electrical transport experiments have provided much valuable data on the electronic properties of materials under high pressure including the discovery of unexpected superconducting phases. However, research progress in this area has been hampered by the fact that setting up an electrical transport experiment can be very difficult and time consuming. Electrical experiments at Mbar pressures are extremely difficult and remain the preserve of just a few talented experimentalists.

In recent years, several research groups have been turning to modern microfabrication technologies to build electrical circuits within the diamond anvils themselves. Efforts thus far have involved either using boron ion implantation to 'write' electrically conductive paths in the diamond anvil, or lithographic techniques to deposit a metal thin-film circuit onto the diamond surface followed by chemical vapor deposition (CVD) of a protective diamond film over the circuit. Among other advantages, these microfabrication approaches offer the experimentalist the convenience of working with anvils which are already pre-wired and electrically insulated, thus overcoming some of the greatest problems associated with setting up high pressure electrical transport experiments. The ability to precisely control the circuit layout may also lead to further advanced development of some of the diagnostic techniques discussed earlier which utilize electrical circuits placed on diamond anvils such as specfic heat, thermoelectric power, magnetic susceptibility, and de Haas – van Alphen.

12 Acknowledgements

This work performed under the auspices of the U.S. Department of Energy by Lawrence Livermore National Laboratory under Contract DE-AC52-07NA27344.

References

[1] R.L.Reichlin, Measuring the electrical resistance of metals to 40 GPa in the diamond-anvil cell, Rev. Sci. Instrum. **54** 1674, (1983).

[2] J.Gonzalez, J.M.Besson and G.Weill, Electrical transport measurements in a gasketed diamond anvil cell up to 18 GPa, Rev. Sci. Instrum. **57**, 106, (1986).

[3] M.I.Eremets, V.V Struzhkin, H.K. Mao and R.J.Hemley, Superconductivity in Boron, Science, **293**, 272, (2001).

[4] T.A.Grzybowski and A.L.Ruoff, Band-overlap metallization of BaTe, Phys. Rev. Lett., **53**, 489, (1984).

[5] H.Hemmes, A.Driessen, J. Kos, F.A.Mul and R.Griessen, Synthesis of metal hydrides and in situ resistance measurements in a high-pressure diamond anvil cell, Rev. Sci. Instrum, **60**, 474, (1989).

[6] H.Bureau, M.Burchard, S.Kubsky, S.Henry, C.Gonde, A.Zaitsev and J.Meijer, Intelligent anvils applied to experimental investigations: state-of-the-art, High Pressure Research, **26**, 251, (2006).

[7] S.T.Weir, J.Akella, C.Aracne-Ruddle, Y.K.Vohra, and S.A.Catledge, Epitaxial diamond encapsulations of metal microprobes for high pressure experiments, Appl. Phys. Lett., **77**, 3400, (2000).

[8] L.J. van der Pauw, L.J., Philips Res. Rep., **13**, 1, (1958).

[9] X.Huang, C.Gao, Y.Han, M.Li, C.He, A.Hao, D.Zhang, C.Yu, G.Zou and Y.Ma, Finite element analysis of resistivity measurement with van der Pauw method in a diamond anvil cell, Appl. Phys. Lett., **90**, 242102, (2007).

[10] K.Shimizu, T.Kimura, S.Furomoto, K.Takeda, K.Kontani, Y.Onuki and K.Amaya, Superconductivity in the non-magnetic state of iron, Nature **412**, 316, (2001).

[11] S.Desgreniers, Y.K.Vohra and A.L.Ruoff, Optical response of very high density solid oxygen to 132 GPa, J. Phys. Chem. **94**, 1117, (1990).

[12] K.Shimizu, K.Suhara, M.Ikumo, M.I.Eremets and K.Amaya, Superconductivity in oxygen, Nature, 393, 767, (1998)

[13] L.Liu, and W.A.Bassett, The melting of iron up to 200 kbar, J. Geophys. Res., **80**, 3777, (1975).

[14] H.K.Mao, P.M.Bell, and C.Hadidiacos, Experimental phase relations of iron to 360 kbar, 1400 °C, determined in an internally heated diamond-anvil apparatus, in High Pressure Research in Mineral Physics, M.H.Manghnani, and Y.Syono, Eds., Terra Scientific, Tokyo, 135, (1987).

[15] R.Boehler, J.Nicol and M.L.Johnson, Internally-heated diamond anvil cell: phase diagram and P-V-T of iron, in High Pressure Research in Mineral Physics, M.H.Manghnani, and Y.Syono, Eds., Terra Scientific, Tokyo,, 173 (1987).

[16] E.Knittle, and R.Jeanloz, High-pressure metallization of FeO and implications for the earth's core, Geophys. Res. Lett., **13**, 1541, (1986).

[17] M.Li, C-X.Gao, Y.Ma, A.Hao, C.He, X.Huang, Y.Li., J.Liu, H.Liu and G.Zou, In situ HPHT resistance measurement of $(Fe_{0.125}, Mg_{0.875})_2 SiO_4$ in a designed laser heated diamond anvil cell, J. Phys.: Condens. Matter, **19**, 1, (2007).

[18] H.Cui, Y.Okano, B.Zhou, A.Kobayashi, and H.Kobayashi, Electrical resistivity of tetramethyltetratelluronaphtalene crystal at very high pressures – examination of the condition of metallization of π molecular crystal, J. Am. Chem. Soc., **130**, 3738, (2008).

[19] D.Patel, T.E.Crumbaker, J.R.Sites, and I.L.Spain, Hall effect measurement in the diamond anvil high-pressure cell, Rev. Sci. Instrum., **57**, 2795, (1986).

[20] S.A.Boye, D.Rosen, P.Lazor and I.Katadjiev, Precise magnetoresistance and Hall resistivity measurements in the diamond anvil cell, Rev. Sci. Instrum., **75**, 5010, (2004.)

[21] S.A.Boye, P.Lazor, and R.Ahuja, Magnetoresistance and Hall effect measurements of Ni to 6 GPa, J. Magn. Magn. Mater.,**294**, 347, (2005).

[22] M.I.Eremets, V.V.Struzhkin, H.K.Mao, and R.J.Hemley, Exploring superconductivity in low-Z materials at megabar pressures, Physica B, **329-333**, 1312, (2003).

[23] J.van Straaten and I.F.Silvera, Observation of metal-insulator and metal-metal transitions in hydrogen iodide under pressure, Phys. Rev. Lett., **57**, 766, (1986).

[24] D.Braithwaite, J.Thomasson, B.Salce, T.Nagata, I.Sheikin, H.Fujino, J.Akimitsu, and J.Flouquet, Coexistence of antiferromagnetic order and superconductivity in the spin ladder system $Sr_{2.5}Ca_{11.5}Cu_{24}O_{41}$, in Frontiers of High Pressure Research II, H.D.Hochheimer, B.Kuchta, P.K.Dorhout, and J.L.Yarger, Eds., Kluwer Academic, Dordrecht,383 (2001).

[25] P.L.Alireza, and S.R.Julian, Susceptibility measurements at high pressures using a microcoil system in a diamond anvil cell, Rev. Sci. Instrum., **74**, 4728, (2003).

[26] S.K.Goh, P.L.Alireza, P.D.A.Mann, A.M.Cumberlidge, C.Bergemann, M.Sutherland and Y.Maeno, High pressure de Haas – van Alphen studies of $Sr_2 RuO_4$ using an anvil cell, Current Appl. Phys.,**8**, 304, (2008).

[27] D.A.Polvani, J.F.Meng, M.Hasegawa, and J.V.Badding, Measurement of the thermoelectric power of very small samples at ambient and high pressures, Rev. Sci. Instrum., **70**, 3586, (1999).

[28] C-S.Zha, and W.A.Bassett, Internal resistive heating in a diamond anvil cell for in situ x-ray diffraction and Raman scattering, Rev. Sci. Instrum., **74**, 1255, (2003).

[29] C-S.Zha, K.Mibe, W.A.Bassett, O.Tschauner, H.K.Mao, and R.J.Hemley, P-V-T equation of state of platinum to 80 GPa and 1900 K from internal resistive heating/x-ray diffraction measurements, J. Appl. Phys., **103**, 54908, (2008).

Chapter 4
Advances in Customized
Diamond Anvils

Samuel T. Weir[1] and Yogesh K. Vohra[2]

[1]Lawrence Livermore National Laboratory and [2]University of Alabama,
United States

1 Introduction

The current state of static high pressure science owes much to the remarkable
physical and mechanical properties of diamond, particularly its extremely high
compressive strength and its transparency to a wide range of optical and x-
ray wavelengths. While a large number of different diamond anvil cell designs
have been developed over the years for different experimental needs, the key
components of these cells, the diamond anvils themselves, have changed very
little for most experimental work. With the increasing availability of advanced
technological tools that can be applied to diamond, more and more efforts are
being directed towards customizing diamond anvils for various high pressure
experiments, with the aim of improving experimental sensitivity and accuracy,
increasing the experimental pressure range, as well as improving the ease of
experimental setup.

For diamond anvil customization, several types of basic fabrication process steps can be identified: diamond removal, diamond growth, deposition of
electrically conductive layers, and lithographic patterning. (Figure 1). Methods for removing diamond include mechanical polishing, laser drilling, and
plasma etching. For diamond growth, chemical vapor deposition of diamond
films from a methane-hydrogen plasma has made impressive advances over
the years, and it is now possible to grow very high quality single crystal diamond films. A very important feature of this technique is its ability to grow
films epitaxially, meaning that the diamond film grows as a matched, crystallographic extension of the original diamond anvil. This ensures excellent
adhesion between the diamond film and the anvil and minimizes film stresses.

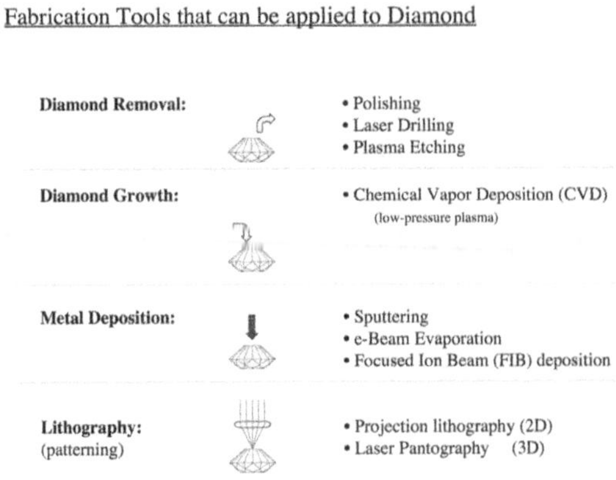

Fabrication Tools that can be applied to Diamond

Diamond Removal:		• Polishing • Laser Drilling • Plasma Etching
Diamond Growth:		• Chemical Vapor Deposition (CVD) (low-pressure plasma)
Metal Deposition:		• Sputtering • e-Beam Evaporation • Focused Ion Beam (FIB) deposition
Lithography: (patterning)		• Projection lithography (2D) • Laser Pantography (3D)

Figure 1. *A selected list of the various tools currently being applied to the customization of diamond anvils.*

Several techniques have become available for depositing conductive layers or paths either on or inside a diamond substrate. Sputter deposition of metal films has been used by a number of research groups for fabricating electrical probe patterns onto diamond anvils for electrical conductivity experiments [1],[2]. More recently, the appearance of focused ion beam (FIB) systems and boron ion implantation techniques have provided experimentalists with new fabrication tools which have the attractive feature that they combine deposition and patterning in one step.

Microlithography is widely used in the semiconductor industry for selecting specific substrate areas for metallization, doping, etching, etc.. Precise patterning of features with a resolution of less than 1 micron is possible with modern lithographic equipment. One complication associated with diamond anvils, however, is that only the relatively flat culet of a diamond anvil can be patterned using standard projection lithography equipment because of their very small depth-of-focus (typically a few microns or less). Therefore, other methods must be used to continue a lithographic pattern down the sides of an anvil such as shadow masking or 3D laser pantography.

This paper will describe the various fabrication tools that are currently being used by experimenters to customize diamond anvils, as well some examples of the use of customized anvils in experiments.

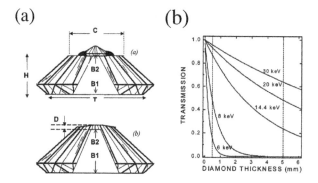

Figure 2. *(a) Laser drilled diamond anvils for x-ray experiments by Dadashev, et. al. [3]. (b) X-ray transmission versus diamond thickness for various x-ray energies. Reprinted with permission from Rev. Sci. Instrum., 72, 2633, 2001. Copyright 2001, American Institute of Physics.*

2 Laser-drilled diamond anvils

Laser drilling diamond anvils in order to improve the transmission of low energy x-rays through the anvils and to reduce background levels for Raman, UV, IR, and Mossbauer experiments has been performed [3]. For example, a dramatic improvement in the transmission of 8 keV x-rays from much less than 1% to approximately 40% can be achieved by reducing the effective anvil thickness from 5 mm to about 0.5 mm. Figure 2a shows two laser drilled diamond anvil designs used by Dadashev, et. al., [3] in their experiments, the first design being a composite of two anvils which are mated together.

3 "Designer" diamond anvils

By combining several fabrication techniques (e.g., metal deposition, lithography, chemical vapor deposition, and plasma etching) in succession, it is possible to produce diamond anvils with embedded electrical circuits for electrical conductivity, magnetic susceptibility, and high-pressure and –temperature experiments. Figure 3 shows a simplified schematic diagram of a designer anvil with thin-film metal electrodes which are embedded in a diamond layer 10-50 μm thick.

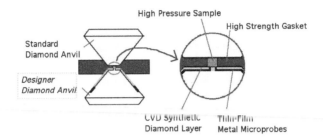

Figure 3. *Simplified schematic of a designer diamond anvil.*

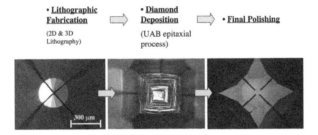

Figure 4. *Summary of the steps involved in designer anvil fabrication.*

4   Designer-anvil fabrication process steps

4.1   Lithography

Fabrication of a designer diamond anvil starts with the lithographic fabrication of electrodes onto the surface of a diamond anvil. Tungsten is an attractive choice for the microprobes for several reasons. First, as a carbide-forming metal, tungsten forms a very strong bond to the diamond anvil substrate. Secondly, since tungsten is a refractory metal (T_{melt}=3410 °C), it is capable of surviving the high substrate temperatures required for epitaxial diamond deposition (∼1000 °C) without melting or diffusing into the surrounding diamond. Tungsten also has a relatively low coefficient of thermal expansion (α_W=4.5x10^{-6} °C^{-1} @ RT), and so thermally induced stresses between the tungsten film and the diamond substrate (α_{DIA}=1.5x10^{-6} °C^{-1} @ RT) are minimized. Finally, interfacial stresses induced by mismatched elastic constants under high pressures are minimized because, like diamond, tungsten has an extremely high bulk modulus (B=308 GPa vs. diamond's B=442 GPa) and has no pressure-induced structural phase transition to at least 4 megabar [4],[5].

Since the electrode pattern must extend about 1 mm down the sides of an anvil, a 3D lithographic process, laser pantography, is used to fabricate the

Lithography on Diamond Anvils

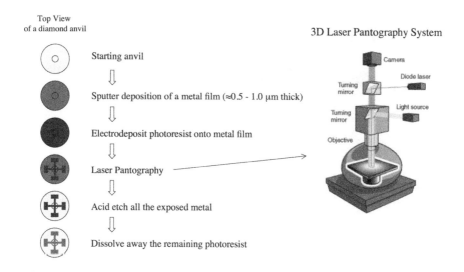

Figure 5. *Summary of the steps involved in patterning a metal film onto a diamond anvil by 3D laser pantography.*

extended circuit lines. Figure 5 summarizes the steps involved. A tungsten film approximately 0.5 μm thick is sputter deposited over the entire anvil, and then a layer of negative photoresist is electrodeposited onto the metal film. The laser pantography system then 'writes' the desired probe pattern onto the photoresist with a focused laser beam. This exposure is accomplished by means of a computer controlled x-y-z stage which moves the anvil relative to the laser's focal point, with movement in the z-direction being used to ensure that the beam is always focused on the photoresist layer.

The photoresist film is polymerized by exposure to light and so exposed areas become relatively insoluble to a chemical developer solution. Thus, after the development step, a photoresist pattern in the shape of the desired probe pattern remains on the metallized anvil surface, with exposed metal surrounding the photoresist pattern. After acid etching the exposed metal and dissolving away the remaining photoresist with acetone, the desired metal probe pattern is left on the anvil. Metal lines with widths as narrow as about 20 μm can be fabricated in this way. This linewidth is satisfactory for metal lines leading from the culet down to electrical contact pads on the sides of the anvil, but for fabricating metal lines on the culet a smaller feature size is desirable.

Standard 2D Projection Lithography used for the Culet

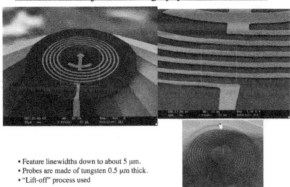

• Feature linewidths down to about 5 μm.
• Probes are made of tungsten 0.5 μm thick.
• "Lift-off" process used

Figure 6. *A tungsten microcoil lithographically fabricated onto the 300 μm diameter culet of a diamond anvil. The tungsten lines are 5 μm wide and 0.5 μm thick.*

Features on the culet, then, are fabricated using a different lithographic process, projection lithography, in which a photoresist layer is deposited onto only the relatively flat culet, and then exposed to UV light through a lithographic mask pattern which is projected onto the photoresist layer using a projection aligner lithography system. In this way, linewidths of less than 5 μm are possible, which is suitable for fabricating more complex patterns such as a tiny microcoils onto 300 μm diameter culets. Figure 6 shows a microcoil pattern fabricated onto a culet using projection lithography.

4.2   Chemical vapor-deposition of diamond

After fabricating a metal circuit onto the surface of a diamond anvil, the next step is to encase the circuit in an epitaxially deposited diamond film using microwave plasma chemical vapor deposition (MP-CVD) (Figure 7). As mentioned previously, epitaxial growth means that the diamond film grows as a matched, crystallographic extension of the starting diamond anvil, thus ensuring excellent film adhesion and strength. Epitaxial diamond was deposited onto the anvil substrate by MP-CVD using a 2% methane in hydrogen gas mixture[6]. The microwave magnetron source operates at a frequency of 2.45 GHz and a power of about 1000-1100 W. Previous experiments have shown that this system is capable of growing homoepitaxial diamond onto a diamond anvil substrate at a substrate temperature of approximately 1000 °C and a chamber pressure of 90 Torr. Typical diamond film growth rates are about 15 μm/hr, and the final as-grown film thickness is normally 40-70 μm.

CVD Diamond Deposition System at the University of Alabama

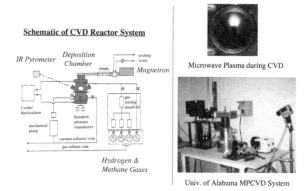

Figure 7. *Simplified schematic of the microwave plasma chemical vapor deposition system (MP-CVD) at the University of Alabama at Birmingham. Reprinted by permission of Taylor & Francis. Copyright 2006.*

Although the presence of a metal circuit pattern on the anvil during MP-CVD growth might be expected to adversely affect the CVD growth process and diamond film adhesion, high pressure experiments performed thus far with designer anvils have found no noticeable weakening of the diamond anvils due to the addition of a metal circuit and a diamond film. Also, the presence of a metal circuit pattern on a diamond anvil appears to have little affect on the epitaxial nature of the MP-CVD diamond film growth. Because the rate of nucleation and growth on clean, smooth metal surfaces is very low, diamond growth proceeds on the exposed diamond surfaces of the anvil and then bridges over the narrow metal circuit lines of the anvil, completely encasing them in diamond. Usually, the diamond film over the circuit lines appears to be completely seamless, although sometimes faint seams can be seen running along some of the length of the circuit lines, possibly indicating the presence of low-angle grain boundaries that formed as a result of the diamond film growing over and meeting from opposite sides of a narrow circuit line.

The low rate of diamond nucleation and growth on clean, smooth metal surfaces also has a practical benefit. By making the electrical contact pads relatively large (e.g., > 250 μm) compared to the thickness of the diamond film on the sides of the anvil, the diamond growth is unable to completely bridge over and cover these contact pads, thus making it easy to make electrical connections to the embedded electrical circuits by attaching wires to the exposed contact pads using silver epoxy.

4.3 Diamond polishing

A final step is to polish the rough as-grown diamond layer on culet of the designer anvil to smooth the surface and shape the culet so that is suitable for high-pressure experiments. Polishing also serves to expose portions of the electrical circuit on the culet so that electrical connections can be made to the sample or to other parts of the designer anvil. This process of exposing and removing excess metal by mechanical polishing is well known in the semiconductor industry as a "Damascene process". The polishing step must be performed very carefully in order to avoid overpolishing the anvil and exposing the metal circuit pattern under the thin diamond layer on the culet.

5 Types of designer anvils

At present, three different types of designer anvils have been fabricated for electrical conductivity, magnetic susceptibility, and high-temperature experiments. Electrical conductivity anvils have been made with four to eight electrodes on their culet, and have been used in experiments up to 2.8 Mbar. For magnetic susceptibility experiments, small 10-turn magnetic sensing coils have been fabricated onto culet, the close coupling of the coil to the sample resulting in a very high signal-to-background ratio.

Efforts have also been made to use the electrical probes to deliver significant amounts of electrical power to the sample region in order to electrically heat high-pressure samples to high temperatures. This was accomplished by taking a designer anvil with eight probes and performing further metallization, patterning, and plasma-etching steps in order to fabricate a thermal insulating layer on the culet for the electrically heated sample [7].

6 "Intelligent" diamond anvils (iDAC)

Another approach to integrating electrical circuits inside diamond anvils is the iDAC method used by H. Bureau and M. Burchard, *et. al* [8]. In this method, high-dose boron implantation (in the range of 10^{16} atoms/cm^2) is used to form conductive paths a few microns beneath the diamond surface. Furthermore, by increasing the boron dose, parts of the diamond can be selectively graphitized in order to form electrical contact pads. From a processing standpoint, the boron implantation method has the attractive feature that it is based around one central tool (the boron implantation system), thus avoiding many of the potential processing, quality control, and compatibility problems which can arise with the designer anvil approach in which multiple tools (e.g., lithography, metallization, CVD diamond deposition, polishing) are required to process each anvil. On the other hand, boron implantation will tend to result in lower conductance circuits than the tungsten lines used in the designer anvil approach, although it has been reported [8] that metallic-

Pressure or Temperature Sensor Anvil

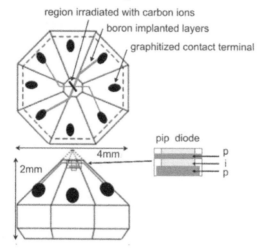

Figure 8. *Schematic of an "intelligent" diamond anvil (iDAC) with a boron implanted PIP diode which can be used as either a pressure sensor or a temperature sensor [8]. Reprinted by permission of Taylor & Francis. Copyright 2006.*

like conductivities can be achieved through boron doping by using extremely high dose levels.

Figure 8 shows an iDAC anvil with an integrated diode center which was fabricated in its culet by boron implantation. The integrated sensor is a PIP diode, in which two boron implanted diamond layers (i.e., "P" or hole doped diamond) sandwich a thin insulating layer of intrinsic, undoped diamond. This anvil also illustrates the excellent depth control of the ion implantation method: By controlling the energy of the boron ion beam, the two boron doped layers could be precisely placed at different depths beneath the diamond surface, forming an overlapping region separated by a thin, undoped diamond layer. Electrical conduction across the thin insulating layer is sensitive to both the pressure and the temperature of the structure, which has obvious applications to high-pressure and –temperature experiments. Figure 9 shows the resistance of the sensor as a function of both pressure and temperature.

7 Integrated circuit technique using alumina films

Other recent work in the fabrication of anvils with integrated circuits includes the technique of Han et al and Gao, et. al. [9, 10] in which a molybdenum

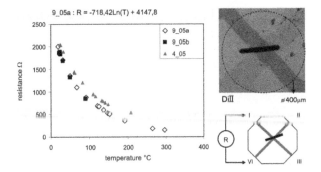

Figure 9. *Resistance versus temperature plot of a PIP diode on an iDAC. Also shown is a photomicrograph of the PIP diode showing two overlapping boron-inplanted lines at different depths [8]. Reprinted by permission of Taylor & Francis. Copyright 2006.*

metal layers are deposited onto anvils with deposited films of alumina used for electrical and thermal insulation. Lithography is used to pattern both the metal and alumina films. Figure 10 shows the process steps used. This technique has been used to perform room-temperature electrical conductivity experiments up to 106 GPa, and has also been used for high-pressure and –temperature conductivity experiments [11].

8   Focused ion beam (FIB) systems

A relatively new tool which has attracted some interest for fabricating metal patterns on diamond anvils is the focused ion beam (FIB) system. In effect, the tool utilizes an ion beam induced chemical reaction to 'write' conductive paths onto a substrate. In this process of ion-assisted chemical vapor deposition (IACVD), an organometallic vapor (e.g., $C_9H_{16}Pt$) is introduced into a deposition chamber containing the substrate, creating a thin, adsorbed layer on the substrate. A focused beam of gallium ions on the substrate causes the adsorbed organometallic layer to locally decompose, depositing a thin metal layer onto the substrate. Figure 11 shows a small platinum x-ray marker measuring 2.8 μm x 4 μm x 0.5 μm which was deposited inside a gasket hole with a FIB system.

Another obvious application of FIB systems is to deposit electrical pathways onto the culets of diamond anvils, for possible use in conductivity experiments. Finally, in addition to depositing metal layers onto substrates, a FIB's ion beam can also be utilized to ion mill and section substrates. This suggests future possibilities involving the use of FIB systems for precision micro-machining of the culet of a diamond anvil in order to, for example,

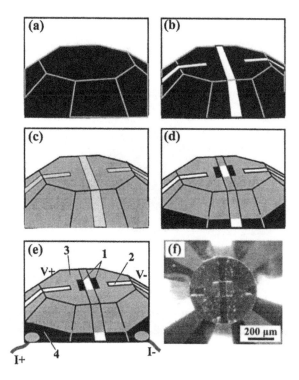

Figure 10. *The lithographic fabrication of electrodes with an alumina insulating film according to the process of Han et al and Gao, et. al [9, 10]. Reprinted with permission from Appl. Phys. Lett., 86, 64104, 2005. Copyright 2005, American Institute of Physics.*

enlarge the sample chamber or perform "gasketless" experiments with hydrothermal diamond anvil cells [12].

9 Further examples of the use of customized anvils in high-pressure experiments

9.1 Electrical conductivity of ferropericlase

The electrical conductivity of ferropericlase was measured to over 1 Mbar using a designer diamond anvil with six tungsten probes [19]. A polycrystalline sample of $(Mg_{0.75}Fe_{0.25})O$ was loaded into a 60 μm diameter sample chamber of a DAC along with a small 5 μm ruby sphere for pressure measurements. Figure 12 shows images of the designer anvil used in this experiment. The amount of electrical probe-to-probe leakage was very low (probe-to-probe re-

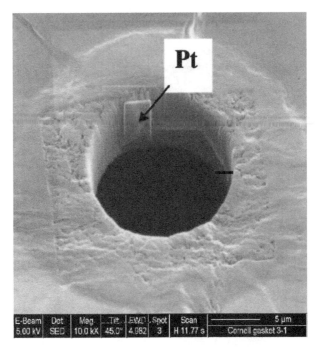

Figure 11. *A platinum pressure marker deposited inside a gasket hole by means of a focused ion beam (FIB) deposition system. [18] The platinum marker measures 2.8 μm x 4 μm x 0.5 μm. Reprinted with permission from Appl. Phys. Lett., 86, 014103, 2005. Copyright 2005, American Institute of Physics.*

sistance > 10 GΩ), as required when making resistance measurements on very high resistivity materials such as ferropericlase. Incidently, the high probe-to-probe resistance of the designer anvil is yet another indicator of the high quality of the CVD diamond film on it.

A plot of the electrical conductivity of the $(Mg_{0.75}Fe_{0.25})O$ sample as a function of pressure is shown in Figure 13. The electrical conductivity increases by an order of magnitude up to 50 GPa, and then decreases by about a factor of three in the pressure range of the high-spin to low-spin transition of ferropericlase. The sample was found to be semiconducting throughout the entire pressure-temperature range studied of up to 101 GPa and 500 K, with the activation energy of the low-spin phase equal to 0.26 eV at 81 GPa and equal to 0.27 eV at 101 GPa.

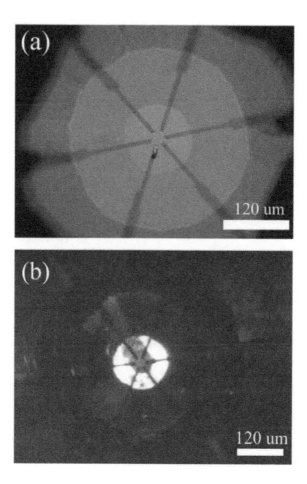

Figure 12. *(a) A 6-probe designer anvil for high-pressure electrical conductivity experiments. (b) The same designer anvil under high pressure with a ferropericlase sample.*

9.2 Magnetic susceptibility of holmium

The heavy lanthanide elements exhibit a wide variety of magnetically ordered structures at low temperatures due to an interplay between strong correlation and indirect exchange effects involving their f-electrons. The very compact lanthanide $4f$ shells result in the effective Mott-localization of these electrons and strong intra-atomic f-electron correlations, resulting in the formation of localized magnetic moments at the ionic sites. These localized magnetic moments interact with each other due to an Ruderman-Kittel-Katsuya-Yoshida (RKKY) indirect exchange mechanism mediated by the surrounding Fermi

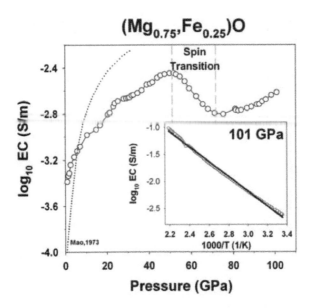

Figure 13. *The electrical conductivity of ferropericlase as a function of pressure to 101 GPa. The inset shows the electrical conductivity versus temperature taken at a pressure of 101 GPa.*

sea of conduction electrons, resulting in the appearance of magnetic order at low temperatures. The application of pressure changes the interatomic distances and can alter the net balance of aligning fields felt by the f-electrons at a given ionic site due all the surrounding ions.

Holmium undergoes two magnetic transitions at zero pressure and low temperatures: an antiferromagnetic transition (basal-plane spiral structure) at T_N=133 K, and a ferromagnetic transition at T_C=20 K. To measure the magnetic properties of this element at high pressures, magnetic susceptibility experiments were performed with a DAC by subjecting the sample to a time-varying ac-magnetic field (typically 3 Oe @ 10 kHz) while monitoring the voltage induced in a tiny magnetic sensing coil located in the culet of a designer diamond anvil next to the sample. Figure 14 shows an image of the culet of the designer anvil with a 10-turn magnetic sensing coil encased in a 10-30 μm thick layer of diamond. An attractive feature of the designer anvil approach to magnetic susceptibility experiments is that the sensing coil can be made very small and placed extremely close to the sample (within a few microns), which translates into a very high signal-to-background ratio and very good signal strength. Because of the high signal-to-background ratio, supplemental background compensation coils and electronics are not needed, and the voltage output from the designer anvil's sensing coil can be directly connected to the

Figure 14. *A multiloop designer anvil for magnetic susceptibility experiments. The outer and inner diameters of the coil are approximately 280 μm and 90 μm, respectively.*

input of a lock-in amplifier [16].

For these experiments, a holmium sample (99.9% purity, 40-mesh flakes, Alfa Aesar) was loaded into the 80 μm diameter sample hole of a gasket made from a high-strength, non-magnetic alloy (MP35N). Magnetic susceptibility data was then taken as a function of temperature by cooling the DAC in a closed cycle He refrigerator (Cryomech ST-15) at several different pressures. Figure 15 shows magnetic susceptibility data for holmium, with both the ferromagnetic and antiferromagnetic transitions clearly visible. With increasing pressure, the transition temperatures associated with both of these transitions steadily decrease. For the antiferromagnetic transition, the decrease in the transition temperature is roughly linear with pressure and drops at a rate of about 4.95 K/GPa. Both magnetic transitions disappear from view at pressures above 12.4 GPa, indicating either the loss of magnetic order or the transition into a new magnetically ordered phase with a much smaller magnetic signature.

10 Future prospects

The increasing availability of a wide variety of fabrication tools for customizing diamond anvils promises exciting possibilities for the future growth and development of new high-pressure experimental techniques and capabilities. Epitaxial diamond growth and the ion implantation of boron into diamond are two tools which are well suited to anvil customization since both of these tools preserve the single-crystal nature of the anvil, which suggests the pos-

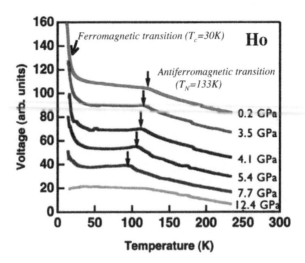

Figure 15. *The magnetic susceptibility of holmium as a function of temperature at various pressures up to 12.4 GPa [17]. The dark arrows point to the ferromagnetic and antiferromagnetic transitions of the sample. Reprinted with permission from Phys. Rev. B 71, 184416 (2005). Copyright 2005 by the American Physical Society.*

sibility of customized anvils having strengths which rival those of standard diamond anvils.

The ability to fabricate Mbar survivable electrical circuits into diamond anvil culets is a significant advance because it opens up many possible paths for future diagnostic exploration and development. Implanted electrical circuits can be used for sensing small voltages (e.g., for conductivity experiments or temperature measurements via small microfabricated thermocouples or resistive sensors), for detecting small magnetic fields (e.g., for magnetic susceptibility), and for transporting electrical power to the sample chamber (e.g., for high-temperature experiments). The work of Bureau et. al. [8] to fabricate tiny pressure and temperature sensors in diamond culets suggests further possible applications involving the measurements of thermal and thermoelectric properties at ultra-high pressures.

What are the current pressing issues for further development? For high pressure experiments such as NMR, de Haas – van Alphen, magnetic susceptibility and high-temperature electrical heating, it would be very desirable to further reduce the resistance of the embedded circuits. The detection sensitivity of a magnetic susceptibility designer anvil, for example, is limited by the Johnson noise generated by magnetic sensing coil itself. This noise competes with the tiny induced voltage signal generated in the coil. The magnitude of the Johnson noise v_n is given by

$$v_n = \sqrt{4k_B T R \Delta f}$$

where k_B is Boltzmann's constant, T is the absolute temperature in Kelvin, R is the resistance of the coil, and Δf is the frequency bandwidth over which the noise is measured. For a typical designer anvil coil resistance of 1 kΩ, the Johnson noise level at room temperature is approximately 1 nV (Assuming $\Delta f \sim 0.1$ sec^{-1}, which corresponds to about a 10 second integration time for a lock-in amplifier.). Lower coil resistances would reduce this noise floor level and enable detection sensitivity to be increased. For NMR and de – Haas van Alphen experiments lower circuit resistances may make it possible to fabricate designer anvils with high quality factor ("high-Q") resonant detection circuits. Finally, for high-temperature electrical heating experiments, lower circuit resistances mean that less waste heat is dissipated along the path lengths of the circuits, and that more power is available for heating the high pressure sample itself. Since metals which have significantly higher electrical conductivities than tungsten (e.g., copper, silver, gold, aluminum) all have melting temperatures which are too low to survive the epitaxial diamond growth process, future designer anvil efforts will likely focus on attempts to increase the thicknesses of the electrical circuit lines in order to reduce resistances.

11 Further development of CVD diamond growth technology

One particularly interesting and ambitious effort to dramatically push forward the frontiers of static high pressure technology is the growth of very large single crystal diamond anvils by Yan, et. al. [13, 14], with the ultimate goal of producing single crystal diamonds of over 100 carats. In their work, diamond is grown by microwave plasma chemical vapor deposition (MP-CVD, the same basic diamond film growth technique used for designer diamond anvils, but the growth rate is boosted from the ~10 μm /hr range used for designer anvil fabrication to the ~100 μm/hr range by introducing relatively large amounts of nitrogen, which acts as a growth catalyst, into the deposition process. The process involves the use of starting synthetic Type Ib diamond plates, with growth on the {100} surface. At present, the growth of single crystal diamonds of over 10 carats has been reported.

An interesting, unexpected characteristic of the CVD diamond grown by this process is the extremely high hardness exhibited by the diamond after high temperature annealing. The Vickers hardness of the annealed CVD diamond was in the range of ~160 GPa, considerably larger than that of annealed or unannealed diamond (~90-140 GPa)[15].

12   Acknowledgements

This work performed under the auspices of the U.S. Department of Energy by Lawrence Livermore National Laboratory under Contract DE-AC52-07NA27344.

References

[1] Grzybowski T.A., Ruoff A.L., Band-overlap metallization of BaTe, Phys. Rev. Lett.**53** 489, (1984).,

[2] Hemmes H., Driessen A., Kos J., Mul F.A., Greissen R., Synthesis of metal hydrides and in situ resistance measurements in a high-pressure diamond anvil cell, Rev. Sci. Instrum., **60**, 474, (1989).

[3] Dadashev A., Pasternak M.P., Rozenberg G.Kh., Taylor R.D., Applications of perforated diamond anvils for very high pressure research, Rev. Sci. Instrum., **72**, 2633, (2001).

[4] Ruoff A.L. , Xia H., Luo H., Vohra Y.K., Miniaturization techniques for obtaining static pressures comparable to the pressure at the center of the earth: X-ray diffraction at 416 GPa,Rev. Sci. Instrum. **61**, 3830, (1990).

[5] Ruoff A.L., Xia J., Xia Q., The effect of a tapered aperture on x-ray diffraction from a sample with a pressure gradient: studies on three samples with a maximum pressure of 560 GPa., Rev. Sci. Instrum. **63**, 4342, (1992).

[6] Catledge S.A., Vohra Y.K., Weir S.T. ,Akella J., Homoepitaxial diamond films on diamond anvils with metallic probes: the diamond/metal interface up to 74 GPa, J. Phys. Cond. Matt., **9**, 67, (1997).

[7] Weir S.T., Jackson D.D., Falabella S., Samudrala G., Vohra Y.K., An electrical microheater technique for high-pressure and high-temperature diamond anvil cell experiments, Rev. Sci. Instrum.,**80**, 13905, (2009).

[8] Bureau H., Burchard M., Kubsky S., Henry S., Gonde C., Zaitsev A., Meijer, J., Intelligent anvils applied to experimental investigations: state-of-the-art, High Pressure Research, **26**, 251, (2006).

[9] Han Y., Gao C., Ma Y., Liu H., Pan Y., Luo J., Li M., He C., Li Y., Li X., Liu J., Integrated microcircuit on a diamond anvil for high-pressure electrical resistivity measurement, Appl. Phys. Lett., **86**, 64104, (2005).

[10] Gao C., He,C., Li M., Han Y., Ma Y., Zou G., In situ conductivity measurement of matter under extreme conditions by film fabrication on diamond anvil cell, J. Phys. Chem. Solids, **69**, 2199, (2008).

[11] Li M., Gao C.-X., Ma Y.-Z., Hao A-M, He C.-Y., Huang X.-W., Li Y.-C., Liu J., Liu H.-W.,Zou G.-T., In situ HPHT resistance measurement of $(Fe_{0.125}Mg_{0.875})_2$ SiO_4 in a designed laser heated diamond anvil cell, J. Phys. Condens. Matter, **19**, 425210, (2007).

[12] Chou I-M, Bassett W.A., Anderson A.J., Mayanovic R.A., Shang L., Containment of fluid samples in the hydrothermal diamond-anvil cell without the use of metal gaskets, Rev. Sci. Instrum., **79**, 115103, (2008).

[13] Yan C-S, Vohra Y.K., Mao H-K,Hemley, R.J., Very high growth rate chemical vapor deposition of single-crystal diamond, Proc. Natl. Acad. Sci. USA, **99**, 12523, (2002).

[14] Ho S-S, Yan C-S, Liu Z., Mao H-K, Hemley, R.J., Prospects for large single crystal CVD diamond,Industrial Diamond Review, **28**, January 2006.

[15] Yan C-S, Mao H-K, Li W., Qian J, Zhao Y., Hemley, R.J., Ultrahard diamond single crystals from chemical vapor deposition,Phys. Stat. Sol. (a),**201**, R25, 2004.

[16] Jackson D.D., Aracne-Ruddle C., Malba V., Weir S.T., Catledge S.A., and Vohra Y.K., Magnetic susceptibility measurements at high pressure using designer diamond anvils, Rev. Sci. Instrum. **74**, 2467 (2003).

[17] Jackson D.D., Malba V, Weir S.T., Baker P.A., Vohra Y.K., Phys. Rev. B. **71**, 184416 (2005).

[18] Ruoff, A.L., Sun. L., Natarajan, S., Zha, C-S, and Stupian, G., Technique for X-Ray Markers at High Pressure in the Diamond Anvil Cell, Rev. Sci. Instrum. **76**, 036102 (2005).

[19] Lin, J-F, Weir S.T., Jackson D.D., Evans W.J., Vohra Y.K., Qiu W., Yoo C-S, Electrical conductivity of the lower mantle ferropericlase across the electronic spin transition, Geophys. Res. Lett., **34**, L16305 (2007).

Chapter 5
Equations of State for Solids over Wide Ranges of Pressure and Temperature

Wilfried B. Holzapfel

Department-Physik, Universität Paderborn, Germany

1 Introduction

Equations of State (EOS) for many reference materials are needed not only for the realization of a reliable *Practical Pressure Scale*, but also for many applications in *Geo- and Planetary-Sciences*. More generally, the EOS of solids are related to all the other thermo-physical properties of solids, which implies, that a reliable formulation of EOS should be based on a complete thermodynamic modelling, which includes not only $p(V,T)$, but also all the other thermo-physical properties, as for instants the *Free Energy*, $F(V,T)$, the *Internal Energy*, $U(V,T)$, the *Entropy*, $S(V,T)$ and their partial derivatives, as for instance the (isothermal) Bulk Modulus, $K_T(p,T)$, *Thermal Volume Expansion Coefficient* or, in other words, the *(thermal) Volume Expansivity*, $\alpha(p,T)$, and the *Heat Capacities*, $C_p(p,T)$ and $C_V(p,T)$. One may notice, that I have represented here these thermo-physical quantities either as functions of volume and temperature or as functions of pressure and temperature. The more convenient thermodynamic variable in applications is usually pressure, and volume is more convenient in theoretical modelling. In any case, the EOS $p(V,T)$ or its inverted form $V(p,T)$, allow to interchange these variables. One must notice however, that $F(V,T)$ is only a *thermodynamic Potential*, which gives a complete definition of the thermodynamic system, if it is represented as a function of volume and temperature. In fact, the Free Energy has here a more fundamental meaning than any of the other thermo-physical quantities, because it is related directly to the quantum statistical modelling, which starts from a calculation of the energies for all the possible quantum states of the solid, $E_n(V)$, and uses this information in the calculation of the *Partition Function*:

$$Z(V,T) = <e^{-H/(kT)}> = \Sigma e^{-E_n(V)/(kT)}$$

The Free Energy is than given by the simple relation:

$$F(V,T) = -kT \ln(Z(V,T))$$

Before I discuss the usual approximations made in the calculation of the partition function (in section 3), I present in section 2 a review of the most commonly used *Parametric EOS Forms*, which represent with temperature dependent "parameters" $V_0(T)$, $K_0(T)$, $K_0'(T)$, $K_0''(T)$ ... the pressure as a function of volume or in reversed form the volume as a function of pressure. These "parameters" denote the volume, the isothermal bulk modulus and its first and higher order isothermal pressure derivatives all for zero pressure. Mostly the difference of these values for zero and ambient pressure is much smaller the experimental uncertainty and therefore not explicitly noted.

2   Parametric EOS forms

When one compares the various analytical forms for parametric EOS given in the literature, one may distinguish the three different types give below:

2.1   Invertible EOS

These forms can be inverted analytically from $p(V)$ to $V(p)$, which is useful in special applications, but results in restrictions on the analytical forms, which limit the range of applicability very seriously. In other word, the extrapolation of these forms beyond the range of the fitted data leads usually rapid to strong divergence with respect to any reasonable extrapolation.
The simplest of these parametric forms was derived by Murnaghan [1], assuming just a linear pressure dependence for the bulk modulus:

$$K(p) - K_0 + K_0' \cdot p$$

or in other words: $K_0'' = 0$.
By integration he obtains

$$p = (K_0/K_0') \cdot ((V_0/V)^{K_0'} - 1)$$

and the inverted form

$$V = V_0 \cdot (1 + (K_0'/K_0) \cdot p)^{-1/K_0'}$$

Since this form includes basically two free parameters, I will refer to this form as **Mu2**.

Freund and Ingalls [2] allowed for a finite value of $K_0'' \neq 0$ and obtained by integration the third order form **FI3**:

$$p = 1/b \cdot [\exp((1/a) \cdot (1 - (V/V_0)^{1/c})) - 1]$$

and the inverted form:

$$V = V_0 \cdot [1 - a \cdot \ln(1 + b \cdot p)]^c$$

with

$$a = (1 + K_0')/(1 + K_0' + K_0 K_0'')$$
$$b = (K_0'/K_0) - K_0''/(1 + K_0')$$

and

$$c = (1 + K_0' + K_0 K_0'')/(K_0'^2 + K_0' - K_0 K_0'')$$

Although **FI3** may fit experimental data better over wide ranges in pressure, it diverges rapidly with respect to more reasonable form on extrapolation to higher pressures, as we will see in the comparison with other forms later. Another more reasonable second order invertible EOS form was proposed by Davis and Gordon [3]. I call this form **DG2** with

$$p = K_o \cdot (1 - (V/V_0))(V/V_0)^{-2}(1 - (1/2)(K_0' - 3)(1 - (V/V_0))$$

and

$$V = V_0 \cdot (K_0' - 1)/(K_0' - 2 \pm \sqrt{1 + 2(K_0' - 1) \cdot (p/K_0)}$$

with $+$ for $V < V_{rp}$, $-$ for $V > V_{rp}$ and the spinodal-volume

$$V_{rp} = V_0 \cdot (K_0' + 1)/K_0'$$

Even more complicated forms can be found in the literature [4], however, there are only very limited applications, where invertible forms are really needed [5].

2.2 Finite strain EOS

Based on the theory of finite strain, Birch [6] introduced a series expansion of the strain energy in terms of Eulerian strain and obtained:

$$p_{BEL} = (3/2) \cdot K_0 \cdot x^{-7} \cdot (1 - x^2) \cdot (1 + \sum_2^L c_k \cdot (x^{-2} - 1)^{k-1})$$

whereby $x = (V/V_0)^{1/3}$ is used for convenience. This Birch-Equation of the order L is called here **BEL**. In most cases, only the second order form **BE2** is used, however in the later comparison with other forms also the third order form **BE3** will be applied.

2.3   Effective potential EOS

When one consider a close packed solid with nearest neighbour interactions
only, one can start from any reasonable interatomic potential to derive a
pressure-volume relation for this idealized solid at zero temperature. Mainly
two different types of interatomic potentials had been used in this approach
initially: On the basis of the Mie-potential [7], Grüneisen [8] proposed a very
flexible three parameter form **MG3**, which is represented by

$$p_{MG3} = (1/n) \cdot K_0 \cdot (V/V_0)^{-m} \cdot (1 - (V/V_0)^n)$$

The two exponents m and n determine $K_0' = 2m - n$. For $m = 7/3$ and $n = 2/3$
this form becomes equivalent to the first order form **BE1** with $K_0' = 3$.
Although various other interatomic potentials had been used to derive some
other reasonable EOS forms as discussed for instance recently [9], only the
form based on the Rydberg potential [10] and presented first by Stacey et al.
[11] will be discussed here in more detail as second order effective Rydberg
form with the label **ER2**, because it was advertised later as "universal" EOS
[12] without reference to Rydberg and with many false statements regarding
its universality. With $x = (V/V_0)^{1/3}$ and $c_{BR2} = (3/2)(K_0' - 1)$ this form is
given by

$$p_{ER2}(x) = 3 \cdot K_0 \frac{1-x}{x^2} \exp(c_{ER2} \cdot (1 - x))$$

Due to the fact that this form shows the wrong asymptotic behaviour at
very strong compression, the exponent in the leading term was later change
[13] from 2 to 5, the parameter $c_0 = -\ln(3 \cdot K_0/p_{FG0}$ with $p_{FG0} = a_{FG} \cdot$
$(Z/V_0)^{5/3}$ for the pressure of a Fermi gas with the electron number Z in the
volume V_0 and the Fermi gas parameter $a_{FG} = 0.02337 GPa \cdot nm^5$ replaced
the previous parameter c_{ER2} of ER2 and an adapted polynomial expansion
(with the order L) was added [14] to increase the range of applicability for
this form APL:

$$p_{APL}(x) = 3 \cdot K_0 \frac{1-x}{x^5} e^{c_0 \cdot (1-x)} \cdot \left[1 + \sum_{k=2}^{L} c_k \cdot x \cdot (1-x)^{k-1} \right]$$

This form shows the correct asymptotic behaviour at very strong compression
and allows for an analytic integration, which results in the corresponding free
energy [15]:

$$E_{APL}(x) = \frac{9V_0 K_0}{2x^5} e^{c_0 \cdot (1-x)} \cdot \{1 - (c_0 + 2 - 2S_L) \cdot x \cdot FE1(c_0 \cdot x) - 2 \cdot x \cdot D_L(x)\}$$

with

$$S_L = \sum_{k=2}^{L} c_k \cdot (1 + k/c_0)$$

$$FE1(x) = 1 - x \cdot \exp(x) \cdot E1(x)$$

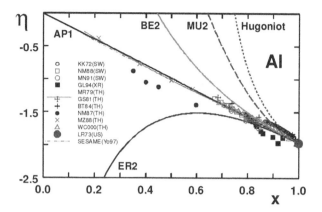

Figure 1. $\eta-x$ *plot for Al at ambient temperature from [16] with reduced shock wave data (SW), theoretical data (TH), x-ray data (XR), and ultrasonic data (US) from different source given in [16]. The curve labelled AP1 corresponds to the first order EOS form AP1. The second order Birch, Murnaghan, and effective Rydberg forms are marked with BE2, MU2, and ER2 respectively. The curve with Hugoniot corresponds to the smoothed original shock wave Hugoniot.*

$$E1(x) = \int_x^\infty \frac{e^{-z}}{z} dz$$

and

$$D_L(x) = \sum_{k=2}^{L} c_k \cdot p_k(x, c_0)$$

The first two polynomials in this form are given by

$$p_2(x) = (2 - x)/c_0$$

and

$$p_3(x, c_0) = 3 - 3 \cdot x + x^2 + (x/c_0)$$

respectively, and the parameter

$$c_2 = (3/2) \cdot (K_0' - 3) - c_0$$

would be zero for $K_0' = 3 + 2/3 \cdot c_0$, which means that the first order form AP1 uses values for K_0', which are always larger than 3 and increase for the heavier elements due to the increase of c_0 with Z. This trend appears to be better than a constant value $K_0' = 4$ implied by the first order Birch equation BE1 as discussed previously in more detail [16].

Due to the limited space allowed for the discussion of these different EOS forms, I can point out here only briefly, that reasonable behaviour of the free energy at largely expanded volume is reproduced by DG2, by all the Birch forms and by the effective potential form. However, the correct asymptotic behaviour under strong compression is only reproduced by APL. This fact has been discussed in detail previously [9, 14, 16] and is illustrated best in Fig. 1. with the use of a *linearization scheme* $\eta(x)$, which plots the logarithm of the scaled pressure

$$\eta(x) = \ln\left(\frac{p}{p_{FG0}} \cdot \frac{x^5}{1-x}\right)$$

over the scaled length x for Al with experimental and theoretical data for ambient temperature from many different sources cited in the original work [16]. The calculated curves for the different EOS forms use all the same (best) values for V_0, K_0 and K_0' . Obviously BE2 and MU2 diverge rapidly to higher values and ER2 to lower values with respect to the theoretical limit $\eta(0) = 0$, whereas all the forms APL meet this limit and even the first order form AP1 fits already perfectly all the data within the experimental and theoretical uncertainty. Similar plots for many other elements and more detailed discussions are given in the original work [13, 14, 16], however for the later thermodynamic modelling, I would like to point out here already, that the use of the APL form appears to give the most robust basis for this modelling of "regular" solids in wide ranges of pressure and temperature [13, 14, 16]. The difference of the smoothed Hugoniot data from AP1 corresponds to the thermal correction and can be use for an estimate of the uncertainty in the calculation of isothermal data from shock wave results.

3   Thermodynamic modeling

As mentioned in the introduction, thermo-physical properties of solids in wide ranges of pressure and temperature can be modeled within a thermodynamic framework, which starts from models for all the quantum states of the solid and uses the partition function for the calculation of a thermodynamic potential (some times also called "fundamental function" or "Gibbs potential"), which is in the present approach the free energy $F(V, T)$. In this approach one can start from the energy of the total ground state, which could be the energy $E_G(V)$ either with or without the contribution from phonon zero point motion. Most rigorously $E_G(V)$ should refer to the energy of the static lattice, since the phonon zero point motion can result in an unusual volume dependence of the free energy for quantum solids like H_2 or He at moderate pressure. With a good model for the static lattice energy one includes in the next step quasi-harmonic phonon excitations, adds anharmonic phonon contributions, defect excitations, conduction electron excitations and also magnetic excitations, if necessary. Usually, all these excitations are treated as independent, which means, that phonon-phonon and electron-phonon couplings are neglected. This procedure results in good models for "regular" solids, but one should keep in mind that strong anharmonicity (near distortive phase transition) or heavy Fermion systems may need more specialized treatments.

3.1 Static lattice

The static lattice energy is modeled in all our evaluations of thermo-physical data for solids under strong compression by the second order APL energy form AP2, discussed in section 2.

3.2 Quasi-harmonic phonons

Rigorous calculations of quasi-harmonic phonon spectra result phonon dispersion curves that are in rather detailed and correspondingly strongly structured phonon densities of states (DOS). However, the main features of the thermo-physical properties are reasonably well reproduced by much simple models. Einstein, for instance [17], replaced the complicated phonon spectrum by just one characteristic frequency. Nernst and Lindeman [18] noticed that two Einstein oscillators give much better results. Debye [19] got better results for the specific heat capacities at low temperatures with a quadratically increasing phonon DOS up to a characteristic cutoff frequency, which correspond to the "Debye temperature" on conversion from phonon frequency to a temperature scale. Although combinations of Debye spectra with several Einstein frequencies have been used for detailed modeling [20], I prefer a simple pseudo-Debye model with one effective Debye temperature and either one or two Einstein frequencies [9, 21, 22] to keep the model sufficiently simple and flexible. To avoid the inconvenient integral form used in the Debye model, I use a "modified pseudo Debye" (MPD) form for the Debye contribution to the internal energy:

$$U_{MPD}(V,T) = 3k_B \cdot \Theta(V) \cdot u_{MPD}(T/\Theta(V))$$

with the volume dependent pseudo-Debye temperature $\Theta(V)$ and the scaled internal energy form $u_{MPD}(t) = g \cdot t^4/(g \cdot a + t^3)$, which depends on the scaled temperature $t = T/\Theta(V)$ and includes a weight factor $g = 0.068$ together with the parameter $a = 0.0434$, which are selected in such a way that the Debye form is best represented at low temperature. Einstein contributions to the internal energy are modelled by the form $u_E(t, f) = f/(\exp(f/t - 1))$ with the frequency scaling factors f. Usually, this factors is determined in best fits of the specific heat at ambient pressure without any pressure dependence. In the case of Diamond [22] however, it was necessary to take into account two frequency factors f_1 and f_2 with a volume dependence of $f_2(V)$ to model the special dispersion of the Grüneisen parameters for Diamond [22]. The total quasi-harmonic phonon contribution to the internal energy is given in this case by

$$
\begin{aligned}
U_{qh}(V,T) \quad = \quad & 3k_B \cdot \Theta(V) \cdot (u_{MPD}(T/\Theta(V)) + (0.5 - g) \cdot u_E(T/\Theta(V), f_1) \\
& + 0.5 \cdot u_E(T/\Theta(V), f_2(V)))
\end{aligned}
$$

In the simple cases like Cu, Ag and Au, one may use just one fixed $f = f_1 = f_2$ only [21]. With partial differentiation and integration, one obtains from $U_{qh}(V,T)$ also the corresponding $C_{Vqh}(V,T)$, $S_{qh}(V,T)$, $F_{qh}(V,T)$ and for the case of no dispersion in the Grüneisen parameters $(\partial f/\partial V = 0)$ the quasi-harmonic thermal pressure is given by

$$p_{qh}(V,T) = \frac{\gamma_{vib}(V)}{V} \cdot U_{qh}(V,T),$$

where V is the atomic volume when $U_{qh}(V,T)$ is the internal energy per atom.

3.3   Mie-Grüneisen approach

Grüneisen noticed in his famous paper [8] that all the thermo-physical properties of solids can be well described, if one assumes that the internal energy and all the other thermo-physical properties depend only on a scaled temperature $t = T/\Theta(V)$ and no other volume dependence except a proportionality to $\Theta(V)$. This assumption implies all the mode Grüneisen parameters $\gamma_v = -\ln v/\ln V|_T$ have the same value as the (average) vibrational Grüneisen parameter $\gamma_{vib} = -\ln \Theta/\ln V|_T$. Within this so called Mie-Grüneisen approximation the "thermal" Grüneisen parameter $\gamma_{thm} = \alpha_V K_T V/C_V$ and the "baric" Grüneisen parameter $\gamma_{bar} = p_{vib} \cdot V/U_{vib}$ are all identical to γ_{vib}. However both dispersion in the mode-Grüneisen parameters as well as anharmonicity lead to explicit (different) temperature dependences in γ_{thm} and γ_{bar} due to effects, which go beyond the Mie-Grüneisen assumption.

3.4   Anharmonicity

While dispersion in the mode Grüneisen parameters leads to a smooth variation in both γ_{thm} and γ_{bar} from their common low temperature value to different high temperature values [23] anharmonicity results in different steady increases of both γ_{thm} and γ_{bar} with increasing temperature. In fact, one may distinguish two different kinds of anharmonicity [23, 24], one related to anharmonicity of the single oscillators due to an anharmonic potential shape and the other due to phonon-phonon coupling. Effects from anharmonic potential can be modeled reasonably on the basis of a classical mean field approach with additional quantum corrections [24]. Since the classical mean field approach has been use long ago [25] to evaluate the volume dependence of γ_{vib}, the same approach has been use for the anharmonic contributions only recently [24]. Basically volume, pressure, bulk modulus and a "correlation parameter" λ determine the average quasi-harmonic force constant of the mean field approach:

$$k2(V) = \left(\frac{6\hbar}{k_B}\right)^2 \cdot \sqrt[3]{\frac{V}{2}} \cdot K(V) \cdot \left(1 - \frac{2}{3}(1+\lambda)\frac{p(V)}{K(V)}\right)$$

From this force constant one obtains the average phonon frequency and the related characteristic temperature

$$\Theta_{MF}(V) = \frac{4\hbar}{3k_B} \sqrt{\frac{k2(V)}{bsf \cdot m}},$$

where m represents the atomic mass and bsf stands for the "bond screening factor", which allows to adjust $\Theta_{MF}(V)$ to the experimental value of the characteristic temperature and takes into account that the mean field potential for the moving atom in the field of the 12 nearest neighbors in a close packed solid is not just the sum of 12 equally stretched bonds but only from the once which are really stretched in a linear motion.

In a similar way a forth order force constant $k4(V)$ is calculated from the cold isotherm and its higher order pressure derivatives K' and K'' to determine an effective anharmonicity parameter

$$a_{MF}(V) = \frac{5\hbar}{24k_B} \frac{k4(V)}{k2(V)^2}$$

and its Grüneisen parameter

$$\gamma_{aMF}(V) = -\frac{\partial \ln a_{MF}(V)}{\partial \ln V}.$$

From the detailed calculation of the vibrational free energy in this classical mean field approximation one can calculate all the other thermo-physical quantities and obtains for the thermal pressure:

$$
\begin{aligned}
p_{vib}(V,T) \;=\;& \frac{\gamma_{vib}(V)}{V} U_{vib}(V,T) \\
&\times \frac{1 - \frac{\gamma_{aMF}(V)}{\gamma_{vib}(V)} a_{MF}(V) \cdot T(1 - 8.8 a_{MF}(V) \cdot T + \ldots)}{1 - 2a_{MF}(V) \cdot T(1 - 13.2 a_{MF}(V) \cdot T + \ldots)}
\end{aligned}
$$

The first order anharmonic contribution to the baric Grüneisen parameter is than within this classical mean field approximation:

$$\gamma_{bar} = \gamma_{vib} \cdot \left(1 + \left(2 - \frac{\gamma_{aMF}}{\gamma_{vib}}\right) \cdot a_{MF} \cdot T\right)$$

In practical applications [21, 22, 23, 24] we replace T by $U_{qh}/3k_B$ to take into account some quantum correction. An additional screening factor for a_{MF} allows adjusting the temperature dependence of γ_{bar} to the experimental data at ambient pressure, which means that only the volume dependence of a_{MF} is used for the extension into the high pressure regime where little is known about anharmonicity.

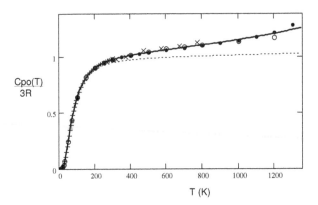

Figure 2. *Heat capacity of Cu at ambient pressure with data +, o, x, • from Ly59, Ma60, Gr72, and Ch98, respectively. The references are given in [28] with details for the fitted curves for C_{p0} and C_V0, solid and dashed curve, respectively.*

3.5   Conduction electrons

Due to the fact that in most metals the change of the conduction bands with pressure is not well known and the thermal pressure of the conduction electrons is very small anyhow, I use the usual free electron approximation with a fixed effective Fermi temperature T_f for the heat capacity of the conduction electrons (per atom) [26]: $C_{el}(T) = k_B \cdot (\pi^2/2) \cdot (T/T_F)$. Since the temperature dependence of the anharmonic phonon contribution is similar at moderately high temperatures to this contribution, the two contributions can be distinguished only at very low temperatures. Uncertainties in the modeling of the contribution from the conduction electrons could be absorbed in the fitted anharmonicity parameter.

3.6   Other contributions

It is clear that defects contribute also to the free energy especially in soft materials near melting [27], however, in our usual evaluations we do not include these contributions and do not consider the model to be very precise near melting due to the fact that our knowledge about defects is limited. Furthermore we have considered so far only "regular" solids without magnetic contribution or electronic interconfiguration crossing and would like to point out here only that these are interesting cases still for further studies.

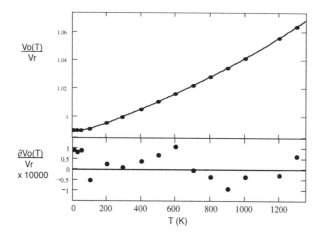

Figure 3. *Relative volume of Cu at ambient pressure with experimental data from TK75. The reference is given in [28] with details for the fitted curve. The lower part of this figure shows the deviation of the data from the fit.*

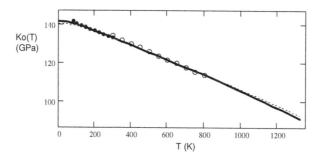

Figure 4. *Isothermal bulk modulus of Cu at ambient pressure with data o and • from CH66 and VT79, respectively. The references are given in [28] with details for the fitted solid and dashed curve representing fits with and without anharmonicity.*

4 Comparison with experimental results

This model has been applied in detail to the evaluation of thermo-physical data for Cu, Ag, Au [21], and Diamond [22]. Earlier versions of this approach without the mean field constraints on anharmonicity had been applied before to Cu, Ag, and Au [28] and also to rare-gas solids [29]. The Figures 2 to 5 illustrate just for the case of Cu the kind of precision obtained within these fits of experimental data. Since all the deviations of the experimental data from the fits are well within the experimental uncertainty, one can conclude

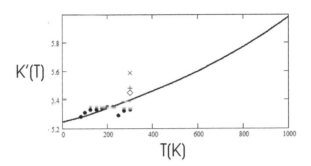

Figure 5. *Pressure derivative of the isothermal bulk modulus for Cu at ambient pressure with data . The references are given in [28] with details for the fitted curve.*

that the modeling is adequate.

5   Comparison of thermodynamic and parametric formulations

Due to the fact that the volume dependence of the thermal pressure is completely determined from the volume dependence of any cold isotherm, one must expect that the analytic form for the high temperature isotherm is different with respect to the low temperature isotherms. This means that isotherms calculated from the correct thermodynamic model may be fitted by the same analytic form as the low temperature isotherms, however, the fitted parameters for $K_0(T)$, $K_0'(T)$ will deviate from the "true" thermodynamic parameters and must be considered as "effective" parameters only. This fact has been discussed in detail recently with respect to the EOS of Cu, Ag, and Au [21], and also for Diamond [22]. I have tried two types of effective parametric forms fitting the thermodynamic data:

- AP2 with the correct $K_0(T)$ and an effective $K_0'(T)$ and

- The correct $K_0(T)$ and $K_0'(T)$ with the next higher order form AP3, which includes than an additional free parameter for an effective $K_0''(T)$.

Since both approaches give small systematic deviations in opposite directions, it was clear that the best approach results from a combination of the two. In other words, the AP3 form with the thermodynamic $K_0(T)$ and effective values for $K_0'(T)$ and $K_0''(T)$ reproduces at temperatures up to the melting curve and pressures up to 300 GPa all the thermodynamic isotherms with a precision of better than 0.1% in pressure, which is much better than the present experimental and theoretical accuracy.

6 Conclusions

I have developed software to model the thermo-physical properties of solids in wide ranges of pressures and temperatures (up to the melting curve) on the basis of experimental data for one cold isotherm at T_r with the parameters V_{0r}, K_{0r}, and K'_{0r}, on the one hand, and temperature dependent data at ambient pressure preferably from zero up to the melting temperature for the volume or thermal expansion, for the constant pressure heat capacity, for the bulk modulus and preferably also for its pressure derivative. Initially [28] we tried to use correlated least square fitting of all the parameters in this thermodynamic model, however, due to strong correlations and specific hierarchy in the parameters we noticed that a successive manual control of the fits on the basis of this hierarchy in the parameters worked best [21]. Problems with Diamond on the basis of one average vibrational Grüneisen parameter only were overcome by the use of two different Grüneisen parameters for the acoustic and optic modes modeling the dispersion in the mode Grüneisen parameters [22]. This software (written in Mathcad) is easy to handle and available on request. It could serve easily for the evaluation of a wide range of literature data on thermo-physical properties to establish a library of robust EOS data for a wide range of regular solids to be used also for the refinement of the "international practical pressure scale" (AIRAPT IPPS) [30].

References

[1] Murnaghan F.D., Finite deformations of an elastic solid, Am. J. Math. **59**, 235, 1937.

[2] Freund J., Ingalls R., Inverted isothermal equations of state and determination of B_0, B'_0 and B''_0, J. Phys. Chem. Solids **50**, 263, 1989.

[3] Davis L.A., Gordon R.B., J. Chem. Phys. **46**, 2650, 1967.

[4] Baonza V.G., Caceres M., Nunez J., Universal compressibility behaviour of dense phases, Phys. Rev. B **51**, 28, 1995.

[5] Holzapfel W.B., Refinement or the ruby luminescence pressure scale, J. Appl. Phys. **93**, 1813, 2003.

[6] Birch F., Finite elastic strain of cubic crystals, Phys. Rev. **71**, 809, 1947.

[7] Mie G., Zur kinetischen Theorie der einatomigen Körper, Ann. d. Phys. **11**, 657, 1903.

[8] Grüneisen E., Theorie des festen Zustandes einatomiger Elemente, Ann. d. Phys. **IV**, 257, 1912.

[9] Holzapfel W.B., Equations of State and Thermophysical Properties of Solids under Pressure, in High-Pressure Crystallography, Katrusiak A., McMillan P., Eds., Kluver Acad. Publ., Netherlands, 2004, p. 217-236.

[10] Rydberg R., Graphische Darstellung einiger bandenspektroskopischer Ergebnisse, Z. Physik **73**, 376, 1932.

[11] Stacey F.D., Brennan B.J., Irvine R.D., Finite strain theories and comparisons with seismological data, Geophysical Surveys **4**, 189, 1981.

[12] Vinet P., Ferrante J., Smith J.R., Rose J.H., A universal equation of state for solids, J. Phys. Condens. Matter **19**, L467, 1986.

[13] Holzapfel W.B., Equations of state for ideal and real solids under strong compression, Europhys. Lett. **16**, 200, 1991.

[14] Holzapfel W.B., Equations of state for solids under strong compression, High Press. Res. **16**, 81, 1998.

[15] Holzapfel W.B., Comment on Energy and pressure versus volume: Equation of state motivated by the stabilized jellium model, Phys. Rev. **B67**, 026102, 2003.

[16] Holzapfel W.B., Physics of solids under strong compression, Rep. Prog. Phys. **49**, 29, 1996.

[17] Einstein A., Die Plancksche Theorie der Strahlung und die Theorie der spezifischen Wrme, Ann. d. Phys. **22**, 180, 1907.

[18] Nernst W., Lindemann F.A., Untersuchungen über die spezifische Wärme bei tiefen Temperaturen. V., Sitzungsber. d. Berl. Akad., 494, 1911.

[19] Debye P., Zur Theorie der spezifischen Wärme, Ann. d. Phys. **39**, 789, 1912.

[20] Kieffer, S.W., Thermodynamics and lattice vibrations of minerals. 3. Lattice dynamics. Rev. Geophys. Space Phys. **17**, 1, 1979 .

[21] Holzapfel W.B., Nicol M.F., Refined Equations of State for Cu, Ag, and Au in the Sub-TPa-region, High Press. Res. **27**, 377, 2007.

[22] Tse J.S., Holzapfel W.B., Equations of State for Diamond in wide Ranges of Pressure and Temperature, J. Appl. Phys. **104**, 043525, 2008.

[23] Holzapfel W.B., Effects of Phonon Dispersion and Anharmonicity on thermophysical Properties of Solids, Z. Naturforsch. **63b**, 718, 2008.

[24] Holzapfel W.B., Equation of State for Solids with Mean-Field Anharmonicity, High Press. Res. **26**, 313, 2006.

[25] Barton M.A., Stacey F.D., The Grüneisen parameter at high pressure: A molecular dynamic study, Phys. Earth Planet. Inter. **39**, 167, 1985.

[26] Tipler P.A., Physik, Spektrum Akad. Verl., Heidelberg, 1994, 1356.

[27] Karasevskii A.I., Holzapfel W.B., Influence of Vibrational Anharmonicity and Vacancies on Thermodynamic Properties of the Rare Gas Crystals, Fizika Nizkikh Temp. **29**, 951, 2003.

[28] Holzapfel W.B., Hartwig M., Sievers, W., Equations of State for Cu, Ag, and Au for wide Ranges in Temperature and Pressure up to 500 GPa and Above, J. Phys. Chem. Ref. Data **30**, 515, 2001.

[29] Holzapfel W.B., Hartwig M., Rei G., Equations of State for Rare Gas Solids under Strong Compression, J. Low Temp. Phys. **122**, 401, 2001.

[30] Bean V. E., Akimoto S., Bell P. M., Block S., Holzapfel W. B., Manghnani M. H., Nicol M.F., Stishov S. M., Another Step Towards an International Practical Pressure Scale, 2nd AIRAPT IPPS Task Group Report, Physica **139 & 140 B**, 52, 1986.

Chapter 6
High-Pressure Crystallography

John B. Parise

Mineral Physics Institute, Department of Geoscience and Chemistry
Department, Stony Brook University, United States

1 Introduction

Crystallographers determine the atomic-level structure of materials. Regardless of the state of the sample, the basic theory of elastic scattering required to determine the atomic arrangement in materials is the same. Particulars of the experiment that allows us to collect reliable data appropriate for structure determination for crystalline, nano-crystalline, liquid, glassy materials etc, will be different. The compromises that accompany collection and interpretation of data when the sample is contained in a high-pressure vessel will be different from those required when the sample to collected at ambient conditions. The basic theory is the same however: diffraction or elastic scattering arises from the interference of radiation scattered from atoms, arranged periodically or not. The x-ray and neutron interference patterns contain amplitude information but not phase. Crystallographers use these patterns and employing chemical intuition, comparisons with related compounds in databases, or more commonly well-known computational techniques, deduce the arrangement of atoms and refine those locations using least squares techniques to a resolution of better than 0.01Å. A century of work by crystallographers has produced a library of structures that are used by condensed matter scientists to interpret structure-property relationship in materials as diverse as minerals, catalysts and biologically active molecules. More than 18 Nobel prizes emphasize the central role crystallography plays in a variety of research endeavors.

1.1 Resources

Several excellent resources are available in print and on the web and these provide advanced introductions to the subject of crystallography. Web resources that are likely to survive a decade or so and that are updated regularly with pointers to more specialized offerings include those of the high-pressure commission (http://www.iucr.org/iucr-top/comm/chp/index.htm) of the International Union of Crystallography (http://www.iucr.org/). Several sites compile the resources needed for modern powder and single crystal scattering, the most comprehensive of which are the Collaborative Computational Projects (CCP; http://www.ccp14.ac.uk/). Excellent textbooks provide introductions and advanced materials for powder [1, 2] and single crystal [3] techniques.

2   Technical developments

Measurement of the weak diffuse scattering in diffraction data dominated by sample Bragg scattering, and parasitic and Compton scattering from the HP cell, presents considerable challenges. Recent technical developments now allow us to study a variety of important disordered materials *in situ*, over an wide range of conditions, with unprecedented precision. These technical developments include protocols for the collection of reliable data from diamond anvil cells [4, 5] and large volume devices [6]; determining the present limits on model discrimination for nano-materials at both ambient [7, 8, 9] and high P [10], the combined use of ultrasonic and scattering techniques at HP to unambiguously reveal transitions in amorphous materials [11] and integrating the use of theory to complement data taken from HP devices [12, 13].

2.1   High energy X-ray scattering

We began using HP at 1-ID-C and 11-ID-B/C beamlines at the APS to investigate glasses, nano-crystalline materials, and other poorly crystalline materials where the application of pressure significantly influences disruptions in long-range translational symmetry [14, 15, 16, 17]. The large Q-range and penetrating power provided by x-rays with E > 60 keV ($\lambda < 0.2$Å) complements our efforts in neutron scattering [8, 30]. The 11-ID-B beamtime has also been invaluable in commissioning and testing the HP cells developed as part of the Spallation Neutron And Pressure (SNAP) beamline construction project at the SNS. The symbiotic development of new classes of pressure cells being produced with the SNS, innovative focusing optics, and more sensitive detectors allowed us to address a broad range of new scientific problems, discussed in more detail below. The XOR beamlines 1-ID-C, 11-ID-B and 11-ID-C at the APS provide a remarkable opportunity since they are, at least for now, unique worldwide in their capabilities for high pressure Quantitative Pair distribution function (QHP-PDF [5]) analysis. The 11-ID-B is the only dedicated HE beamline in the U.S. focusing solely on application of the PDF technique. An upgrade at 11-ID to an optimized undulator is complete and significantly increases the photon flux available. This is particularly advantageous for the study of disordered materials at high pressure which is a flux and brightness limited experiment [4, 5, 10]. Performing these measurements at high pressure is challenging and is only possible using strategies designed specifically for these QHP-PDF measurements, including:

1. A combined use of the Bragg scattering (Rietveld refinement) with PDF analysis on both x-ray and neutron sources provides the greatest chance of discrimination between competing models [12, 13].

2. Measurement of spectroscopic and physical properties in combination with scattering on recovered samples provides significant insights [11, 18].

3. Background corrections are critical/challenging. Care in measuring realistic blanks is essential [6].

4. For good data measurement times are necessarily long (hours at 11-ID-B and C at APS).

5. Measurements to high Q ($> 35\text{Å}^{-1}$) are required to reduce truncation ripples and to improve real-space resolution. Signals are weak at higher Q, hence the need for longer counting times. SNAP/NOMAD at SNS and 11-ID-B/C at APS are optimized for these measurements.

6. Ideally, single crystals are used in these studies [19] (Figure 1); we have utilized these whenever available and the postdoc on this proposal, Lauren Borkowski, has an ideal background in HP single crystal research, powder scattering and synthetic chemistry.

7. Theoretical approaches, Rietveld refinement and the use of Reverse Monte Carlo to analyze data, by fitting not only to the total scattering and PDF but also to the Bragg peaks, allows the recovery of partial 3D information.

2.2 Background to the pair distribution function technique

The determination of crystal structure at high pressure from Bragg scattering alone, or taking into account the total scattering, is, in principle, no more difficult than the determination of crystal structure from data collected at ambient conditions. Many of the methodologies developed at 11-ID [20, 21] have been transferred essentially unchanged, including the use of 2-D detectors, to the high-pressure studies. The pressure cell required to maintain high-P conditions usually imposes a number of compromises. Typically, parasitic scattering from thePcell, peak broadening, asymmetry and peak-shifts due to deviatoric stresses, and several other systematic errors, all compromise data quality. Obtaining reliable data requires minimizing parasitic scattering. We have made a start on this problem in a series of recent experiments using high brightness high-energy x-ray scattering we [14, 4, 5] and several other groups [22, 23] are obtaining exciting results. In the examples below, we review some of our recent work in high-pressure devices aimed at collecting data suitable for the study of the total scattering at high pressure.

The PDF, $G(r)$, gives the probability of finding an atom at a given distance r from another atom and can be considered as a bond length distribution [24]:

$$G(r) = 4\pi r \left[\mu(r) \quad \rho_0\right] = \frac{2}{\pi} \int_0^\infty Q[S(Q) - 1]sin(Qr)dQ$$

where $\rho(r)$ is the microscopic pair density, ρ_0 is the average number density, and Q is the magnitude of the scattering vector $(Q = (4\pi sin\theta)/\lambda)$. Experimentally it is not possible to measure data up to infinite Q, and the cutoff at finite values of Q_{max} decreases the real space resolution of the PDF. This causes some aberrations in the form of 'termination ripples' which propagate through $G(r)$ as high frequency noise. For both x-ray and neutron scattering experiments, high energies are required in order to access high values of Q_{max} to obtain the most accurate Fourier transform of the reduced structure function $F(Q)$.

The PDF is obtained from the Powder Diffr. (x-ray or neutron) via a Fourier transform of the normalized total scattering intensity, $S(Q)$:

$$S(Q) = 1 + \left[I^{coh}(Q) - \sum c_i |f_i(Q)|^2\right] / \left|\sum c_i f_i(Q)\right|^2$$

Where $I^{coh}(Q)$ is the coherent scattering intensity per atom, c_i is the atomic concentration and f_i is the x-ray scattering factor for species i. Conventional high real-space resolution measurements typically make use of energy resolving point detectors, such as high-purity germanium, that are scanned over wide angular ranges. These measurements are very slow and generally are unsuitable for studies in the diamond anvil cell where sample volumes are typically less than $10^{-2}mm^3$. For large volume devices such as the Paris-Edinburgh cell (Fig 1) the use of a point counter with tight collimation provides excellent signal-to-noise discrimination, at the expense of 8-hour data collection times. Recent advances in measurement strategies for quantitative PDF analysis reduced data collection times by several orders of magnitude over the conventional scanning approach based on point by point detection [4, 20], by combining high energy x-rays $(> 60keV)$ with imaging plate (IP) or amorphous Si (a-Si) area detectors to measure the scattered intensity to moderately high values of momentum transfer $(Q_{max} < 30\text{Å}^{-1})$. In combination with modified high pressure cell designs [4, 20] IP and a-Si detectors provide moderate resolution PDF data in tens of minutes for large volume devices and hours for the DAC, rather than hours or not at all. This combination of focusing, detectors and HP cells enables previously impractical parametric and in-situ studies of complex functional materials. Consequently, such studies based on this approach have become more routine/widespread and this approach has rapidly become the method of choice for HP PDF measurements

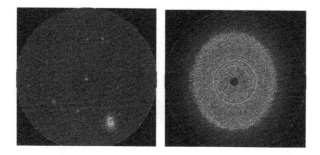

Figure 1. *Typical two-dimensional sample exposure illustrating the single crystal diamond spots alongside sample Powder Diffr. These spots, including the surrounding diffuse scatter, must be carefully masked prior to data integration. To avoid saturation of the detector and obtain optimal counting statistics, the bright spots from the diamond anvils demonstrate the need to average many short exposures. The intensity of parasitic DAC Compton scatter is best observed when comparing the central beamstop shadow with adjacent background intensity. The diamond anvil Compton scatter and sample diffuse scatter cannot be distinguished by eye.*

for a growing number of users at the APS. The practicality of high-resolution measurements ($Q_{max} > 35\text{Å}^{-1}$) and high pressure/variable temperature measurements is still limited primarily by the availability of a properly configured beamline with optics, pressure devices, detectors and methodology matched to user needs. This was very much the situation the US high-pressure community found itself in as the SNS was constructed. At that time it was decided, based on user needs, to build a high- pressure beamline configured with flexibility in mind and with a "synchrotron" philosophy with the optics, cells and detectors integrated and optimized. We very much desire the same outcome at the PDF beamline, 11-ID-B; a beamline integrated and optimized for measurements of the QHP-PDF in either large volume or diamond anvil cells, and with a user community engaged in setting priorities.

2.3   Quantitative total scattering studies at high PT

For our initial trial measurements (Figures 1 and 2) we found scattering to 20Å^{-1} is sufficient for high quality PDFs of crystalline gold reference in a relatively standard diamond anvil cell [4]. Diffraction patterns were collected using a MAR345 imaging plate detector. Data treatment included subtraction of background, determined from exposures at ambient-P without the sample in position, and exclusion of single-crystal diamond spots (Figure 2), followed by integration with Fit2D [25]. To avoid saturation of the detector on any single exposure, and to obtain optimal counting statistics, data were obtained

John B. Parise

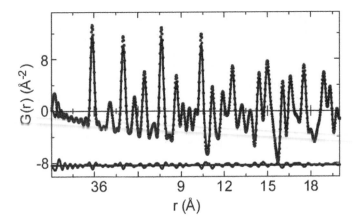

Figure 2. *A plot comparing the PDF of gold taken from a 1mm capillary (solid black) vs. that taken in the DAC at 1 bar (dash black). The difference plot below (solid black) shows no asymmetry about PDF Au-Au correlations suggesting our DAC background correction is valid.*

by averaging many short exposures. Typically, data were collected for ten 5-second exposures, which were averaged to attain optimum counting statistics. Integrated data was processed to PDFs using the program PDFgetX2 [26], where standard corrections as well as those unique to the image plate geometry were applied [20]. Full profile fitting of the PDF was performed using program PDFFIT [27].

The DAC is versatile, portable and easily interfaced to LT and resistive heating apparatus. For selected problems, large samples in the gas cell or Paris-Edinburgh cell will be required. The use of area detectors usually precludes the use of diffracted beam collimation. This is not so problematic for diamond cell studies where contributions by Bragg scattering from single crystal diamond anvils can be excluded ex post facto using software such as Fit2D [25]; the loss of information on the area detector is not so severe since integration around the Debye rings provides sufficient statistics. Some difficulty arises in the case of large volume high pressure devices (LVHPD) were components in the beam tend to give rise to parasitic scattering around the whole Debye ring [28, 29]. When measurement of the diffuse elastic contribution to the pattern is important, as they are for total scattering, subtraction techniques may eliminate the diffuse scattering component. For this reason, use of a radial collimator [30, 31, 32] along with a large sample size in a LVHPD is advantageous for quantitative studies of glassy and cryptocrystalline materials.

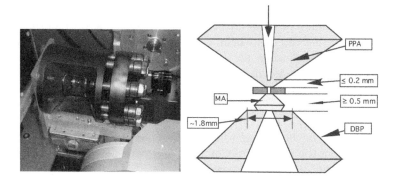

Figure 3. *(Left) moissonite (SiC) anvil cell of the type being developed by the Carnegie Institution for SNAP at the SNS and capable of pressure > 35GPa and large volume. (Right) schematic drawing of a DAC with perforated anvils. DBP: Diamond Backing Plate made of 0.25 carat diamond. Upper conical hole cut to accommodate diameter of incoming beam. PPA: Partial Perforated Anvil made of ~0.25-carat diamond with a conical hole. MA: Miniature Anvil made of ~0.05 carat diamond.*

2.4 Quantitative high pressure pair distribution function (QHP-PDF) analysis: current practice

Comparison of parameters derived from both the Rietveld technique [33, 34] and PDF refinement were consistent and demonstrated the corrections used for background and Compton scattering were valid (Figures 2 and 3; [4]). Several potential pitfalls are avoided by looking at heavy scatterers such as gold including the dominance of sample scattering compared to the coherent scattering contributions from the methanol:ethanol (4:1) pressure-transmitting medium at $80keV$.

The correlations from the alcohol pressure medium must exist however and will become more obvious when light elements are used. This will require some combination of samples being loaded in He, elimination of the pressure medium, perhaps along with heating of the sample to eliminate deviatoric stress, and the collection and proper normalization of blanks for subtraction of the pressure-medium contribution. The installation of a gas-loading (high pressure helium) device at GSECARS at the APS will greatly improve this situation. The protocol described above works well for heavy scatterers, and for lighter scatterers it is essential to properly correct for background to remove long-wavelength errors in $S(Q)$, which will Fourier transform into physically meaningless peaks in the low-r(Å) region of the PDF. Tests of the validity of our background correction included comparison of PDFs obtained from gold in a capillary vs. gold in the DAC at ambient-P; we found the difference in PDF between the capillary and DAC is minimal (Figure 3). The significant contribution to the background from diamond will need to be addressed [4] for studies of glasses, melts and light elements, since in these cases Compton scattering can overwhelm the signal of interest. One solution is to maximize sample size while not compromising P capabilities and in cases where $P > 35GPa$ are required, larger volume gem anvil cells are currently under development for the SNAP project (Figure 5) at the Spallation neutron source [35, 36]. These new devices allow a large sample volume to be taken to pressures above $35GPa$ [10] and have been used at the 11-ID beamline without focusing. In those cases where the sample is sufficiently large to allow tight collimation, the problem of Compton and parasitic scattering from high P cell components is greatly reduced. Another alternative that we adopted, and is being used by other APS users at 1-ID, is to increase signal-to-noise discrimination by removing some of the diamond from the beam path using perforated diamond anvils with the geometry shown in Figure 4. In this case, a conical hole of about $0.5mm$ maximum and $80\mu m$ minimum diameter is perforated into the diamond to within $200\mu m$ of the $350\mu m$-culet. A similar hole is made into the diamond located towards the detector and a miniature anvil set upon it. Providing a beam can be introduced down the hole, in the direction of the arrow in Figure 3, Compton scattering can be significantly reduced. It is important in this case to focus the incident beam rather than to use beam slits to define the beam size. This maximizes the x-ray flux on the sample. Beamline optics at beamline 1-ID at the APS are well matched to the studies of nano-crystalline and glassy materials at high PT. Focused x-rays with energies in the $80 - 120keV$ range provide data to $Q > 20\text{Å}^{-1}$ with standard imaging plate geometries, while minimizing background from the DAC. The high-energy x-rays at beamline 1-ID are delivered by a bent double-Laue monochromator followed by vertically focusing refractive lenses. The liquid nitrogen cooled monochromator [37] consists of two bent Si(111) Laue crystals arranged to sequential Rowland conditions and provides high flux in a beam of preserved source brilliance (divergence and size). The focusing refractive lenses, placed immediately after the monochromator, are of either the cylindrical aluminum or saw-tooth silicon [38] types, giving line-foci of 16-80 microns in vertical size at the end-station, with flux density gains in the range 6-20. The combination of focusing, perforated diamonds and area detectors have allowed QHP-PDF studies of nano-crystalline and glassy materials composed of first row transition and second main group elements. Some recent results are discussed below and typical of the quality of QHP-PDF fits is a study of nano-crystalline FeS. The composition FeS crystallizes in two modifications, the mackinawite and troilite structures [39]. The first step in the formation of iron sulfides under hydrothermal conditions is the nucleation of a reduced, short-range ordered iron monosulfide (FeSfresh) that is generally believed to be a precursor to crystalline mackinawite. Using the total scattering technique we recently confirmed [39] FeSfresh is nano-crystalline with a particle size of about $4nm$, that it is single phase and that its PDF can be modeled using the mackinawite structure. The high-P behavior of FeSfresh was unknown. Nano-FeSfresh is single phase, can be made reproducibly in the nano-crystalline form and does not require the use of capping agents for stabilization. We thought it an excellent candidate material to explore the differences in high-P behavior between bulk crystalline and nano-crystalline materials. The high-P behavior of the troilite phase is well known [40] transforming at about 4.6 and 7.2 GPa to the so-called MnP and FeS-III phases, respectively. The 3 HP phases are easily distinguished in the case where highly crystalline troilite is used as a starting material. All three polymorphs possess structures related to the NiAs-type and are distinguished by the presence of superlattice reflections resulting from atomic displacements from positions in the aristotype NiAs-related phase. In the case of neutron scattering [40] the differences are even more obvious as the transitions to MnP and FeS-III are accompanied by changes in the long range magnetic order resulting in large changes in magnetic scattering at low Q.

For nano-crystalline mackinawite, sharp features in the diffraction pattern collected at $9.1 GPa$ using the protocols described above, occur at positions expected for the sub-lattice reflections of the NiAs-related phases, troilite, MnP-type and FeS-III, suggesting the coordination number of iron has increased from 4 to 6. As expected, attempts to fit mackinawite- and troilite-related models to these data were unsuccessful. At $P < 7GPa$ we would expect FeS-III to be the stable polymorph of FeS and indeed this structure [40, 41] provides a better fit to the data than either troilite or the MnP-related structure (Figure 5). The differences between the fits for MnP-type and FeS-III models however are subtle and underscore the need for the collection of the highest possible quality data and to combine HEX data with total neutron scattering.

2.5 Quantitative high pressure pair distribution function (QHP-PDF) analysis: next steps

Results obtained using focused high-energy beams, and modified HP cells demonstrate quantitative data suitable for PDF analysis and the derivation of refined structure models can be obtained. The current set-ups (Figs. 1 and 4) for studies with focused beams at 1-ID and unfocused beams at 11-ID cover a very broad range of important science. Greater gains are possible, especially in the area of low Z-materials, with a modest investment in personnel to implement improvements in experimental design that will improve signal-to-noise discrimination by

1. increasing sample volume at mega-bar pressures and

2. drastically reducing the Compton background from the diamonds by employing energy discrimination diffraction optics.

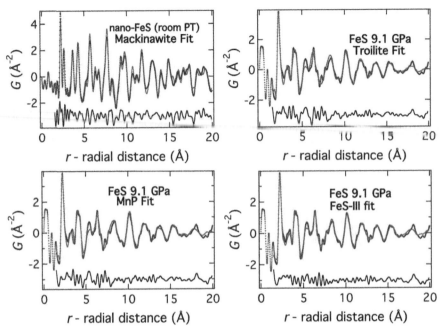

Figure 4. *Fits to the $G(r)$ for nano-crystalline FeS (mackinawite) at room P and T (top left) and the fit to the $G(r)$ obtained from data collected at 9.1 GPa (Figure 7) using the model for troilite (top right), the stable phase at ambient P for bulk FeS shown in Figure 5a, and using the model for the MnP-related form (bottom left) and FeS-III the stable phase above about 4.6 GPa and 7.2 GPa, respectively ([41, 40]). The experimentally determined $G(r)$ in each case is shown as a dotted line and the model as a continuous line. The difference curve (black line) is plotted below and on the same scale as the experimental and model-derived curves.*

2.5.1 Increasing sample volume

We have developed and successfully tested a two-phase composite gasket that allows nearly twice the volume of sample at pressures approaching a megabar [42]. The gasket utilizes well-sintered and preferentially oriented graphite produced for use as neutron monochromators. The graphite, machined into very small disks with a specialized electronic discharge machine, is loaded into the diamond anvil cell, reinforcing the metallic gasket by increasing its shear-strength. In initial tests we find the sample volume increase by about a factor of 2- 4, compared with samples loaded in metal gaskets above about 30 GPa. The gasket will assist studies examining the total scattering of materials with low electron densities, such as, $SiO2$, $CaSiO3$, and $MgSiO3$ at $P > 50 GPa$.

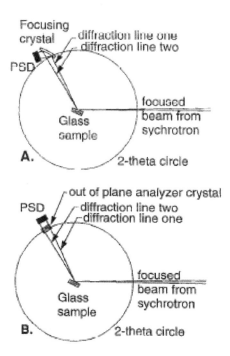

Figure 5. *Schematic diagrams of (A) focusing crystal, position sensitive detector (PSD) method. (B) Out-of-plane analyzer crystal, PSD method. A Suitable PSD include IP and a-Si detectors used at APS.*

2.5.2 Energy discrimination without sacrificing maximum pressure

For QHP-PDF studies of important weakly scattering systems, such as H_2O in the DAC, a new approach may be advantageous. Compton background from the diamonds is a serious problem, and can only be partly offset with the use of perforated diamonds (Figure 4). Another possibility is to use approaches pioneered by Beno [43] (Figure 6) in which a curved perfect crystal analyzer is used to simultaneously diffract multiple powder lines into a position sensitive detector. This technique [43] has been shown to possess high resolution, low background, and very high counting rates. This data measurement technique provides excellent energy resolution, discriminating against Compton scattering. A variant [43] of this technique uses a flat analyzer crystal to deflect multiple diffraction lines out of the equatorial plane (Figure 6). These techniques have not previously been tested with high-pressure devices. However the data reported by Beno et al. [43] were collected on a bending magnet at the NSLS. Using a focused high-energy beam at APS will provide, we believe, superior data on materials containing light elements, especially when coupled with the composite gasket discussed above. Carrying out trial measurements to access the utility of composite gaskets and the analyzer crystal diffraction geometry to obtaining reliable QHP-PDF will be a priority of the first year of funding. Studies of materials containing lighter atoms, such as silica, ice and glassy materials will be challenging.It will be important to optimize the experiment by balancing Q-space resolution, minimizing Compton and other parasitic scattering from the pressure vessel, and maximizing signal from the sample. Most of the equipment required for these experiments is already in place and a modest investment in personnel time adapting the "Beno approach" to HP studies will provide revolutionary new data on liquids and glasses at very high pressures.

References

[1] Klug H.P., Alexander L.E., X-Ray Diffraction Procedures: For Polycrystalline and Amorphous Materials, Wiley-VCH, 1974.

[2] Pecharsky, V. and Zavalij,P. Fundamentals of Powder Diffraction and Structural Characterization of Materials. , Kluwer, Boston, (2003) pp713.

[3] Giacovazzo, C., ed. Fundamentals of Crystallography. IUCr, Oxford University Press: Oxford., (1985) pp 654.

[4] Martin C.D., Antao S.M., Chupas P.J., Lee P.L., Shastri S.D., and Parise J.B., Quantitative high-pressure pair distribution function analysis of nanocrystalline gold. Appl. Phys. Lett., **86**, 061910, (2005).

[5] Parise J.B., Antao S.M., Michel F.M., Martin C.D., Chupas P.J., Shastri S.D., and Lee P.L., Quantitative high-pressure pair distribution function analysis. J. Synchrotron Radiat., **12**, 554, (2005).

[6] Chapman K.W., Chupas P.J., Kurtz C.A., Locke D.R., Parise J.B., and Hriljac J.A., Hydrostatic low-range pressure applications of the Paris-Edinburgh cell utilizing polymer gaskets for diffuse X-ray scattering measurements. J. Appl. Crystallogr., **40**, 196, (2007).

[7] Michel F.M., Debnath S., Strongin D.R., Ehm L., Lee P., Chupas P.J., Tarabrella C., Schoonen M.A.A., and Parise J.B., Ferrihydrite, the Iron Core in Native Horse-Spleen Ferritin. Science, submitted, (2008).

[8] Michel, F.M., Ehm L., Liu G., Han W.Q., Antao S.M., Chupas P.J., Lee P.L., Knorr K., Eulert H., Kim J., Grey C.P., Celestian A.J., Gillow J., Schoonen M.A.A., Strongin D.R., and Parise J.B., Similarities in 2-and 6-line ferrihydrite based on pair distribution function analysis of X-ray total scattering. Chem. Mater., **19**, 1489, (2007).

[9] Michel F.M., Ehm L., Antao S.M., Lee P., Chupas P.J., Liu G., Strongin D.R., Schoonen M.A.A., Phillips B.L., and Parise J.B., The Structure of Ferrihydrite, a Nanocrystalline Material. Science, **316**, 1726, (2007).

[10] Ehm, L., Antao S.M., Chen J., Locke D.R., Michel M.F., Martin C.D., Yu T., Parise J.B., Chupas P.J., Lee P.L., Shastri S.D., and Guo Q., Studies of local and intermediate range structure in crystalline and amorphous materials at high pressure using high-energy X-rays. Powder Diffr., **22**, 108, (2007).

[11] Antao, S.M., Benmore C.J., Bychkov E., and Parise J.B., Network rigidity in GeSe2 glass at high pressure. Phys. Rev. Lett., **100**, 115501, (2008).

[12] Yang L., Tulk C.A., Klug D.D., Ehm L., Martin D., Chakoumakos B.C., Molaison J.J., and Parise J.B., A New Hexagonal Phase for Pressure-Quenched Xe Clathrate Hydrate. in American Crystallographic Association Book of Abstracts (http://neutrons.ornl.gov/conf/aca2008/abstracts.shtml). Oak Ridge Tennessee: ACA Volume: Pages (2008 of Conference).

[13] Chakoumakos B.C., Yang L., Klug D.D., Tulk C.A., Martin D., Ehm L., and Parise J.B., Guest Atom Disorder in sII and sH Krypton Clathrate Hydrates. . in American Crystallographic Association Book of Abstracts (http://neutrons.ornl.gov/conf/aca2008/abstracts.shtml). Oak Ridge Tennessee: ACA Volume: Pages, (2008 of Conference).

[14] Mei Q., Benmore C.J., Hart R.T., Bychkov E., Salmon P.S., Martin C.D., Michel F.M., Antao S.M., Chupas P.J., Lee P.L., Shastri S.D., Parise J.B., Leinenweber K., Amin S., and Yarger J.L., Topological changes in glassy GeSe2 at pressures up to 9.3 GPa determined by high-energy x-ray, and neutron diffraction measurements. Phys. Rev. B, **74**, 014203, (2006).

[15] Tulk C.A., Benmore C.J., Urquidi J., Klug D.D., Neuefeind J., Tomberli B., and Egelstaff P.A., Structural studies of several distinct metastable forms of amorphous ice. Science, **297**, 1320, (2002).

[16] Guthrie M., Tulk C.A., Benmore C.J., Xu J., Yarger J.L., Klug D.D., Tse J.S., Mao H.K., and Hemley R.J., Formation and structure of a dense octahedral glass. Phys. Rev. Lett., **93**, 115502, (2004).

[17] Mei Q., Benmore C.J., and Weber J.K.R., Structure of liquid SiO2: A measurement by high-energy x-ray diffraction. Phys. Rev. Lett., **98**, 057802, (2007).

[18] Martin C.D., Chaudhuri S., Grey C.P., and Parise J.B., Effect of A-site cation radius on ordering of BX6 octahedra in (K,Na)MgF3 perovskite. Am. Mineral., **90**, 1522, (2005).

[19] Wilhelm C., Boyd S.A., Chawda S., Fowler F.W., Goroff N.S., Halada G.P., Grey C.P., Lauher J.W., Luo L., Martin C.D., Parise J.B., Tarabrella C., and Webb J.A., Pressure-induced polymerization of diiodobutadiyne in assembled co-crystals. J. Am. Chem. Soc., **130**, 4415, (2008).

[20] Chupas P.J., Qiu X.Y., Hanson J.C., Lee P.L., Grey C.P., and Billinge S.J.L., Rapid-acquisition pair distribution function (RA-PDF) analysis. J. Appl. Crystallogr., **36**, 1342, (2003).

[21] Chupas, P.J., Chaudhuri S., Hanson J.C., Qiu X.Y., Lee P.L., Shastri S.D., Billinge S.J.L., and Grey C.P., Probing local and long-range structure simultaneously: An in situ study of the high-temperature phase transition of alpha-AlF3. J. Am. Chem. Soc., **126**, 4756, (2004).

[22] Sheng H.W., Liu H.Z., Cheng Y.Q., Wen J., Lee P.L., Luo W.K., Shastri S.D., and Ma E., Polyamorphism in a metallic glass. Nat. Mater., **6**, 192, (2007).

[23] Sen S., Gaudio S., Aitken B.G., and Lesher C.E., Observation of a pressure-induced first-order polyamorphic transition in a chalcogenide glass at ambient temperature. Phys. Rev. Lett., **97**, 025504, (2006).

[24] Egami T. and Billinge S.J.L., Underneath the Bragg Peaks: Structural Analysis of Complex Materials. Pergamon Materials Series, ed. R.W.Cahn. Vol. 7. 2003, Kidlington: Elsevier. 316.

[25] Hammersley A.P., Svensson S.O., Hanfland M., Fitch A.N., and Hausermann D., Two-dimensional detector software: from real detector to idealised image or two-theta scan. High Pressure Res., **14**, 235, (1996).

[26] Qiu X., Thompson J.W., and Billinge S.J.L., PDFgetX2: a GUI-driven program to obtain the pair distribution function from X-ray Powder Diffr. data. J. Appl. Crystallogr., **37**, 678, (2004).

[27] Proffen T. and Billinge S.J.L., PDFFIT, a program for full profile structural refinement of the atomic pair distribution function. J. Appl. Crystallogr., **32**, 572, (1999).

[28] Chen J., Weidner D.J., Vaughan M.T., Li R., Parise J.B., Koleda C., and Baldwin K.J., Time resolved diffraction measurements with an imaging plate at high pressure and temperature. Reviews in High Pressure Science and Technology, **7**, 272, (1998).

[29] Chen J., Weidner D.J., Vaughan M.T., Parise J.B., Zhang J., and Xu, Y. A Combined CCD/IP Detection System for Monchromatic XRD Studies at High Pressure and Temperature, in Science and Technology of High Pressure, M.H. Manghnani, W.J. Nellis, and M.F. Nicol, Editors. Universities Press Ltd.: Hyderabad, India pp. 1035, (2000).

[30] Parise, J.B., Antao S.M., Martin C.D., and Crichton W., Diffraction studies of order-disorder at high pressures and temperatures. Powder Diffr., **20**, 80, (2005).

[31] Mezouar M., Faure P., Crichton W., Rambert N., Sitaud B., Bauchau S., and Blattmann G., Multichannel collimator for structural investigation of liquids and amorphous materials at high pressures and temperatures. Rev. Sci. Instrum., **73**, 3570, (2002).

[32] Crichton W.A., Mezouar M., Monaco G., and Falconi S., Phosphorus: New in situ powder data from large-volume apparatus. Powder Diffr., **18**, 155, (2003).

[33] Toby B.H., EXPGUI, a graphical user interface for GSAS. J. Appl. Crystallogr., **34**, 210, (2001).

[34] Larson A.C. and Dreele R.B.V., GSAS Manual, in Secondary GSAS Manual, Secondary Larson, A.C. and R.B.V. Dreele, Editors. Publisher: Place Published, (2000).

[35] Xu J.A., Mao H.K., Hemley R.J., and Hines E., Large volume high-pressure cell with supported moissanite anvils. Rev. Sci. Instrum., **75**, 1034, (2004).

[36] Xu J.A., Mao H.K., Hemley R.J., and Hines E., The moissanite anvil cell: a new tool for high-pressure research. J. Phys.: Condens. Matter, **14**, 11543, (2002).

[37] Shastri S.D., Fezzaa K., Mashayekhi A., Lee W.K., Fernandez P.B., and Lee P.L., Cryogenically cooled bent double-Laue monochromator for high-energy undulator X-rays (50-200 keV). J. Synchrotron Radiat., **9**, 317, (2002).

[38] Shastri S.D., Combining flat crystals, bent crystals and compound refractive lenses for high-energy X-ray optics. J. Synchrotron Radiat., **11**, 150, (2004).

[39] Michel F.M., Antao S.M., Chupas P.J., Lee P.L., Parise J.B., and Schoonen M.A.A., Short- to medium-range atomic order and crystallite size of the initial FeS precipitate from pair distribution function analysis. Chem. Mater., **17**, 6246, (2005).

[40] Marshall W.G., Nelmes R.J., Loveday J.S., Klotz S., Besson J.M., Hamel G., and Parise J.B., High-pressure neutron-diffraction study of FeS. Phys. Rev. B, **61**, 11201, (2000).

[41] Nelmes R.J., McMahon M.I., Belmonte S.A., and Parise J.B., Structure of the high-pressure phase III of iron sulfide. Phys. Rev. B: Condens. Matter, **59**, 9048, (1999).

[42] Martin C.D., Meng Y., Prakapenka V., and Parise J.B., The post-perovskite structure of MgGeO3 and geometrical constraints inherent to the perovskite structure-type. J. Appl. Crystallogr., submitted, (2007).

[43] Beno M.A., Knapp G.S., Armand P., Price D.L., and Saboungi M.L., Application of New Synchrotron Powder Diffr. Techniques to Anomalous Scattering from Glasses. Rev. Sci. Instrum., **66**, 1308, (1995).

Chapter 7
Optical Spectroscopy at High Pressure

Mario Santoro

LENS, European Laboratory for Non-linear Spectroscopy, Sesto Fiorentino, Florence, Italy and CNR-INFM CRS-SOFT, c/o Universita di Roma La Sapienza, Rome, Italy
Email: santoro@lens.unifi.it

1 Introduction

Optical, vibrational spectroscopies, e.g. Raman scattering and Infrared (IR) absorption spectroscopy, have been major tools, through the years, for investigating the physical and chemical properties of molecular materials under high pressures, along with X-ray diffraction (XRD). While the direct outcome of XRD techniques is the microscopic static structure, Raman and IR spectroscopies probe molecular dynamics, and are the leading techniques in providing direct information on the inter-atomic(molecular) interactions, which drive molecular motions. On the other hand, these interactions depend on the interatomic distances, and therefore vibrational spectroscopies also provide an indirect probe of the microscopic structure of materials, which can constrain the assignment of XRD patterns. More crucially, in some cases where the application of XRD is extremely challenging, such as those involving very thin samples of liquid and amorphous materials made of low Z elements confined in the diamond anvil cell (DAC), vibrational spectroscopy has provided a unique tool to unveil the unknown structure of new materials (see the case of amorphous, non molecular CO_2 in section 5). Also, Raman and IR spectroscopies are commonly used in the investigation of phase diagrams of materials, since the intensity, frequency and linewidth of the vibrational peaks are extremely sensitive to the fluid-solid and solid-solid phase transitions. Indeed, optical spectroscopy techniques coupled to low, room and high temperatures DACs allow one, at the state of the art, to investigate the pressure-temperature (P-

T) phase diagram of molecular systems in the 0.1-300 GPa and 4-2500 K P-T range.

In this brief review, I will try to give some basic notions on Raman and IR spectroscopies, connected to DAC techniques. Some general, and theoretical aspects will also be mentioned. Then I will describe some typical spectroscopy set-ups, used with DACs. Finally, I will discuss a few highlights on two archetypal simple systems of increasing complexity, O_2, and CO_2.

2 General aspects

The general, theoretical treatment of Raman and IR spectroscopy can be found in many textbooks (ref. [1], [2]). Here we mention a few points which are of particular relevance in discussing the application of optical spectroscopy to molecular systems at high pressures.

In the common nomenclature of solid state physics, the vibrational modes involving the internal and external degrees of freedom of the molecules, in molecular crystals, are usually classified as vibrons and lattice phonons, respectively. Raman scattering probes the time fluctuations of the total electric polarizability driven by the different modes, e.g. the volume spanned by the oscillating molecules, to which the polarizability fluctuations are proportional. In the same way, the IR absorption process probes the fluctuations of the total electric dipole, which are bounded to the instantaneous charge separation. This view provides a tool for predicting the Raman and IR activity in molecular crystals, at least in the most simple cases. As a basic example we can consider the simple X_2 molecule (X=O, N, H, I, etc.). In the stretching mode this molecule does span a volume, which supports Raman activity; on the other hand, the molecule always retains the inversion symmetry, e.g. it exhibits null instantaneous charge separation, which makes the mode IR inactive. We now consider two such molecules, mutually oriented as the opposite edge of a rectangle (D_{2h} dimer). In the *in phase* stretching mode the dimer experiences a net spanned volume and, as a consequence, Raman activity, while it always retains the inversion symmetry and the mode is IR inactive. In the *out of phase* mode, the net volume fluctuation is zero, since the extension of one molecule is exactly compensated by the contraction of the other, and the mode is Raman inactive; on the other hand, the dimer does not have instantaneous inversion symmetry any longer, and the mode is IR active. Finally, we consider the case of a molecular crystal having one such dimer in each primitive cell. Raman and IR first-order processes deal with crystalline modes where all primitive cells fluctuate exactly in phase with one another. Therefore, we only need to consider a single cell, and we simply recover the case of the single dimer. The analysis of more complex modes requires group theory methods, which are beyond the scope of this brief review. It should be clear now that Raman and IR spectroscopy can provide important indications about the structure of molecular solids, and can be used extensively in constraining the variety of new, unknown structures which are found at high-pressure conditions.

We now consider the effect of pressure on the intermolecular interaction in a very simplified model [3]. We consider the dimer described above. The internal modes are driven by strong intramolecular forces which can be modelled by hard springs. These modes are also perturbed by weak intermolecular interactions, modelled by a soft spring with elastic constant G, which depends on the intermolecular distance R. The interaction potential (vibrational coupling) energy of the system, V_{int}, is then: $V_{int} = Gx_1x_2$, where x_1 and x_2 are the internal, stretching coordinates of the two molecules, respectively. It is an easy exercise to show that the vibrational coupling is responsible for the frequency splitting of the *in phase*, Raman active mode (wavenumber ν_R) and the *out of phase*, IR mode (ν_{IR}). It results that: $G = 2\pi^2\mu c^2(\nu_R^2 - \nu_{IR}^2)$, where μ is the reduced mass of the X_2 molecule and c is the velocity of light. When a molecular crystal made of the simple dimers is under pressure, the vibrational coupling constant G in increased and, through the G, also the Raman and IR frequency splitting, because the intermolecular distance is reduced. The parameter G can be measured vs. pressure, e.g. vs. R, once the equation of state of the crystal is known. The knowledge of the function $G(R)$ provides physical insight into the nature of the coupling parameter itself (see the case of O_2, in section 4). At very high pressure, the intermolecular distances tend to approach the values of the intramolecular lengths. When this case occurs, the chemical bonds can reconstruct in order to lower the total energy of the solid, and new molecular species, or even non molecular structures, will ultimately result. This extreme circumstance is of great appeal in high-pressure science, as will be shown below in a few real cases, and optical spectroscopy is very useful in investigating the molecular transformations through the changes they produce in the vibrational spectra. Even when molecular transformations are still far from being achieved, the tendency of the molecules to rearrange their configurations can lead to phase transitions between different molecular phases. Discontinuities in the pressure (temperature) shift at constant temperature (pressure) of the Raman and IR frequencies, or in the slope of these shifts, will be observed, depending on whether the transition is first- or second-order. Therefore, optical spectroscopy has been used extensively in investigation of the high-pressure phase diagram of molecular solids.

3   The Raman and IR spectroscopy set-up

In this section we will briefly treat the main aspects of Raman and IR spectroscopy techniques devoted to DACs. First, we discuss the Raman apparatus. Figure 1 shows a schematic diagram of a typical micro-Raman set up employed with the DAC. Ion lasers are commonly used as the excitation source. These lasers provide different lines, spanning from the violet to the deep red spectral range. Ti:sapphire lasers are also employed, with tunable wavelength in the

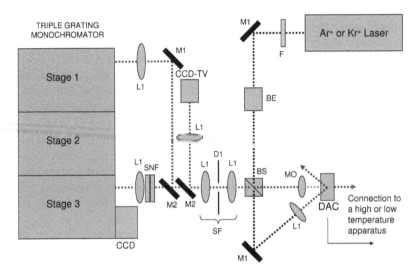

Figure 1. *Schematic of a typical micro-Raman set-up devoted to DACs. An Ar$^+$ (488.0 nm, 514.5 nm) or a Kr$^+$ (647.1 nm, 752.5 nm) ion laser is used as the Raman excitation source. The scattered beam is dispersed by a grating monochromator and detected by a CCD detector. F: laser bandpass filter; M1 and M2: aluminium coated or dielectric mirrors (M2s are revovable); BE: 2X-5X beam expander; BS: beamsplitter (removable); MO: long working distance micro-objective (10X-50X); L1: lenses (50-100 mm focal length); D1: pin-hole (30-100 μm diameter); SF: spatial filter assembly; CCD-TV: CCD video camera for visual inspection of the sample; SNF: one or two holographic super plus notch filters (Kaiser Optics).*

red/near infrared region. The availability of different excitation lines is crucial for a number of reasons. It allows one to discriminate between Raman and fluorescence peaks when new, unknown materials or phases are produced at high pressure. In contrast to the fluorescence bands, the frequency shift of the Raman peaks, with respect to the laser line, does not depend on the excitation line. Therefore, new Raman peaks can be checked by changing the laser line. Alternatively, it can be checked whether the new peaks in the Stokes side of the spectrum exhibit their anti-Stokes counterpart. Also, the blue/green laser lines excite fluorescence signals from the diamond anvils, which make a background to the Raman signal of the sample. This background is typically much broader than the Raman peaks, and is strongly reduced when a red excitation line is used, although this line can still induce some intense, narrow fluorescence peaks. On the other hand, blue/green laser lines are always to be used in very high temperature experiments (> 1000 K), where the thermally excited radiation overwhelms the spectra in the red wavelengths region.

Another crucial part of the apparatus consists of the objective lenses which focus the laser beam on the sample and collect the scattered light. The choice of these parts is constrained by the small sample size, whose typical diameter and thickness are equal to 10-200 μm and 5-50 μm, respectively. In order to optimally couple the small sample to the Raman set-up, one needs an objective with a large numerical aperture (NA). Also, the laser beam diameter should be expanded to illuminate the entire entrance lens of the objective. As a simple case we can consider a thin, focusing lens with diameter and focal length equal to 5 mm and 10 mm (NA=0.24), respectively. This lens focuses an expanded, gaussian laser beam (λ=500 nm) into an about 1 μm focal spot, with depth of focus equal to about 6 μm, which is inside the constrains posed by the gasket hole size of DACs. Actually, micro-objectives are used rather than simple thin lenses; these objectives are made of a cascade of many lenses properly designed to minimize the geometrical and chromatic optical aberrations. An additional constraint posed by the DAC is that the working distance, e.g. the distance between the sample and the external surface of the first lens of the objective, needs to be large, in the range of 15-30 mm. Mitutoyo and Nikon provide long working distance micro-objectives with NA=0.28-0.5.

We consider now which is the best scattering geometry to be adopted. In the back-scattering geometry, the incident and the scattered beam are focused and collected (collimated) by the same objective. A beam splitter is needed in this case, as indicated in the figure. This configuration allows the best transverse resolution. Alternatively, one can use a quasi-backscattering configuration, also indicated in the figure, where the incident beam is focused by a different lens. We do not need the beam-splitter now. In this configuration the Raman and fluorescence background signal coming from diamond anvils are greatly reduced. On the other hand, the limited aperture of the DAC, and the actual size of the collecting micro-objective, often reduce the maximum numerical aperture of the focusing lens, which in turn decreases the transverse resolution and the Raman signal of the sample. Whatever scattering configuration is adopted, a spatial filter on the collected optical path should then be used. In this filter, two lenses produce and collect (collimate), respectively, an image of the sample magnified by a typical factor of 10. A pinhole is posed onto the image plane for selecting that very part of image which is illuminated by the laser spot. This procedure results in a greatly reduced background signal from the diamonds. The cleaned, collimated beam is finally focused either onto a CCD-video camera for visual inspection of the sample, or onto the entrance slit of a grating monochromator.

As far as the spectrometer is concerned, we have several options. A single-grating monochromator (stage 3 in Figure 1) can be used in connection with one or two super plus notch filters (for an extensive treatment of the grating monochromator, see ref. [4]). This configuration exhibits high throughput, because the optical elements included in the monochromator are very few: only two mirrors and one diffraction grating. The notch filters are mandatory for removing the quasi-elastic Rayleigh component of the scattered signal which, otherwise, overwhelms the inelastic signal. Unfortunately, the filters have an important drawback: they completely remove the signal within a spectral region as wide as 200-300 cm^{-1} around the zero Raman frequency. This prevents measurements of low-energy Raman modes, such as low-energy phonons, or the continuous, collision-induced spectrum of fluid materials. In order to access the low frequency region of the spectrum (down to a few wavenumbers), a triple monochromator can be used in subtractive-mode, in connection with the CCD detector. The scattered signal enters stage 1, which is a single monochromator, and is dispersed by the diffraction grating; the dispersed light enters stage 2, through a slit which cuts out the Raileigh component of the spectrum. Stage 2 is a single monochromator identical to stage 1 with the grating rotated in the opposite phase, and its job is to counter-disperse the signal, e.g. to recompose it into a white beam. Finally, the cleaned beam goes to stage 3 which disperses it onto the CCD plane, thereby resolving the spectrum. The total throughput of the triple monochromator is lower than that of the single device by about one order of magnitude. This makes the single monochromator still preferable when the measurement has to be fast, such as in challenging very high-temperature experiments in DACs, where the high-temperature conditions can be supported by the diamond anvils only for short times before breaking occurs. The best frequency resolution of the spectrometer is only determined by the third stage. It depends on the focusing length of the mirrors, typically 500-750 mm, the number of lines per millimeter of the grating, typically 300-2500 lines/mm, the wavelength, and the pixel size of the CCD, typically 20 μm. An overall instrumental linewidth as low as a few tenths of wavenumbers can be achieved at 750-800 nm. The spectrometer schematically reported in figure 1 can be used either as a single or as a triple monochromator by collecting the scattered light onto the entrance slit of stage 3 or stage 1, respectively. This kind of device is provided by different companies, such as Roper-Princeton (TriVista555) or Jobin-Yvon.

As a final step of the Raman set up we consider the CCD detector. Many of these devices are available on the market (Princeton-Roper, Andor), and are either Peltier or liquid-nitrogen cooled. The detector noise is typically determined by the read-out device, and it can amount to only a few electrons per pixel. The spectral range of CCDs cover the 200-1000 nm interval, and the maximum quantum efficiency can be higher than 90 percent. CCDs with enhanced efficiency in the near UV or near IR spectral range are available, which is relevant for very high temperature or low fluorescence measurements, respectively.

Figure 1 also indicates that the DAC can be used in connection with high- or low-temperature apparatus such as a cryostat, an oven, or a laser-heating set-up, and Raman spectra can be measured *in situ*. The high-temperature techniques in DACs constitute a broad and rapidly developing field, which is beyond the scope of this brief review and are treated elsewhere in these proceedings.

We now discuss the IR spectroscopy apparatus. The Fourier Transform Infrared Spectroscopy (FTIR) is the most common techniques for measuring

IR absorption and reflection spectra of materials. The basic principles of the FTIR are very well treated in ref. [4]. The FTIR spectrometer usually employs the Michelson interferometer, or some modification of it. In this device a collimated light beam is produced by a broad band source such as a graybody lamp or a synchrotron source. The beam is split by a beam splitter, BS, into two partial beams, which are reflected by two mirrors and are again superimposed at the BS, where they interfere each other before they reach the detector in the observation plane. If one of the mirrors is translated by Δx, the path difference between the two interfering beams changes by $\Delta s = 2\Delta x$. The intensity of the transmitted light is measured as a function of the changing path difference, while one of the mirrors moves along a straight line. The power spectrum of the source is then obtained as the Fourier transform of the interferogram, which is the changing component of the detector signal, e.g. the time correlation function of the two beams. The spectral resolution of the interferometer is roughly equal to the inverse of Δs. FTIR set-ups with maximum Δs equal to 5-10 cm are commonly available, with resolution equal to 0.1-0.2 cm^{-1} (Bruker, Nicolet), which is enough for high-pressure applications. The Bruker IFS-125 is a much more expensive set-up, where the arm of the moving mirror is about 3.0 m long; this set-up is really mandatory when one is interested in measuring the rotational, fine structure of vibrational transitions in gaseous systems. In real instruments, and in contrast to the schematic description given above, many additional mirrors are employed, both flat and concave, along with diaphrams and optical filters. The resulting measured power spectrum is modulated by the spectral reflectivity of all the optical components, and the bulk absorption of the filters and the beam splitter. The sample is inserted in the region where the two beams are superimposed, after the two separated paths, e.g. between the BS and the detector; the bulk absorption of the sample modulates the measured power spectrum, I_s, as any other optical component. Once the reference spectrum, I_r, is known, which is measured after having removed the sample from the cell, the absorbance A of the sample is obtained as: $A = -\log_{10}(I_s/I_r)$. We recall that $A = \alpha d \log_{10} e$, where α is the absorption coefficient and d is the sample thickness.

The FTIR apparatus, in principle, performs well in a very broad spectral range extending from the far IR to the near UV, e.g. from 10 cm^{-1} to 50000 cm^{-1}, allowing measurements of vibrational spectra and low energy electronic spectra of materials. We note that the lowest accessible frequency with DACs is actually limited by the diffraction losses due to the small gasket hole, and is in the best conditions equal to 30-50 cm^{-1}. Different lamps (mercury, Globar, tungsten), beam splitters (mylar, KBr, Si or Al coated quartz) and detectors (bolometer, DTGS, MCT, InSb, Si) are available, and have to be optimally combined depending on the frequency region one is dealing with. The whole spectrometer has to be operated under vacuum or N_2 atmosphere, in order to remove the strong absorption lines of gaseous CO_2 and water vapor which are always present in the normal atmosphere.

When one employs DACs, an IR microscope has to be used in connection

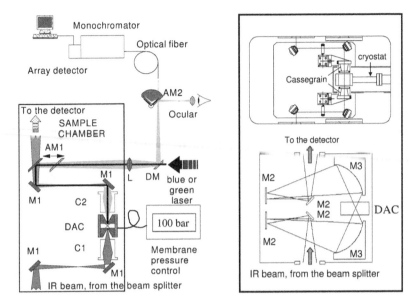

Figure 2. *Schematic of two typical micro-IR set-up devoted to DACs, inserted in the sample chamber of a FTIR intererometer (not shown). The apparatus on the left side is based on Cassegrain micro-objectives (the picture is not in scale and the objectives are only schematically drawn). This set-up is also reported on the right side, at the top of the box, where the DAC is clamped to a cryostat. The set-up on the right side (bottom) is based on ellipsoidal mirrors. M1 and M2: aluminium coated, flat mirrors; C1 and C2: aluminium or gold coated Cassegrain objectives; DM: dichroic mirror; L: lens; AM1 and AM2: removable mirrors; M3: aluminium coated, ellipsoidal mirrors.*

with the standard FTIR apparatus, in order to fit the IR beam to the small gasket hole as well as possible. The size of the beam spot that can be achieved at the sample plane is determined by the properties of the light source. With synchrotron radiation, an almost diffraction-limited small spot is obtained [5]; on the other hand, when an extended source (lamp) is used, the spot size is determined by the size of the source image through the microscope. In Figure 2, I show two typical optical configurations implementing the IR microscope, which are routinely used (High Pressure lab. at LENS) with DACs in a FTIR apparatus employing extended sources [6], [7]. In the set-up on the left hand side of the Figure, two Cassegrain micro-objectives are used for symmetrically focusing the incident IR beam on the DAC and collecting the transmitted radiation, respectively. An image of the source is produced in the sample chamber, with typical linear size of a few millimiters. This is re-focused onto the sample in the DAC through a demagnification ratio of 0.25, ending up into a final spot of about 1 mm. Ealing Cassegrain micro-objectives are used with large numerical aperture and long working distance (WD) equal to 0.28 and 24.5 mm, respectively. Objectives with NA=0.5 and WD=23.2 mm are also available. We notice that the small mirror of the Cassegrain produces an obscuration of the beam amounting to about 20 percent; on the other hand, the quality of the demagnified image is very good. Alternatively, ellipsoidal mirrors are used, as shown in the right hand side (bottom part) of the figure. In this case, the first image of the lamp, and the demagnified one onto the DAC sample, correspond to the two focal points of the ellipsoid. The demagnification ratio is 0.22. With this set-up the drawback of the obscuration is avoided, but the system is affected by relevant geometrical aberrations. It is also possible to use diaphragms for limiting the size of the source. The standard FTIR apparatus is provided with a set of diaphragms of decreasing diameter from 12.0 to 0.5 mm. This allows one to produce a final spot on the sample as small as 100-150 μm, thereby providing the capability of measuring different regions of inhomogeneous samples, confined in large gasket holes. The obvious drawback here is the intensity decrease of the transmitted beam.

Several additional set-ups to be used with the DAC can be inserted into the sample chamber of the FTIR apparatus. On one side of the Figure 2 is shown a cryostat which clamps the body of a DAC. Also, heaters of different type can be used; we have measured *in situ* the IR absorption spectrum of samples in DACs, while heating the cell up to at least 700 K. Of course, at some high-temperature point, the strong IR thermal emission from the hot DAC will saturate the detector. In Figure 2 is also shown an optical set-up for pressure measurement, based on the pressure shift of the fluorescence lines of ruby and other fluorescent materials such as Sm:YAG or SrB_4O_7:Sm^{2+}. A blue or green laser excitation source is focused onto the sample thorough a lens with low numerical aperture and one of the two Cassegrain micro-objectives. The red fluorescence light is back-collected along the same path, and coupled either to an ocular or to a monochromator+CCD apparatus, through an optical fiber.

4 Oxygen

As a simple molecular system with unique features by virtue of its spin ($S = 1$), dense oxygen is a benchmark for condensed matter theories [8]. The high-pressure phase diagram of oxygen includes up to seven solid phases (Figure 3), which are affected by spin-spin interactions. One relevant effect of these forces is that all the phases, with the exception of γ (the structure of ζ is not resolved yet), are formed by molecular layers with the molecules orthogonal to the layers. In the γ phase the molecules are affected by dynamical, orientational disorder. A variety of physical phenomena occur in the solid at high pressure, such as the anti-ferromagnetic ordering (α and δ), molecular clustering (ϵ), and metallization (ζ). These phenomena have been quantitatively investigated by means of optical spectroscopies. These techniques have been very useful in the extensive study of the phase diagram. One phase, η, was recently discovered

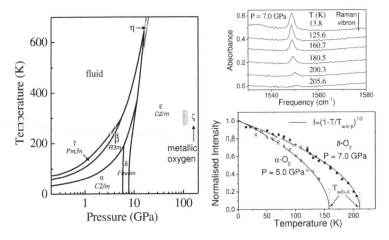

Figure 3. *Left panel: high pressure phase diagram of oxygen. The symmetry of the solid phases is indicated. The boundaries of the η phase are still debated [9], [22]. The knowledge of the true $\epsilon - \zeta$ phase boundary is limited, due to metastability. Right panel (top): absorption spectra of the IR vibron versus T along an isobaric run in δ-O_2, up to the $\delta - \beta$ phase boundary [10]. Vertical stick: position of the Raman vibron peak. Right panel (bottom): integrated, normalised intensity of the IR vibron peak measured versus T, along two isobaric runs up to the $\alpha(\delta) - \beta$ phase boundary.*

by means of Raman spectroscopy alone [9].

In the α and δ phases, beyond the Raman vibron peak, was measured an IR vibron peak, which is due to the long-range anti-ferromagnetic order (Figure 3) [10]. The magnetic, primitive cell of these strutures contains two molecules on the vertex and on the basis center of the conventional, crystallographic cell, with antiparallel spins. The Raman and the IR vibrons are the in- and out-of-phase stretching modes of the pair of molecules, respectively. Although a theoretical treatment of the interaction induced intensity of the IR vibron is still lacking, a direct, experimental probe of the anti-ferromagnetic origin of this peak is represented by the temperature dependence of the intensity. Indeed, measurements along isobaric runs have shown that the IR intensity continuously decreases upon increasing the temperature, and vanishes at the transition temperature to the β phase (Figure 3), which does not exhibit long-range order of the spins. This behavior suggests that the intensity of the IR peak is an order parameter. Also, it has been shown that the spin-spin interaction is the source of the strong vibrational coupling, which produces the large and strongly pressure dependent frequency separation between the IR and Raman vibron peaks [10]. The anti-ferromagnetic interaction between the pair of molecules (1 and 2) in the primitive cell is described by the Heisenberg Hamiltonian: $H = -J\widehat{S}_1 \cdot \widehat{S}_2$, where J (< 0)

is the exchange integral, and $\widehat{S}_1$ and $\widehat{S}_2$ are the molecular spin operators. The J depends on the intermolecular distance R and on the intramolecular lengths r_1 and r_2. If the J is expanded up to the second order in the intramolecular, stretching coordinates x_1 and x_2, we obtain a vibrational coupling term in the spin averaged intermolecular potential whose force constant G is given by: $G = - < \widehat{S}_1 \cdot \widehat{S}_2 >_T \partial^2 J/\partial x_1 \partial x_2|_{x_1=0;x_2=0}$. From measurements of the pressure shift of the Raman and IR vibron peaks, one can obtain the $\partial^2 J/\partial x_1 \partial x_2|_{x_1=0;x_2=0}$ parameter as a function of pressure, and then as a function of R (exponential behavior) since the equation of state is known (see Section 2). Also J was obtained, as a function of R, on the basis of optical spectroscopy [11]. The near IR absorption bands due to the transitions from the ground state $^3\Sigma_g^-$ to the lowest electronic excited states, $^1\Delta_g$ and $^1\Sigma_g^+$, were measured in the α, δ and β phases. The Heisenberg exchange interaction adds a negative contribution to the ground state energy, which shifts to higher values the frequency of the absorption bands with respect to the ideal, spin disordered configuration. Therefore, measurements of the frequency of these bands allows one to obtain the exchange integral J.

The ϵ or "red" phase of solid oxygen is stable in a large pressure range, up 96 GPa at room T, where metallization starts to occur. Because of dramatic color changes from deep red to brown, this phase was thought to be characterized by relevant changes in the electronic structure, since its first discovery [8]. The structure of the ϵ phase was found to be monoclinic [8], but the exact position of the molecules and the local (site) symmetry was unknown, for long time. Raman and IR spectroscopy studies provided, for the first time, evidence of molecular clustering [7], [11]. The model of a molecular crystal made of O_4 units (strong D_{2h} dimer of parallel O_2 molecules) seemed to provide a proper interpretation of spectroscopic data. The IR vibron peak discontinuously softens at the $\delta - \epsilon$, first-order phase transitions, and its intensity increases by a few orders of magnitude (Figure 4), becoming comparable to that observed in molecules with allowed IR activity such CO_2 and N_2O. This strong peak was assigned to the out-of-phase stretching mode (b_{3u}) of the two constituent O_2 molecules of the O_4 unit, which is indeed IR active. The in-phase stretching mode (a_g) was assigned to the Raman vibron, which also exhibits discontinuous softening at the $\delta - \epsilon$ phase transition. The vibron frequencies softening is a mark of the increased intermolecular interaction between the two molecules, supported by the charge transfer. This mechanism shifts some electronic charge from the intramolecular to the intermolecular region between the interacting O_2, thereby weakening the intramolecular O_2 bonds. A new, rather intense peak was then observed in the far IR region of the absorption spectrum, whose intensity increase vs. pressure is very steep between 10 and 20 GPa and levels off at higher pressures, paralleling that of the IR vibron peak. For this reason the new band was suggested to have a vibron nature itself and was assigned to the other IR active, internal mode of the O_4, which is the out-of-phase stretching mode (b_{2u}) of the new O_2-O_2 bonds. The O_2-O_2 in-phase stretching (a_g) and the O_4 in-plane bending mode (b_{1g})

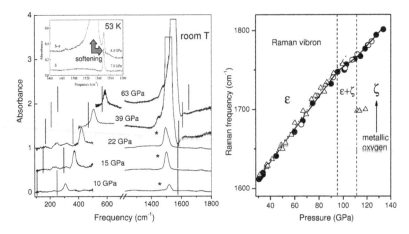

Figure 4. *Left panel: IR absorption spectra of oxygen in the ε phase [7]. The spectra marked with the asterisks are obtained on samples as thin as a few tenths of microns, while the other spectra refer to samples about 50 μm thick. Sticks: position of the Raman peaks. Inset: IR spectra in the vibron region at the δ − ε phase transition, showing coexistence and a frequency jump. Right panel: pressure shift of the Raman vibron peak in the ε and ζ phases. Full and empty dots are measured upon increasing and decreasing pressure, respectively [20]. Triangles are from ref. [19]. The lines are guides for the eye.*

are assigned to the two low frequency peaks in the Raman spectrum, which are reinterpreted as vibron peaks, while the out-of-plane bending mode (a_u) is both Raman and IR inactive. Beyond the strong changes in the vibrational spectra at the δ − ε phase transition, remarkable changes of the electronic excitation bands were also observed in the near IR/visible frequency region of the absorbtion spectrum [11]. These changes were consistently interpreted on the basis of the O_4 molecule, and the anti-ferromagnetic coupling in solid oxygen was suggested to be the driving force leading to the formation of the diamagnetic O_4 molecule. A magnetic collapse in the ε phase was also shown in a recent neutron diffraction study [12]. Only recently, structural studies based on high quality single-crystal [13] and powder x-ray diffraction [14] data have determined the true structure of the ε phase. These studies have directly demonstrated that the ε phase is actually made of O_8 units, exhibiting a vibrational spectrum which is more complex than that described above [13]. Details on the weak chemical bond holding together the O_8 cluster were recently provided by an x-ray spectroscopy study [15], where the closed shell, diamagnetic electronic structure of this unit was confirmed. We notice that *ab initio* molecular dynamics calculation have not yet been able to predict the observed structure of the ε phase to be the thermodynamically stable one [16], [17]. The ε phase is still a challenge for computational techniques.

Further, even more profound modifications of the electronic structure in solid oxygen are encountered above 96 GPa, where the insulator-to-metal, $\epsilon-\zeta$ phase transition occurs. The high-pressure metallization of oxygen was first reported on the basis of optical transmission and reflectivity measurements in the near-IR/visible frequency region [18]. In the transmission measurements, the ϵ phase exhibited optical anisotropy, since different optical density spectra were obtained with light propagating along direction perpendicular or parallel to the molecular axis. This is also observed by means of simple visual observation, showing light and deep colored facets when white light is transmitted thorough the sample. Measurements of the transmission spectra have also been showing that the optical band gap decreases upon increasing pressure and vanishes at about 110 GPa, which indicates the insulator-to-metal transition. Also, it was found that the sample reflectivity abruptly increases at pressures above 90 GPa, which was taken as an indication for pressure-induced creation of nearly free electrons, in agreement with the results from transmission measurements. Raman spectroscopy measurements from different groups [19], [20], although not quantitatively agreeing with each other, have found that the metallic, ζ phase still exhibits a vibron peak in the spectral region of the intra-O_2, O-O stretching mode, whose frequency pressure behavior is down shifted with respect to that of the ϵ phase (Figure 4). These results suggest that the ζ phase is a molecular crystal made of O_2 molecules with increased intermolecular interaction with respect to the ϵ phase. What kind of molecular cluster, or molecular polymer, if any, forms the ζ phase still remains to be determined. The full structure of this phase has not been experimentally resolved so far, while *ab initio* calculations predict monoclinic $C2/m$ or $C2/c$ structures [17], [21]. An even more intriguing open issue, and a challenge for optical spectroscopy investigations, is whether the further compression of metallic oxygen will finally end up into an atomic solid.

5 Carbon dioxide

The high-pressure phase/kinetic diagram of carbon dioxide includes up to seven solid phases and an amorphous form [23], [25], [24] (Figure 5). Although CO_2 is still a very simple molecule, the solid-solid phase transitions in this system are affected by extremely large metastabilities, which has prevented the identification of the true phase boundaries above 10-15 GPa. In this pressure range strikingly different kinetic boundaries have been found in the various studies, depending on the different P-T paths and analysis techniques. It is remarkable that beyond the well known phase-I (dry ice), all the other transformations, including the high P-T decomposition of CO_2 into carbon and oxygen [26], have been discovered, over the past two decades, by means of optical spectroscopy. Experimental and theoretical studies have shown that all the phases below 30 GPa are molecular crystals with orientational order of the intramolecular axes, although an open discussion persists on the interpretation of the XRD data on phases II and IV [23]. The most striking high-pressure phenomenon in CO_2 is the reversible transformation to non-molecular, extended solids above 30 GPa, where the thermodynamic phase boundaries are severely confused. This confusion is most likely due to the energetic competition of many extended, meta-stable structures, as can be inferred from the results of *ab initio* calculations.

The discovery of non-molecular, crystalline carbon dioxide, indicated as phase V, was achieved by means of Raman spectroscopy [27]. The vibrational data identified this phase as an extended solid with carbon-oxygen single bonds, similar to the quartz polymorph of SiO_2. In this picture, the carbon atom is tetrahedrally coordinated by oxygen atoms, forming a three-dimensional network of corner shearing CO_4 tetrahedra. The main point supporting the analogy between phase-V and quartz was the sharp peak observed at about 800 cm^{-1} (40 GPa), which was assigned to the symmetric, inter-tetrahedral stretching of C-O-C single bonds. The Raman spectrum of CO_2-V was subsequently confirmed by other experimental investigations [26], [28] (see figure 5). Only one XRD study has been performed on phase V so far [29], where it was shown that this material is indeed made of CO_4 tetrahedral units, arranged in an orthorhombic, trydimite-like structure. Also, phase V was found to exhibit a very high bulk modulus, equal to 365 GPa, which is higher than that of all the SiO_2 crystals. Theoretical studies based on *ab initio* calculations have predicted the tetrahedral structure of phase V, although a tetragonal, β-cristobalite arrangement of the CO_4 units was predicted to be much more stable than the trydimite-like one [30], [31]. Also, the predicted bulk modulus for a variety of tetrahedrally-coordinated phases ranges between 1/2 and 1/3 of the measured one. The Raman spectrum of β-cristobalite CO_2 was also calculated [30] (figure 5), and it was found to well reproduce the most intense experimental Raman peaks of phase V. On the other hand, the experimental and the *ab initio* calculated [31] XRD patterns of phase V and β-cristobalite differ. It appears that a lot of work has still to be done for reconciling all these findings and definitely solving the structure of CO_2-V. Surely, additional experimental XRD data should be concluded. It could also be the case that the Raman spectrum is not suitable for discriminating between the trydimite-like and β-cristobalite structures. Anyway, the IR absorption spectrum should be investigated as well, for providing more constrains to the interpretation of XRD data. Finally, it should be considered whether the experimental samples of CO_2-V could actually be made of a metastable mixture of different phases.

Recently, a non-molecular, extended, glassy form of carbon dioxide, which was predicted to exist by Serra et al. [32] was experimentally discovered [33], and its structure was resolved by means of optical spectroscopy [34]. The IR spectrum of the new material (a-carbonia, or a-CO_2) exhibits three prominent, broad bands: A, B, and C (figure 5). Based on the comparison between the experimental and the *ab initio* calculated spectra, it resulted that band B is mainly contributed by the antisymmetric C-O-C stretching modes of silica-like

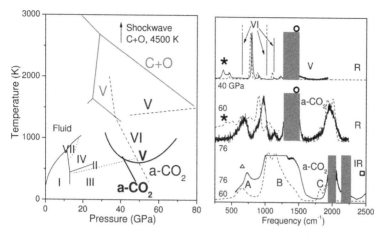

Figure 5. *Left panel: high pressure phase/kinetic diagram of CO_2 [23], [25], [24]. Continuous lines below 25 GPa are phase boundaries. Dotted lines below 30 GPa and 1000 K: kinetic boundaries for forming phase IV from VII, and II from III. Lines above 25 GPa: kinetic boundaries for forming the non molecular phases V and VI, the form a-CO_2, and the dissociated material, C+O, from molecular CO_2 (dashed [25], thick grey [26], and thick black lines [28], [33], [34] correspond to different studies). Right panel: room T experimental [28], [25], [33], [34] (continuous lines and sticks) and theoretically predicted [30], [36], [34] (dashed lines and sticks) Raman (R) and IR (IR) spectra of phase V, VI, and of a-CO_2. Stars and open symbols: approximate frequency of the lattice phonons and the internal modes, respectively, of molecular, solid CO_2. Blocks: frequency ranges of diamond bands.*

CO_4 units. Also, it results that band C is entirely due to the stretching mode of the carbonyl (C=O) units of partially reacted sites, where carbon is in a three-fold coordination. The picture is that of an extended, three-dimensional network of interconnected CO_4 and CO_3 units, where the amount of silica-like sites is equal to about 60 percent. This picture is also supported by the comparison of the experimental and theoretical Raman spectra (figure 5), which leads to the interpretation of the broad peak around 900 cm^{-1} as due essentially to the symmetric C-O-C stretching mode of the CO_4 coordination sites, while the broad peak at about 2000 cm^{-1} is entirely due to carbon-oxygen stretching mode of the CO_3 sites, analogously to case of the IR spectrum. Carbon in a higher coordination, as five and/or six, was ruled out in a-carbonia, at variance with the case of silicon in dense SiO_2 structures. Indeed, the transformation to six-fold coordinated carbon dioxide is predicted to occur in the multimegabar pressure range [35]. Enthalpic considerations suggested that a-carbonia is a metastable intermediate state of the transformation of molecular CO_2 into fully tetrahedral phases. The amorphous, extended character of a-carbonia was confirmed by XRD [33]. On the other hand, due to experimental limitations, the direct structural information was rather qualitative and inadequate to univocally assess the local structure of a-CO_2. Therefore, this is one important case when optical spectroscopy is the crucial key for determining the microscopic structure of a new material.

Another non-molecular, crystalline phase, CO_2-VI, was recently found, in coexistence with a-CO_2 [25], (figure 5). The Raman spectrum of this phase mixture exhibited three new sharp lines, with the strongest one at about 1020 cm^{-1}, superimposed on the broad bands of a-carbonia. Although CO_2-VI was identified as an extended structure of carbon in sixfold coordination, subsequent *ab initio* calculations have shown that non-molecular, octahedral crystals of CO_2 are mechanically unstable at these experimental conditions, decomposing into molecular CO_2 [34]. The calculations also tried to identify the actual structure of CO_2-VI on the basis of hypothetical, crystalline structures which include both CO_4 and CO_3 coordination sites, similarly to a-carbonia. These crystals, which are meta-stable in the P-T range of phase VI, exhibit layered structures and differ from each other in the relative amounts of σ and π bonded oxygen atoms. The main (noncarbonylic) Raman peak of the predicted crystal structures, whose origin can be traced to the stretching of C-O single bonds, lies close to the observed main peak of phase VI, while its precise frequency depends strongly on the local environment. It was then concluded that the structure of CO_2-VI could, indeed, consist of some combination of CO_3 and CO_4 sites. Another study, based on *ab initio* calculations, considered hypothetical, meta-stable, layered structures composed of a two-dimensional network of corner-sharing CO_4 tetrahedra [36]. Two out of the four calculated Raman frequencies for this crystals remarkably agree with the most intense peaks of phase VI, thereby providing and alternative route for determining the structure of this phase. On the other hand, the interpretation of the XRD data is controversial.

6   Concluding remarks

In this review I have briefly discussed the application of optical spectroscopy to high-pressures studies. I have illustrated some crucial, technical aspects of the Raman and IR set-ups, such as different ways to fit the light beams to the small samples in DACs; efficient methods to collect light signals from the samples; and the relevant characteristics of the spectrometers. All of this has been discussed in connection with the case of simple molecular systems, but the discussion is quite general and is appropriate to all kind of samples in DACs. I then considered, as concrete examples, the study of two case systems of increasing complexity, O_2 and CO_2. I have shown how Raman and IR spectroscopy have been the leading tool, along with XRD, in the investigation of different physical aspects of these systems such as: (i) the phase/kinetic diagram (discovery of new solid phases and determination of phase and kinetic boundaries); (ii) the antiferromagnetic spin ordering; (iii) the molecular clustering; (iv) the metallization in solid O_2; and (v) the reversible transformations into non-molecular, extended, crystalline and amorphous solids in CO_2. It is worthwhile to stress one again that, while XRD is the crucial technique for unveiling the microscopic structure of materials, it is rather limited in investigating the nature of the forces holding together solid and fluid materials. On the other hand, optical spectroscopies, although indirect in the study of the structure, are the direct probe of these forces, thereby representing the main route to add physical insights to the investigation of phenomena at extreme P-T conditions.

7   Acknowledgements

My research at LENS is supported by the European Union under Contract RII3-CT2003-506350, and by "Firenze Hydrolab" through a grant by Ente Cassa di Risparmio di Firenze.

References

[1] Hayes, W. and Loudon, R., *Scattering of Light by Crystals*, Dover Pubblications, Inc., Mineola, New York, 2004.

[2] Turrell, G., *Infrared and Raman Spectra of Crystals*, Academic Press, London and New York, 1972.

[3] van Kranendonk, J., *Solid Hydrogen*, Plenum, New York, 1983.

[4] Demtröder, W., *Atoms, Molecules and Photons*, Springer-Verlag, Berlin-Heidelberg, 2006, Chapeter 11.

[5] Goncharov, A.F. et al., in *Science and Technology of High Pressure*, Manghnani, M.H., Nellis, W.J. and Nicol, M., Eds. Universities Press, Hyderabad, India, 2000, 90.

[6] Bini, R. et al., Experimental setup for Fourier transform infrared spectroscopy studies in condensed matter at high pressure and low temperatures, *J. Chem. Phys.*, 68, 3154, 1997.

[7] Gorelli, F.A. et al., The ϵ phase of solid oxygen: evidence of an O_4 molecule lattice, *Phys. Rev. Lett.*, 83, 4093, 1999.

[8] Freiman, Y.A., and Jodl, H.J., Solid Oxygen, *Phys. Rep.*, 401, 1, 2004.

[9] Santoro, M. et al., New phase diagram of oxygen at high pressures and temperatures, *Phys. Rev. Lett.*, 93, 265701, 2004.

[10] Gorelli, F.A. et al., Antiferromagnetic order in the δ phase of solid oxygen, *Phys. Rev. B*, 62, 3604, 2000.

[11] Santoro, M. et al., Antiferromagnetism in the high-pressure phases of solid oxygen: low-energy electronic transitions, *Phys. Rev. B*, 64, 064428, 2001.

[12] Goncharenko, I.N., Evidence for a magnetic collapse in the epsilon phase of solid oxygen, *Phys. Rev. Lett.*, 94, 205701, 2005.

[13] Lundegaard, L.F. et al., Observation of an O_8 molecular lattice in the ϵ phase of solid oxygen, *Nature*, 443, 201, 2006.

[14] Fujihisa, H. et al., O_8 cluster structure of the epsilon phase of solid oxygen, *Phys. Rev. Lett.*, 97, 085503, 2006.

[15] Meng, Y. et al., Inelastic x-ray scattering of dense solid oxygen: evidence for intermolecular bonding, *Proc. Natl. Acad. Sci. U.S.A.*, 105, 11640, 2008.

[16] Neaton, J.B., and Ashcroft, N.W., Low-energy linear structures in dense oxygen: implications for the ϵ phase, *Phys. Rev. Lett.*, 88, 205503, 2002.

[17] Ma, Y., Oganov, A.R., and Glass, C.W., Structure of the metallic ζ-phase of oxygen and isosymmetric nature of the $\epsilon - \zeta$ phase transition: *ab initio* simulations, *Phys. Rev. B*, 76, 064101, 2007.

[18] Desgreniers, S., Vohra, Y.K., and Ruoff, A.L., Optical response of very high density solid oxygen to 132 GPa, *J. Phys. Chem.*, 94, 1117, 1990.

[19] Weck, G., Loubeyre, P., LeToullec, R., Observation of structural transformations in metal oxygen, *Phys. Rev. Lett.*, 88, 035504, 2002.

[20] Goncharov, A.F. et al., Molecular character of the metallic high-pressure phase of oxygen, *Phys. Rev. B*, 68, 100102, 2003.

[21] Tse, J.S. et al., Electronic structure of ϵ-oxygen at high pressure: GW calculations, *Phys. Rev. B*, 78, 132101, 2008.

[22] Weck, W. et al., Melting line and fluid structure factor of oxygen up to 24 GPa, *Phys. Rev. B*, 76, 054121, 2007.

[23] Santoro, M. and Gorelli, F.A., High pressure solid state chemistry of carbon dioxide, *Chem. Soc. Rev.* , 35, 918, 2006.

[24] Giordano, V.M., Datchi, F., Molecular carbon dioxide at high pressure and high temperature, *Europ. Phys. Lett.* 77, 46002, 2007.

[25] Iota, V. et al., Six-fold coordinated carbon dioxide VI, *Nature Mat.*, 6, 34, 2007.

[26] Tschauner, O., Mao, H.K., and Hemley, R.J. New transformations of CO_2 at high pressures and temperatures, *Phys. Rev. Lett.* 87, 075701, 2001.

[27] Iota, V., Yoo, C.S., and Cynn, H., Quartzlike carbon dioxide: an optically nonlinear extended solid at high pressures and temperatures, *Science*, 283, 1510, 1999.

[28] Santoro, M. et al., *In situ* high P- T Raman spectroscopy and laser heating of carbon dioxide, *J. Chem. Phys.*, 121, 2780, 2004.

[29] Yoo, C.S. et al., Crystal structure of carbon dioxide at high pressure: "superhard" polymeric carbon dioxide, *Phys. Rev. Lett.* 83, 5527, 1999.

[30] Dong, J., Tomfohr, J.K., and Sankey, O.F., Rigid intertetrahedron angular interaction of nonmolecular carbon dioxide solids, *Phys. Rev. B*, 61, 5967, 2000.

[31] Dong, J. et al. Investigation of hardness in tetrahedrally bonded nonmolecular CO_2 solids by density-functional theory, *Phys. Rev. B*, 62, 14685-14689 (2000).

[32] Serra, S. et al., Pressure induced solid carbonates from molecular CO_2 by computer simulation, *Science* 284, 788, 1999.

[33] Santoro, M. et al., Amorphous silica-like carbon dioxide, *Nature*, 441, 857, 2006.

[34] Javier, A.M. et al., Mixed threefold and fourfold carbon coordination in compressed CO_2, *Phys. Rev. Lett.*, 100, 163002, 2008.

[35] Holm, B. et al., Theoretical investigation of high pressure phases of carbon dioxide, *Phys. Rev. Lett.*, 85, 1258, 2000.

[36] Togo, A., Fumiyasu, O., and Tanaka, I., Transition pathway of CO_2 crystals under high pressures, *Phys. Rev. B*, 77, 184101, 2008.

Chapter 8
Inelastic X-Ray Scattering on High-Pressure Fluids

Federico A. Gorelli

Research Center SOFT-INFM-CNR, Università di Roma "La Sapienza", I-00185, Roma, Italy and LENS, Via N. Carrara 1, I-50019 Sesto Fiorentino, Firenze, Italy

1 Introduction

The knowledge of thermo-physical properties of materials at elevated pressure (P) and temperature (T) conditions is important in applied thermodynamics and geophysical/planetary science. Of particular interest is the study of fluid materials which, in the case of gaseous systems, are in the supercritical fluid state where not much is known about the evolution of physical properties. Properties such as elastic moduli, dispersion relations and damping of acoustic waves, viscosity and relaxation times are connected to the dynamics of density fluctuations [1], and can be determined, for example, by means of light scattering techniques [2, 3, 4, 5, 6, 7, 8]. The derived dynamical properties pertain to the hydrodynamic realm since the probe wavelength (a few hundred nanometer) is much larger than the coarse grained microscopic structure of any molecular system. Length scales comparable to the mean inter-particle distances can be assessed by coherent inelastic neutron (INS)[9, 10]and x-ray scattering (IXS)[11, 12, 13, 14]. The large size of neutron beams, however, limits such studies to relatively large samples, thus preventing the use of diamond anvil cells. These limitations can be overcome in the case of IXS, since undulator-based synchrotron X-rays can be focused down to small spot sizes, in the micrometer range. A few remarkable efforts have been performed in this direction using moderate pressures (<0.2 GPa) for liquid metals[15, 16]. Indeed, while studies on crystalline systems are routinely performed up to pressures of several tens of GPa [17], experiments on liquids are scarce [18, 19] and a quantitative visco-elastic analysis of the IXS spectra

has not been attempted, mainly due to problems of parasitic scattering from
the sample environment and the diamonds, which is particularly critical in
the study of light elements.

Here we present a basic introduction to the IXS technique on fluids and the
apparatus for performing quantitative IXS measurements on fluid and solid
samples in a diamond anvil cell (DAC), which can be used to reach pressures of
the order of 50 GPa and temperatures as high as 1000 K by resistive heating.
In the final part the comparison of experimental spectra on fluid Ar with
those obtained from molecular dynamics (MD) simulations, together with the
generalized heat capacity ratio and the longitudinal viscosity obtained from
the spectral analysis will be presented, demonstrating the state of the art of
this technique applied to samples in the diamond anvil cell.

2   General aspects

When a perturbation is applied on a fluid, its effect is damped by dissipation
phenomena: diffusions, viscous flows and thermal exchanges. Even without
an external perturbation, spontaneous microscopic fluctuations are anyhow
present in the fluid. These naturally occur in a broad band of wavelengths and
frequencies. Spontaneous fluctuations, according to the dissipation-fluctuation
theorem [20], are dissipated in the same way as the applied perturbation. For
this reason, studying the response of the system as a function of frequency and
momentum of the induced perturbation, one can obtain basic information on
the structure and the dynamics of the unperturbed system at different length
and timescales.

Formally, the response of a fluid to an applied perturbation is known only in
two limiting cases:

- At low momentum and frequency: the hydrodynamic limit. Here information on the macroscopic properties of the system can be obtained,
  while microscopic information can be provided only indirectly.

- At very high momentum: the single-particle limit. Here the system behaves as a non-interactive ensemble of particles; therefore information
  on collective dynamics cannot be retrieved.

At intermediate momentum, which means wavelengths comparable to intermolecular distances (mesoscopic region) both local structures and dynamics become important. This is the most interesting region because information
on the local arrangement of molecules and the interactions among them can
be obtained. Unfortunately, neither the hydrodynamic nor the single-particle
limit can describe the response of the system in this range. Up to now, a
formally exact description of the fluid dynamics in the mesoscopic region is
missing. Nevertheless, there are some phenomenological approaches, such as
the one based on the memory function, able to describe sufficiently well the
observed phenomenology.

The main role in the description of fluid dynamics is played by the time correlation functions. A time correlation function is defined as the thermodynamic average of the product of two dynamical, i.e. time-dependent, variables. Each one represents an instantaneous deviation (fluctuation) of a physical quantity, $A(\vec{r}, t)$, with respect to its equilibrium value, $\langle A \rangle$:

$$\delta A(\vec{r}, t) = A(\vec{r}, t) - \langle A \rangle \tag{1}$$

The average, $\langle \ldots \rangle$, is carried out over the phase coordinates of all molecules in the fluid with an equilibrium ensemble as weighting function. Considering the spatial and temporal invariance of the liquid, the time correlation function, $C_{A,B}(\vec{r}_1, \vec{r}_2, t_1, t_2)$, of the dynamical variables $A(\vec{r}_1, t_1)$ and $B(\vec{r}_2, t_2)$ is therefore:

$$
\begin{aligned}
C_{A,B}(\vec{r}_1, \vec{r}_2, t_1, t_2) &= V \langle \delta A(\vec{r}_1, t_1) \delta B(\vec{r}_2, t_2) \rangle \tag{2} \\
&= V \langle \delta A(\vec{r}, t) \delta B(0, 0) \rangle \tag{3} \\
&= C_{A,B}(\vec{r}, t) \tag{4}
\end{aligned}
$$

where $\vec{r} = \vec{r}_2 - \vec{r}_1$, $t = t_2 - t_1$, and V is the volume. For $t = 0$, $C_{A,B}(\vec{r}, t)$ assumes its maximum value, $\lim C_{A,B}(\vec{r}, t) = 0$ for $t \to \infty$, indicating the loss of any correlation between the two variables when they are evaluated at very different times.

Among all possible fluctuating variables describing the dynamics of fluids, a crucial role is played by density fluctuations since they are directly probed by a large number of spectroscopic techniques. The density function can be expressed as follows:

$$n(\vec{r}, t) = \frac{1}{\sqrt{N}} \sum_{i=1}^{N} \delta(\vec{r} - \vec{R}_i(t)) \tag{5}$$

where N is the number of particles and $\vec{R}_i(t)$ are their positions. The density-density correlation function is:

$$
\begin{aligned}
G(\vec{r}, t) &= V \langle \delta n(\vec{r}_1, t_1) \delta n(\vec{r}_2, t_2) \rangle \tag{6} \\
&= \frac{V}{N} \left\langle \sum_{i,j=1}^{N} \delta\left(\vec{r}_1 - \vec{R}_i(t_1)\right) \delta\left(\vec{r}_2 - \vec{R}_j(t_2)\right) \right\rangle - \frac{N}{V} \tag{7}
\end{aligned}
$$

In inelastic spectroscopic measurements the experimental observable is the time and space Fourier transform of $G(r, t)$, which is usually called the dynamical structure factor, $S(\vec{Q}, \omega)$:

$$S(\vec{Q}, \omega) = \int_V d\vec{r} \int_{-\infty}^{\infty} \left(G(\vec{r}, t) - \frac{N}{V} \right) e^{i(\vec{Q} \cdot \vec{r} - \omega t)} dt \tag{8}$$

It is convenient to introduce the intermediate scattering function, $F(\vec{Q}, t)$, which is the spatial Fourier transform of $G(\vec{r}, t) - \rho$ or, equivalently, the inverse time Fourier transform of $S(\vec{Q}, \omega)$:

$$F(\vec{Q}, t) = \frac{1}{N} \left\langle \sum_{i,j=1; i \neq j}^{N} e^{i\vec{Q} \cdot \vec{R}_i(t)} e^{-i\vec{Q} \cdot \vec{R}_j(0)} \right\rangle = \int_{-\infty}^{\infty} e^{i\omega t} S(\vec{Q}, \omega) d\omega \quad (9)$$

where $\rho = N/V$ is the microscopic density.

The intermediate scattering function evaluated at $t = 0$ gives the well known static structure factor $S(\vec{Q})$:

$$F(\vec{Q}, 0) = \left[\int_{-\infty}^{\infty} e^{i\omega t} S(\vec{Q}, \omega) d\omega \right]_{t=0} = \int_{-\infty}^{\infty} S(\vec{Q}, \omega) d\omega = S(Q) \quad (10)$$

This quantity may be obtained from a diffraction experiment using x-rays or neutrons and by Fourier transforming it; then $g(\vec{r}) = G(\vec{r}, 0)$ can be derived.

In the following, considering an isotropic fluid the vector label will be dropped and only scalar quantities will be considered. In an isotropic fluid, the intermediate scattering function satisfies the Langevin equation:

$$\frac{\partial^2 F(Q, t)}{\partial t^2} + \omega_0^2 F(Q, t) + \int m(Q, t - t') \quad (11)$$

where $m(Q, t)$ is the second-order memory function for density fluctuations and

$$\omega_0^2(Q) = \frac{k_B T Q^2}{M S(Q)} \quad (12)$$

is the square of the isothermal sound frequency, where k_B is the Boltzmann constant, T is the temperature, M is the molecular mass and $S(Q)$ is the static structure factor. Through some algebra the expression for $S(Q, \omega)$ reads:

$$\frac{FT\left\{F(\vec{Q}, t)\right\}}{F(Q, 0)} = \frac{S(Q, \omega)}{S(Q)} = \frac{1}{\pi} \frac{\omega_0^2(Q) m'}{\left(\omega^2 - \omega_0^2(Q) - \omega\tilde{m}\right)^2} + (\omega m') \quad (13)$$

where m' and $\tilde{m}$ are the real and imaginary parts of the Fourier transform of the memory function [21, 22]. In order to give a more explicit physical insight for the expression of the dynamic structure factor given by eq. 10, and in particular of the quantities m' and $\tilde{m}$, the linear response theory will be introduced now [23].

The equation of motion of the damped harmonic oscillator (DHO) forced by an external force is:

$$\frac{d^2 x}{dt^2} + 2\gamma(Q, \omega) \frac{dx}{dt} + \omega_0^2 x = F(t) \quad (14)$$

where $\gamma(Q, \omega)$ is the generalised damping parameter:

$$\gamma(Q, \omega) = \gamma'(Q, \omega) + i\gamma''(Q, \omega) \quad (15)$$

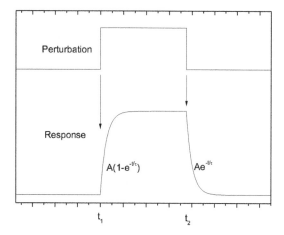

Figure 1. *A time-dependent perturbation is applied at time t_1 and is turned off at time t_2. The response of the system where a relaxation process with a relaxation time τ exists, is shown at the bottom.*

By Fourier transforming eq. (12), the linear susceptibility can be obtained:

$$\frac{\tilde{\chi}}{\tilde{F}} = \chi(Q,\omega) = \chi'(Q,\omega) + i\chi''(Q,\omega) \tag{16}$$

The classical version of the fluctuation dissipation theorem gives:

$$\chi''(Q,\omega) = \frac{\omega}{2k_B T} S(Q,\omega) \tag{17}$$

where k_B is the Boltzmann constant.

This approach for $S(Q,\omega)$ and the one described above on the basis of the Langevin equation coincide once the following identities are assumed, which leads to reinterpret the memory function as a damping function:

$$lm' = Q^2 \gamma' \tag{18}$$

$$m'' = \gamma'' \tag{19}$$

When a time-dependent perturbation is applied on a system, it goes locally out of equilibrium and manifests a response which is characterized by a relaxation process. This is summarized in Figure 1, where a hypothetical perturbation, switched on at time t_1, induces a response in the system which grows with an exponential behaviour with a characteristic relaxation time τ. In a similar way, when the perturbation is switched off at time t_2, the response decays exponentially with the same time τ.

An acoustic wave travelling with frequency ω_L and momentum Q is a time-dependent perturbation which creates regions which are compressed and

expanded in time and which are periodic in space with a period $2\pi/Q$ and
time with a period $T = 2\pi/\omega_L$. The relative magnitude of the two times T
and τ, characteristic of the acoustic wave and of the system, respectively,
determine two limiting cases. When the period of the acoustic wave is much
longer than the relaxation time ($T \gg \tau$), the system has enough time to relax
before the successive perturbation. It is then a fully relaxed regime from the
system point of view. In the opposite case, when the period of the acoustic
wave is much shorter than the relaxation time ($T \ll \tau$), the system does not
have enough time to relax before the successive perturbation. This condition
then corresponds to a fully unrelaxed regime.

The physics of a fluid system, and in particular the description of the
relaxation processes which drive its dynamics, appears in the memory function. For a fluid in which three relaxation processes are present the memory
function can be written as:

$$m(Q,t) \qquad = \left(\omega_S^2(Q) - \omega_T^2(Q)\right) e^{-D_T(Q)Q^2 t} \qquad (20)$$

$$+ \left(\omega_\infty^2(Q) - \omega_S^2(Q)\right) e^{-t/\tau_\alpha(Q)} + 2\Gamma_\mu \delta(t) \qquad (21)$$

where:

$$\omega_T^2(Q) = \omega_0^2(Q) \qquad (22)$$

is the square of the isothermal sound frequency, [see eq. (15)], and $\gamma(Q)$ and
$D_T(Q)$ are the Q-dependent generalizations of the heat capacity ratio and of
the thermal diffusivity:

$$\gamma = \frac{C_P}{C_V} \qquad (23)$$

and

$$D_T = \frac{k}{\rho C_V} \qquad (24)$$

where k is the thermal conductivity.

$$\omega_S(Q) = \omega_T(Q)\sqrt{\gamma(Q)} \qquad (25)$$

is the adiabatic sound frequency of the velocity and $\omega_L(Q)$ is the apparent
sound frequency, which can be obtained from the maxima of the longitudinal
current spectra:

$$\omega_L(Q) = \max\left\{\frac{\omega^2 S(Q,\omega)}{Q^2}\right\} \qquad (26)$$

$\omega_\infty(Q)$ is the infinite frequency sound speed, $\tau_\alpha(Q)$ is the structural relaxation
time and Γ_μ is the strength of the fast (microscopic) relaxation process.

The structure of $m(Q,t)$, characterized by the relaxation times, determines
the details of the apparent dispersion curve $\omega = \omega(Q)$. As a matter of fact, the
presence of a structural and a thermal relaxation process induces deviations
from the linear evolution of the dispersion curve at small values of Q: $\omega = CQ$.

The structural relaxation process is responsible for the so called viscoelastic transition, which occurs as the fluid behaves as a "viscous" system at

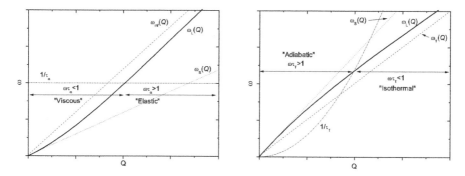

Figure 2. *Plots of frequency Omega versus wavevector Q. The left panel illustrates the viscoelastic crossover. When $\omega\tau_\alpha < 1$ the system is the system is fully relaxed, meaning that the relaxation time is shorter than two successive perturbations. As ω increases the system becomes increasingly less able to relax and the dispersion curves upwards. Eventually when $\omega\tau_\alpha > 1$ the system cannot relax and behaves as a solid. The right panel illustrates the adiabatic to isothermal transition. The crossover between the apparent sound dispersion and the inverse of the thermal relaxation time. The sound dispersion bends downwards with increasing Q and changes slope (sound velocity) passing from the adiabatic towards the isothermal sound dispersion. In the "adiabatic" regime, when $\omega\tau_t > 1$, the system is fully unrelaxed, meaning that the relaxation time is longer than two successive perturbations and the system is not able to dissipate energy through this channel. On the contrary the "isothermal" regime, when $\omega\tau_t < 1$, is the one where the system is fully relaxed.*

frequencies smaller than $1/\tau_\alpha$ and "elastic" for frequencies larger than $1/\tau_\alpha$. The transition is located at the crossover between the two regimes, i.e. where $\omega\tau_\alpha = 1$. In the viscous regime, the frequency is reduced with respect to the intrinsic frequency for the effect of the relaxation process, which is able to dissipate energy. At the visco-elastic transition the dispersion curve switches from one linear behaviour to another corresponding to two different sound velocities, which are the slopes of the dispersion curve. As discussed above the "viscous" regime is the one where the system is fully relaxed, meaning that the relaxation time is shorter than two successive perturbations. On the contrary, the "elastic" regime is the one where the system cannot relax. The structural relaxation time can have a Q dependence, but for simplicity it has been considered constant in this case.

Analogously, the thermal relaxation process can also produce a deviation of the dispersion curve from its linear behaviour. In this case the relaxation time has a well-defined Q dependence, as $1/\tau_T = D_T(Q)Q^2$, so that $\omega > 1/\tau_T$ at small values of Q, corresponding to the fully unrelaxed regime, and

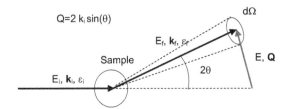

Figure 3. *A schematic diagram of the scattering process.*

then "adiabatic" as the system is not able to dissipate energy between two successive perturbations. In contrast, at higher Q values, when ω is smaller than $1/\tau_T$, the system is able to dissipate energy (fully relaxed) and the regime is then "isothermal". The relaxed and unrelaxed regimes are then inverted as a function of Q with respect to the visco-elastic transition, and in this case the sound speed shows a decrease passing from the adiabatic to the isothermal regimes. These two effects just described can be present simultaneously and can then interfere each other. At higher Q values the dispersion curve first shows a maximum and then a minimum. For Q values similar to Q_m, the Q value of the first maximum in the static structure factor. This is due to the structural correlation at high Q values, i.e. small distances, see eq. (15).

3 Instrumentation

The dynamic structure factor can be obtained from an inelastic x-ray scattering experiment. It can be shown[9] that in the case in which

- the scattering process is dominated by the Thomson term (resonant and spin-dependent terms are negligible),

- the Born-Oppenheimer approximation is valid and

- there are no electronic excitations in the energy transfer range,

the x-ray inelastic scattering cross section is related to the dynamic structure factor through the following formula:

$$\frac{\partial^2 \sigma}{\partial \omega \partial \Omega} = N(e^2/mc^2)^2 \frac{k_f}{k_i} \epsilon_f \cdot \epsilon_i \, |f_m(Q)|^2 \, S(Q,\omega) \qquad (27)$$

where N is the number of scatterers, e is the electron charge and m its mass, c the light velocity, k_i, k_f, ϵ_i, ϵ_f are the momenta and the polarizations of the incident and scattered photons (see Fig. 3). $f_m(Q)$ is the atomic form factor, which modulates the scattered intensity as a function of the exchanged momentum Q.

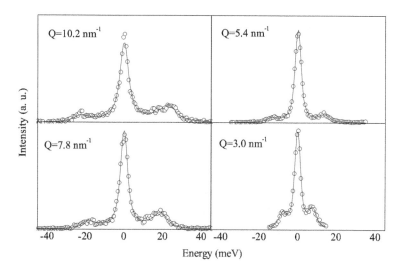

Figure 4. *The inelastic x-ray scattering spectra of supercritical fluid oxygen at room temperature and 5.35 GPa at 4 different values of the exchanged momentum. The spectrum is made of three main features: the central quasi-elastic peak and the two acoustic excitations.*

In an x-ray inelastic scattering experiment it is possible to measure the number of photons which are scattered with an energy

$$E = \frac{h}{2\pi}\omega \tag{28}$$

and have exchanged a momentum Q with the sample:

$$I(Q, E) = \epsilon(Q)N_0\Delta E\Delta QL\rho\frac{\partial^2\sigma}{\partial E\partial Q} \propto S(Q, E) \tag{29}$$

where $\epsilon(Q)$ is the "optical efficiency", N_0 is the number of incident photons, ΔE is the collecting energy width, ΔQ is the collecting angular width, L is the sample length, ρ is the sample density. Typical IXS spectra measured at different Q values on fluid oxygen at 5.35 GPa and room temperature are shown in Fig. 4.

The measured spectra are the result of the convolution of the experimental resolution with the dynamic structure factor:

$$I(Q, \omega) = \int R(\omega') \tag{30}$$

where $R(\omega)$ is the instrumental resolution.

The experimental spectra can be analyzed by using a dynamic structure factor as the one reported in eq. (16), obtained from a theoretical model which

uses the Langevin equation for the classical density fluctuations with memory
function formalism as described above. This classical dynamic structure factor
(symmetric) is related to the quantum dynamic structure factor through the
detailed balance factor:

$$S_q(Q,\omega) = S_{cl}(Q,\omega)\frac{\hbar\omega}{2\pi k_B T}\frac{1}{1-e^{-\frac{\hbar\omega}{2\pi k_B T}}} \tag{31}$$

The parameters present in the memory function [eq. (27)] can be adjusted
by means of a fitting procedure providing relevant information such as the
relaxation times and relaxed and unrelaxed velocities. These values can be
used to determine for example generalised hydrodynamic quantities such as
the generalized longitudinal viscosity which is given by:

$$\eta_L(Q) = \frac{\rho}{Q^2}\left[\left(\omega_\infty^2(Q) - \omega_S^2(Q)\right)\tau_\alpha(Q) + \Gamma_\mu(Q)\right] \tag{32}$$

The dynamic structure factor can be measured also by means of the inelastic neutron scattering technique (INS). In this case the scattering cross
section is

$$\frac{\partial^2\sigma}{\partial\omega\partial\Omega} = N(b)^2\frac{k_f}{k_i}S(Q,\omega) \tag{33}$$

where b is the neutron scattering length which depends on Z and on the
total (electron+nuclear) magnetic moment J. The neutron scattering length
can then be very different for different isotopes of the same element, and also
for the same isotope but with a different value of J. Neutrons have a big
advantage with respect to x-rays as for wavelengths of the order of 1 Å, they
have energies in the range 10–100 meV which are very similar to those of
inelastic excitations. It is important to notice that the accessible dynamical
range is very different between the two techniques. As a matter of fact it results
that for neutrons there is a maximum energy value, typically smaller than 100
meV, while for x-rays, there is practically no limit of the energy range, so that
they can probe high values of the energy transfer even at low Q. Furthermore
for neutrons, in some cases it is difficult to separate the coherent from the
incoherent contributions while x-rays allow a higher Q resolution.

In the last decade, the IXS technique was successfully applied in the study
of collective dynamics in those cases where INS is difficult to apply: kinematic
limitations, large incoherent scattering, multiple scattering, very high momentum resolution, or small samples as in the case of samples in the diamond anvil
cell.

Figure 5 shows a schematic of the triple axis spectrometer present at ID28
at the ESRF. The first "axis" is the high-energy-resolution monochromator,
the second "axis" is the sample goniometer and the third "axis" is the crystal
analyzer. The x-ray photons emerging from the pre-monochromator are highly
monochromated by the backscattering monochromator, which consists of a
Si crystal operating at a Bragg angle of 89.98° ensuring the minimization

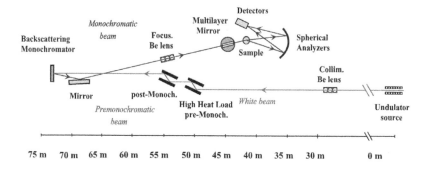

Figure 5. *A schematic diagram of the ID 28 beamline at the ESRF.*

of geometrical contributions to the energy resolution. By using high order Bragg reflections ((hhh) with $h = 7, 8, 9, 11$ or 13) and perfect crystals [24], it is possible to obtain an energy resolution of $\Delta E/E = 10^{-7}-10^{-8}$. The energy of the monochromated photons can be varied by changing the temperature of the Si crystal which is stabilized in the mK range. The x-ray beam is then focussed in the vertical and horizontal planes by a toroidal mirror down to $250 \times 80 \ \mu m^2$. The beam spot can be further decreased in size down to $30 \times 40 \ \mu m^2$ by using a multilayer mirror and/or diffractive Be lenses, making it ideal for measurements on samples inside the diamond anvil cell.

The scattered radiation is then analysed by the spherical analyzers made of 12000 of $0.6 \times 0.6 \ mm^2$ Si single crystals aligned on a spherical surface. The detectors are Peltier-cooled Si diodes working in photon-counting mode (1 count in 30 min). The 9 analyzers are positioned each 0.75° (2θ) apart and are mounted on a 7-m long arm which can rotate from 0 to 55° allowing the simultaneous measurement of energy spectra at 9 different values of the exchanged momentum. The energy scans are made by temperature scans of the high-resolution monochromator Si crystal: $\Delta E/E = -\Delta d/d = \alpha T$ with $\alpha = 2.58 \ 10^{-6} \ K^{-1}$, where d is the lattice spacing and α is the thermal expansion coefficient.

The instrumental resolution can be obtained by measuring the IXS spectrum of a sample which provides a contribution only in the quasi-elastic region. For this purpose a disordered sample, such as Plexiglas, at low temperature and at a Q value corresponding to the maximum of its first sharp diffraction peak can be used. This allows to maximise the elastic contribution to the scattering and to obtain a spectrum which represents the instrumental resolution. In Fig. 6 a resolution spectrum measured using the (999) reflection is shown.

From an experimental point of view there is also an uncertainty on the exchanged momentum at which the spectrum is measured deriving from the acceptance angle of the spherical analyzers. This is at maximum, i.e. for small

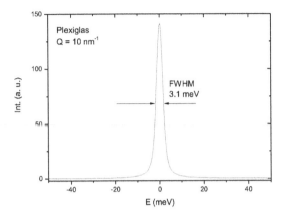

Figure 6. *Instrumental resolution function of ID28 beamline at ESRF using the (999) reflection of the high-resolution monochromator. This was measured using a Plexiglas sample at a Q value where the static structure factor reaches its maximum, so that the intrinsic line width is negligible with respect to the instrumental one.*

values of the scattering angle:

$$\delta Q(\theta) \cong k_0 \cos \frac{\theta}{2} \delta\theta \cong k_0 \delta\theta \cong 0.4 \text{ nm}^{-1} \tag{34}$$

It is worth to notice that the sample scatters and also absorbs the x-rays. It is then recommended, if possible, to have a sample length as long as the x-ray scattering length $1/\mu$ of the sample material in order to have the maximum scattering efficiency. This quantity may vary in a very wide range: from about 50 μm in the case of liquid metals to much larger values in the case of gases or fluids, depending on their density.

4   Systems

Of the various IXS experiments on high-pressure fluids in the recent literature, I will describe here only some selected experiments on liquid and supercritical fluid neon, nitrogen, water and ammonia up to 3 kbar and on liquid water using the diamond anvil cell. In the final part I will describe a recently published paper which shows an experimental procedure that allows to measure quantitatively the dynamic structure factor on samples in the diamond anvil cell.

Liquid and supercritical fluid neon has been studied with IXS using a large-volume high-pressure cell. This cell can sustain pressures up to 3 kbar and can be cooled or heated (up to 700 K). The sample length along the x-ray beam is 10 mm and the windows are made of 1-mm-thick diamond. Room temperature measurements [25] on dense supercritical neon at 3 kbar have provided the dynamic structure factor at a density which is comparable to the one of the liquid at ambient pressure. The dispersion curve shows a linear behaviour at small Q values (1–8 nm^{-1}) as expected in the hydrodynamic regime, see Fig. 7 (left panel). The solid line represents the adiabatic sound velocity in the macroscopic $Q = 0$ limit (1050 m/s) and the dispersion curve follows perfectly that line. This linear behaviour indicates the absence of a structural relaxation process, which would produce a viscoelastic transition as discussed above and then an increase of the sound velocity (change of slope) with Q (see left panel of Fig. 7). At higher Q values, the energy values start to bend and deviate from the linear behaviour and show a minimum around 22 nm^{-1}, i.e. at $Q \approx Q_m$, the Q value of the first maximum in the static structure factor. Such a minimum has already been observed in various liquid systems, and can be explained as a manifestation of the interference between the density fluctuations and the pseudoperiodicity responsible for the sharp feature in the static structure factor at Q_m.

In the same figure the dispersion curve of liquid neon at 80 bar and 35 K, obtained from INS measurements is also reported [26]. This liquid has a density similar to that of the supercritical fluid neon at room temperature and 3 kbar. The energy values of the liquid reported in the left panel of Fig. 7 have been multiplied by the factor 1.99, which corresponds to the ratio between the adiabatic sound velocities at the two thermodynamic states. Differently from the supercritical fluid, the energy values are higher than the solid line representing the adiabatic sound velocity. This indicates the presence of positive sound dispersion, which is related to a structural relaxation process. The effect of the positive dispersion is generally observed in liquid systems and can be considered a fingerprint of the liquid state. Furthermore the minimum observed at $Q \approx Q_m$ in the fluid state, becomes much deeper in the liquid state and the dispersion curve almost reaches zero, resembling solids.

More recently, Cunsolo *et al.* [27] have shown that the amount of positive dispersion decreases on going from the liquid to the supercritical fluid state. This has been evidenced by performing three measurements at the same density, but different pressures and temperatures in the liquid and in the supercritical fluid neon, see right panel of Fig. 7. A more systematic study of the evolution of the positive sound dispersion has been performed by Bencivenga *et al.* [28] on liquid and supercritical fluid nitrogen. The measurements have been performed along an isobar at 400 bar at different temperatures. Here the density changes, but also in this case it has been observed that the positive sound dispersion vanishes on going from the liquid state to the supercritical fluid (see Fig. 8). The spectra have been analysed using a line shape model derived within the framework of the memory function formalism as described above. Figure 8 shows the adiabatic, the isothermal and the apparent sound velocities, and the inverse of the structural and thermal relaxation times.

The apparent sound velocity shows a positive sound dispersion in the liquid state (87 K) which decreases with increasing temperature and disappears

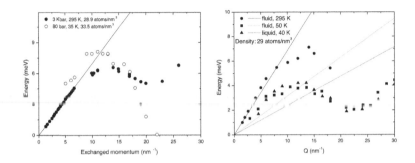

Figure 7. *Left panel: Sound dispersion in liquid and supercritical fluid neon at a similar density. The liquid state (open dots) shows a positive sound dispersion, as the values are higher than the solid line representing the adiabatic sound velocity. On the contrary in the supercritical fluid state (full dots) the sound dispersion values lie on the solid line, indicating the absence of a positive sound dispersion. The positive sound dispersion is a fingerprint of the liquid state and is related to a viscoelastic transition (see text). Right panel: Three isochoric measurements in liquid and supercritical fluid neon at different temperatures. The amount of positive dispersion, decreases on going from the liquid to the supercritical fluid state with increasing temperature.*

on reaching supercritical conditions, i.e. temperatures higher than 126.2 K. However, the results at 128 K still show a small deviation of the apparent sound velocity from the adiabatic one, indicating positive dispersion in the supercritical state, even if very close to the critical temperature. The viscoelastic transition generating the positive sound dispersion takes place when the frequency of the apparent sound velocity crosses or is close to the inverse of the structural relaxation time τ_α. Increasing the temperature and reaching supercritical conditions $1/\tau_\alpha$ also increases and does not cross the dispersion curve, thereby the viscoelastic transition does not occur any longer. The values of the apparent sound dispersion follow the adiabatic sound dispersion at low Q values. The absence of the positive sound dispersion allows putting in evidence the adiabatic to isothermal transition that takes place at Q values for which the inverse of the thermal relaxation time is similar to the apparent sound frequency. In this case the adiabatic to isothermal transition produces a downward bending of the apparent sound dispersion, reaching the isothermal sound dispersion.

An analogous systematic study has been performed on Ne, N_2, NH_3 and H_2O along isobaric measurements at different temperatures, ranging from the liquid up to the supercritical state [29]. This has allowed verifying that the structural relaxation time follows Arrhenius behaviour in the liquid state,

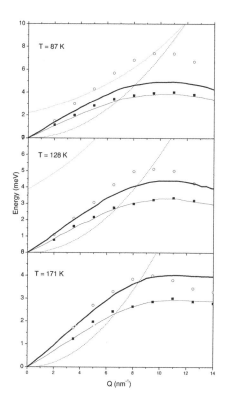

Figure 8. *Dispersion curves of liquid and supercritical fluid nitrogen at the same pressure (400 bar) and different temperatures, ranging from the liquid state (87 K) to the supercritical fluid state at 128 K (which is about the critical temperature, 126 K) and at 171 K. The positive sound dispersion, which is present in the liquid state at 87 K has almost disappeared at 128 K and is completely absent in the supercritical fluid state at 171 K. The dotted and dashed lines represent the inverse of the structural and thermal relaxation times respectively (the first is not reported at 171 K as it is outside the frequency range of interest). The thick and thin solid lines represent the adiabatic and isothermal dispersions respectively derived from S(Q) measurements and on the basis of eq. (19). The open dots represent the apparent sound dispersion obtained from the maxima of the longitudinal current spectra. The squares represent the best fit values of the isothermal dispersion.*

while it deviates on entering the supercritical state. The slope of the linear behaviour of the structural relaxation time in the Arrhenius plot gives the activation energy of the structural relaxation process itself (about the intermolecular bond energy). This is clearly much higher in the case of hydrogen-bonded liquids such as ammonia and water than for non-hydrogen-bonded

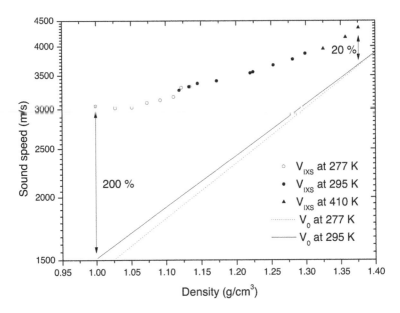

Figure 9. *Sound velocity (logarithmic scale) in liquid water as a function of density. Solid and dotted lines represent the adiabatic sound velocity. Open dots represent the values measured at 277 K using a large volume cell. Full dots represent the values measured at 295 K using the diamond anvil cell. Up triangles represent the data measured at 410 K using the diamond anvil cell. The positive sound dispersion is maximum at ambient conditions and decreases on increasing density and reaches a value which is typical of simple liquids.*

liquids such as nitrogen.

The first experiment on fluids using the DAC has been performed on liquid water[15]. Liquid water exhibits a strong positive dispersion as at ambient conditions the adiabatic sound velocity is about the half of the apparent sound velocity. With increasing pressure, the amount of positive dispersion decreases and reaches values typical of a simple liquid, i.e. about 20% (see Fig. 9). This indicates that under high pressure the hydrogen bond network tends to disappear and water tends to be a simple liquid.

Unfortunately, the quality of the measured spectra could provide only the apparent speed of sound, obtained from the maxima of the spectra (see above), but did not allow to perform a full visco-elastic analysis as in the case described above of the large volume cell. What precludes such a possibility is the reliability of the measured intensity in the quasi-elastic region of the spectra. As a matter of fact, the parasitic scattering generates a spurious quasi-elastic peak which is difficult to remove *a posteriori*. In order to have the possibility of performing a full visco elastic analysis it is then necessary to reduce all the possible sources of parasitic scattering and make the residual one very reproducible. This is obviously a general issue, which becomes crucial using the diamond anvil cell.

In general, the parasitic scattering from air before and after the sample has to be made negligible. This is a potentially huge signal which might completely cover the sample signal and is generally removed by inserting the sample (and the cell that contains it) inside a vacuum chamber. The parasitic scattering from air is so nominally eliminated but it still remains the parasitic scattering from the entrance and exit windows. As the depth of field of the IXS spectrometer is very large, the entrance window has to be placed very far from the sample and the exit one very close to the beamstop, which is generally in air, outside the vacuum chamber. Last but not least the parasitic scattering from the cell windows (which fall within the depth of field) has to be made negligible. This is possible in the case of large-volume cells where the sample thickness is higher than that of the windows. These constrains become critical in the case of experiments on samples inside the DAC and the parasitic scattering becomes more difficult to eliminate. When using the DAC, a magnified image of the sample using a microscope to inspect the sample and measure the pressure *in situ* (optical sensor) is necessary. For this reason the entrance window of the vacuum chamber cannot be placed at a large distance (outside the depth of field of the IXS spectrometer) as this would be incompatible with the microscope working distance (30-40 mm). Furthermore, the relationship between sample and window thickness is the least favourable, as the diamond windows are thick ($\approx$5 mm) while the sample is thin (=100 μm). For this reason the parasitic scattering from the diamond windows cannot be in general considered as negligible and has to be subtracted from the raw data in order to obtain a reliable measurement of the spectrum in the quasi-elastic region once all the other possible sources of parasitic scattering have been eliminated.

Recently it has been demonstrated that it is possible to obtain a reliable IXS spectrum from measurements on samples in the diamond anvil cell [30]. The study has been made on fluid supercritical argon and has benefited from a carefully designed vacuum chamber dedicated to the DAC and from the minimization and control of sources of parasitic scattering.

The dedicated vacuum chamber is shown in Figure 10. The main features are the presence of an entrance slit (80 μm wide) and of the beamstop inside the vacuum chamber. The entrance window is placed at less than 30 mm from the sample, in order to allow a visual inspection and/or pressure measurement using an optical sensor. The slit allows the elimination of the parasitic scattering from air, which in this case would be otherwise very intense as the entrance window is placed very close to the sample. The beamstop blocks the beam inside vacuum, avoiding that the direct beam would pass through the exit window with a consequent generation of parasitic scattering. The entrance

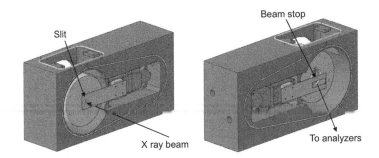

Figure 10. *Different views of the vacuum chamber designed for IXS experiments on samples placed inside a DAC. The "entrance" slit and the beam stop are placed inside the vacuum chamber and can be finely positioned by means of two motorised translation stages. This setup allows to completely eliminate the parasitic scattering (<%1 count/100 sec) apart from the ones originated by the DAC and by the sample. The DAC holder is surrounded by a ceramic heat shield in order to allow resistive heating of the DAC itself.*

slit and the beam stop are motorized in order to allow for a remote fine adjustment. Furthermore the entrance slit motorization has a travel of 25 mm, which allows the complete removal of it from the optical access in order to allow the visual observation and/or pressure measurement. The DAC holder allows the use of a resistive heater and is surrounded by a ceramic heat shield in order to permit heating of the DAC.

Figure 11 shows measurements of the "empty" cell with a standard vacuum chamber and with the dedicated one. Obviously a big improvement is present at low Q's, which means at low scattering angle, where the parasitic scattering described above is more important.

The quasi-elastic peak shown in the right panel of Fig. 11 is only relative to the diamonds of the DAC as the vacuum chamber itself shows that the intensity of the parasitic elastic scattering is completely negligible, i.e., less than 1 count/100 s. A careful study of this spurious residual signal has demonstrated that is correlated to the diamond quality and for reducing this signal at maximum, synthetic diamonds are the best candidates.

Furthermore, it is important to avoid the excitation of a Bragg peak as this may produce artefacts in the spectrum, see left panel of Fig. 12. This undesirable condition can be avoided by measuring the intensity profile of the transmitted direct beam as a function of the DAC rotation around one axis: the minima observed correspond to activations of Bragg peaks (see right panel of Fig. 12).

In order to allow quantitative measurements, the spurious empty cell sig-

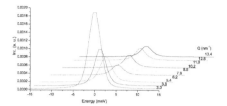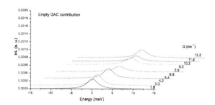

Figure 11. *Empty DAC measurement using a "standard" vacuum chamber, which is not able to eliminate completely the parasitic scattering at small Q values. The quasi-elastic peak at high Q values is only due to the scattering originating from the DAC. Right panel: Empty DAC measurement using the vacuum chamber described in the text (see Fig. 10). In this case the parasitic scattering has been completely eliminated and the quasi-elastic peak originating from the DAC is almost independent of the Q value. The DAC was equipped in both cases with type=Ia diamonds.*

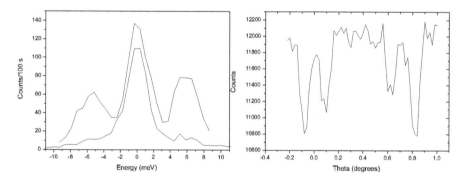

Figure 12. *Left panel: Effect of the activation of a diamond Bragg peak on the empty-DAC spectrum. The spurious peaks centred around 6 and −6 meV have been produced when the DAC orientation was such that a diamond Bragg peak was activated producing these artefacts for some unknown multiple scattering event. This condition can be avoided (see the lower spectrum in the left panel) by orienting the DAC in order not to activate a Bragg peak, corresponding to avoiding the minima of the transmitted direct beam intensity as a function of the DAC orientation (see right panel).*

nal should be reproducible. It has been observed that the quasi-elastic signal strongly depends on the sample purity. Unfortunately, when working with fluids, even if starting with a clean sample, some dust particles, probably present on the gasket surface in contact with the sample, may fall on the diamond culets, especially during unloading cycles. This is illustrated in Fig. 13 where the same sample at the same pressure and temperature, but before and after a

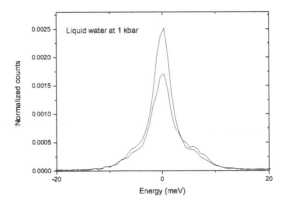

Figure 13. *Effect of dust particles inside the sample chamber. The two spectra have been measured on the same sample and at the same pressure and temperature, but before and after a pressure cycle. The dust particles which have fallen in the scattering volume during the decompression cycle have increased the intensity of the quasi-elastic peak, making impossible a subtraction of the empty cell a posteriori.*

pressure cycle, shows different intensities in the quasi-elastic region. In order to allow for a subtraction of the empty cell *a posteriori*, a good advice would be to measure the empty cell spectrum before and after the experiment and also to verify that during the experiment the quasi elastic peak does not show abrupt changes. This last point is generally ensured by avoiding decompression cycles during the experiment.

In conclusion, the study described in this work[28] shows that if you make parasitic scattering from air negligible, minimize the possibility of multiple scattering from diamonds by orienting the DAC and take extreme care in DAC cleaning (and avoid loading/unloading cycles), then quantitative IXS measurements on low-Z fluids in the DAC are possible. This is illustrated in the Fig. 14 where the spectra measured on fluid Ar have been super imposed on the spectra obtained from molecular dynamics simulations. The empty cell contribution (green line) has been subtracted from the raw spectra producing the red spectra. The agreement between experiment and simulation is excellent, demonstrating that the measured spectrum is reliable even in the quasielastic region and that after the empty cell subtraction, all the spurious effects originating from parasitic scattering in the quasi-elastic region have been eliminated.

The spectra have then been analysed using a viscoelastic model based on the memory function approach as described above and the results are reported in Fig. 15. The apparent dispersion ω_L (full dots) is always higher than the adiabatic sound frequency (black line) indicating the presence of positive dispersion, as obtained in a Brillouin scattering experiment [31]. The related

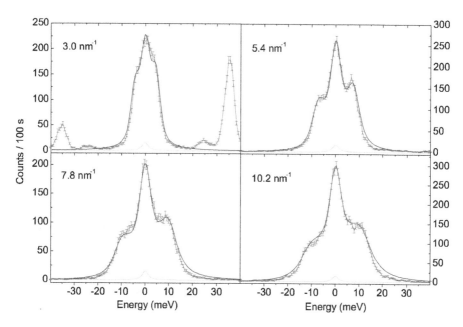

Figure 14. *IXS spectra at the indicated Q values on supercritical fluid argon at 1.2 GPa and room temperature. The red, blue and green lines represent the experimental (after the empty cell subtraction), the simulated and the empty-cell spectra. The agreement between experiment and simulation is excellent demonstrating that quantitative IXS measurements are possible using the DAC.*

viscoelastic transition takes place at Q values around 1.5 nm^{-1} (blue arrow) where the inverse of the structural relaxation time has values comparable to the frequency of the apparent sound dispersion. The adiabatic-to-isothermal transition, indicated by a red arrow, on the contrary is not producing a relevant effect on the apparent dispersion. The expected decrease of the frequency and the consequent passage from the adiabatic towards the isothermal dispersion, as observed in nitrogen (see above) is here compensated by the increase of the apparent dispersion for the effect of the positive dispersion due to the viscoelastic transition.

The positive sound dispersion, occurring in a supercritical fluid, such as argon at 1.2 GPa and room temperature, is not a trivial result. As seen above, the positive sound dispersion is a fingerprint of the liquid state and it progressively disappears on going towards supercritical conditions. However, it is important to notice that the experiments described above are performed along isobars close to the critical pressure, while in this case argon is in deeply supercritical conditions: $P/P_c \approx 245$ and $T/T_c \approx 2$.

A similar observation was for the first time reported on fluid supercritical

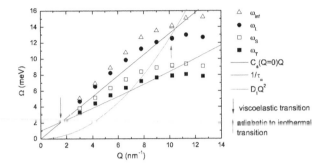

Figure 15. *Sound dispersions and evolution of the relaxation times in supercritical fluid Ar at 1.2 GPa and room temperature as obtained from the viscoelastic analysis. Open triangles, full dots, open squares and full squares represent the infinite, apparent, adiabatic and isothermal sound dispersions respectively. The blue and the red lines represent the inverse of the structural and thermal relaxation times respectively. The black line represents the adiabatic sound velocity as measured in a Brillouin scattering experiment (see text).*

oxygen, where a positive dispersion was observed at high pressure and room temperature using the DAC[16]. Also in this case oxygen is in deep supercritical conditions ($P/P_c \approx 1000$ and $T/T_c \approx 2$) and the presence of positive dispersion has been explained on the basis of a redefinition of the phase diagram in the supercritical region, which is further divided into a liquid-like and a gas-like region by the Widom line, an extrapolation of the liquid-vapour coexistence line, which separates the gas from the liquid state. The pressure and temperature conditions of the measurements performed on fluid argon and oxygen lie in the liquid-like region, and supercritical fluids in this region show the presence of the positive dispersion, a fingerprint of the liquid state.

References

[1] Molecular Hydrodynamics, J.P.Bo and S.Yip, McGraw-Hill New York, 1980.

[2] Dynamic Light Scattering, John Wiley and Sons, New York, 1976.

[3] E.H.Abramson *et al.*, Speed of sound and equation of state for fluid oxygen to 10 GPa, J. Chem. Phys., **110**, 10493 (1999).

[4] E.H.Abramson, L.J.Slutsky and Brown J. M., Thermal diffusivity of fluid oxygen to 12 GPa and 300 °C, J. Chem. Phys., **111**, 9357, (1999).

[5] E.H.Abramson, J.M.Brown and L.J.Slutsky L, The thermal diffusivity of water at high pressures and temperatures, J. Chem. Phys., **115**, 10461 (2001).

[6] F.Decremps et al., Sound Velocity and Absorption Measurements under High Pressure Using Picosecond Ultrasonics in a Diamond Anvil Cell: Application to the Stability Study of AlPdMn, Phys. Rev. Lett. **100**, 035502, (2008).

[7] K.Matsuishi et al., Equation of state and intermolecular interactions in fluid hydrogen from Brillouin scattering at high pressures and temperatures, J. Chem. Phys. **118**, 10683, (2001).

[8] F.Li, Brillouin scattering spectroscopy for a laser heated diamond anvil cell, Apl. Phys. Lett. **88**, 203507, (2006).

[9] S.W.Lovesey, Theory of neutron scattering from condensed matter, Clarendon Press, Oxford, (1987).

[10] P.A.Egelstaff, Neutron scattering studies of simple fluids, Il Nuovo Cimento D, **12**, 403, (1990).

[11] T.Scopigno, G.Ruocco and F.Sette, Microscopic dynamics in liquid metals: The experimental point of view, Rev. Mod. Phys. **77**, 881, (2005).

[12] F.Sette, Dynamics of Glasses and Glass-Forming Liquids Studied by Inelastic X-ray Scattering, Science, **379**, 1550, (1998).

[13] G.Ruocco et al., Equivalence of the sound velocity in water and ice at mesoscopic wavelengths, Nature, **379**, 521, (1996).

[14] T.Scopigno et al., Is the Fragility of a Liquid Embedded in the Properties of Its Glass?, Science, **302**, 849, (2003).

[15] D.Ishikawa *et al.*, Fast Sound in Expanded Fluid Hg Accompanying the Metal-Nonmetal Transition, Phys. Rev. Lett., **93**, 097801, (2004).

[16] K.Tamura *et al.*, Structural instability and the metal–non-metal transition in expanded fluid metals, Non-Crystalline Solids, **353**, 3348, (2007).

[17] M.Krisch and F.Sette in Light Scattering in Solids IX, Novel Materials and Techniques, Eds. M. Cardona and R. Merlin, Springer Verlag, Berlin Heidelberg, (2007).

[18] M.Krisch *et al.*, Pressure Evolution of the High-Frequency Sound Velocity in Liquid Water, Phys. Rev. Lett. **89**, 125502, (2002).

[19] F.A.Gorelli *et al.*, Liquidlike Behavior of Supercritical Fluids, Phys. Rev. Lett., **97**, 245702, (2006).

[20] R.Kubo, The fluctuation-dissipation theorem, Rep. on Progr. Phys., **29**, 255, (1966).

[21] U.Balucani and M.Zoppi, Dynamics of the liquid state, Clarendon Press, Oxford, (1994).

[22] U.Bafile, E.Guarini and Barocchi, Collective acoustic modes as renormalized damped oscillators: Unified description of neutron and x-ray scattering data from classical fluids, Phys. Rev. E, **73**, 061203, (2006).

[23] P.A.Egelstaff, An Introduction to the Liquid State, Oxford University Press, (1992).

[24] R.Verbeni et al., X-ray Monochromator with 2×10^8 Energy Resolution, J. Synchrotron Radiat., **3**, 62, (1996).

[25] A.Cunsolo *et al.*, Dynamics of Dense Supercritical Neon at the Transition from Hydrodynamical to Single-Particle Regimes, Phys. Rev Lett., **80**, 3515, (1998).

[26] A.A.van Well and L.A de Graaf, Density fluctuations in liquid neon studied by neutron scattering, Phys. Rev. A, **32**, 2396, (1985).

[27] A.Cunsolo *et al.*, Microscopic relaxation in supercritical and liquid neon, J. Chem. Phys., **114**, 2259, (2001).

[28] F.Bencivenga *et al.*, Adiabatic and isothermal sound waves: The case of super-critical nitrogen, Europhys. Lett., **75**, 70, (2006).

[29] F.Bencivenga *et al.*, Structural and Collisional Relaxations in Liquids and Supercritical Fluids, Phys. Rev. Lett., **98**, 085501, (2007).

[30] F.A.Gorelli et al., Inelastic x-ray scattering from high pressure fluids in a diamond anvil cell, Appl. Phys. Lett., **94**, 074102, (2009).

[31] M.Grimsditch, P.Loubeyre and A.Polian, Brillouin scattering and three-body forces in argon at high pressures, Phys. Rev. B, **33**, 7192, (1986).

Chapter 9
Optical Spectroscopy in the Diamond Anvil Cell

Alexander F. Goncharov

Geophysical Laboratory, Carnegie Institution of Washington, United States

1 Introduction

The study of materials under extreme pressures and temperatures is developing at an accelerating rate, with the observation of numerous new phenomena and implications that span physical, chemical, Earth and planetary, and biological sciences. An array of probes have been integrated recently to perform *in situ* measurements of materials properties under high static pressure including Raman [1, 2], Brillouin [3, 4], infrared, IR [5, 6], x-ray diffraction [7, 8], x-ray radiography [9] and x-ray spectroscopy [10], and electromagnetic methods [11]. Under extreme pressure-temperature $(P - T)$ conditions [12, 13] created in a diamond anvil cell, DAC (see an example of the laser heated DAC in Figure 1) these techniques have led to myriad discoveries with important implications, including new materials [14], phase transformations [15], anomalous physical properties [16], and superconductivity [17].

Optical spectroscopy is one of the most informative techniques for high pressure studies. The importance of the technique arises from power of spectroscopic methods for studying a wide variety of phenomena and transparency of the diamond windows of the high-pressure cell over a wide spectral range. In many cases, optical methods are unsurpassed, as in case of materials containing low-Z elements, disordered materials, studies of phase transitions, pressure calibration and lattice dynamics (e.g., [1]). Attractive features of optical spectroscopy methods include their non-contact nature, ability to provide high spatial resolution and to obtain sufficiently strong signals from microscopic samples. The use of lasers as light sources revolutionized optical spectroscopy; continuing development of laser technology (e.g., invention of powerful fiber and ultra-short pulse lasers) opens new possibilities (yet not fully realized) in

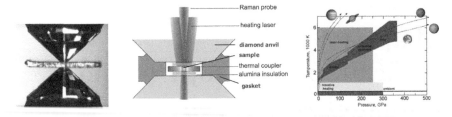

Figure 1. *Diamond anvil cell: left photograph (courtesy of S. Jacobsen), middle - schematic for laser heating experiment (courtesy of J. Crowhurst) and right - its applicability range.*

Table 1. *Electromagnetic radiation: frequency and wavelength ranges.*

Type	Frequency Range	Wavelength
γ-ray	$> 10^{21}$ Hz	< 0.003 nm
x-ray	$3 \times 10^{10} - 10^{21}$ Hz	$10 - 0.003$ nm
UV	$750 - 30000$ THz	$400 - 10$ nm
Visible	$400 - 750$ THz	$750 - 400$ nm
IR	300 GHz $- 400$ THz	1 mm $- 750$ nm
Microwave	$1 - 300$ GHz	30 cm $- 1$ mm
Radio	3 KHz $- 300$ GHz	100 km $- 1$ mm

the field of optical spectroscopy at high pressures. In fact, until the advent of extremely powerful synchrotron sources some 15 years ago, optical techniques were the only available for probing materials in the ultrahigh pressure regime. Optical probes provide important information about elastic, vibrational, and electronic states of materials and their dynamics. Here I give an overview of optical spectroscopy techniques, examples of recent studies, and a prospect of future developments.

2   Spectroscopy units, spectral ranges and dimension constraints

Optical spectroscopy covers a broad frequency range from 300 GHz to 30,000 THz that includes the infrared (IR), visible and ultraviolet (UV) spectral ranges (Table 1).

The relation between the frequency, f of the electromagnetic radiation, its

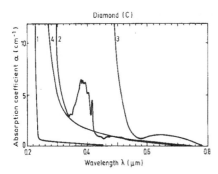 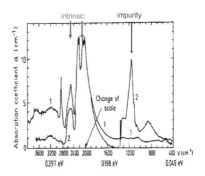

Figure 2. *Left: UV absorption of diamond. Right: IR absorption of diamond* [19, 68].

wavelength λ and the energy E of a light quantum, is given by

$$\frac{E}{h} = f = \frac{c}{\lambda}$$

where c is the speed of light in vacuum and h is the Planck constant. Traditionally, spectroscopy measures energy in wavenumbers $(1/\lambda)$, the number of wavelengths which can be accommodated in the length of 1 cm. Thus, optical spectroscopy spans the 10–10^6 cm^{-1} range. Diamond (used in anvils to create high-pressure conditions, Figure 1) is essentially transparent in this spectral region except for a moderate to strong absorption related to the 2- and 3-phonon intrinsic processes in the 2000–3000 cm^{-1} spectral range and nitrogen impurity bands near 1200 cm^{-1} (Figure 2).

Due to limitations of the sample dimensions in the DAC (normally the characteristic dimensions of the sample do not exceed 100 μm, Figure 1) and the wave properties of light, IR reflectivity and absorption measurements are severely affected by diffraction, which deteriorates the system throughput and reduces its special resolution. These conditions (diffraction limit) can be expressed as follows:

$$\lambda \leq \frac{d}{2NA}.$$

Where d is the characteristic linear dimension of the sample and $NA = n\sin\alpha$ is the numerical aperture of the lens used (determined through the half-angle of the maximum cone of light, α, and the refractive index, n). With common parameters such as $\alpha = 30°$, $n = 1$ (air) and $d = 100$ μm, the diffraction limit is near 100 cm^{-1}. Thus, far and mid-IR measurements would benefit from use of intense light sources (see below).

Table 2. *Elementary excitations probed by optical spectroscopy.*

Type	Typical Spectral Range
Phonons (lattice vibrations)	0–1600 cm^{-1}
Librons (restricted molecular rotations)	10–500 cm^{-1}
Rotons (free molecular rotations)	10–1000 cm^{-1}
Vibrons (intramolecular vibrations)	700–4250 cm^{-1}
Crystal field (electronic)	10000 cm^{-1}
Magnons (spin)	3000 cm^{-1}
Plasmons (free electron)	10000 cm^{-1}
Intra- and interband transitions (electronic transitions)	10000 cm^{-1}
Gap excitations (electronic or quasi- particles)	10–500 cm^{-1}

3   Basic principles

The main principle of the optical spectroscopy is based on the interaction of electromagnetic radiation with elementary excitations in the matter. Probing of the optical spectral range allows to tackle a broad range of elastic, vibrational (lattice), electronic, and magnetic excitations and their dynamics (Table 2). Those excitations can be probed either by matching their energies and polarization (reaching resonance conditions) with the probe light, or via inelastic scattering processes (Figure 3). In the latter, the system is being excited first to a virtual electronic state (or a real excited level in the case of resonance scattering or fluorescence) by an incident (pump) light quantum, and the emitted (scattered) quantum is analyzed for the change in energy and polarization (Figure 3). In the case of elastic (Rayleigh) scattering, the energy of the scattered photon is the same as the incident one, and no excitation is involved in the process. In inelastic light scattering (Raman, Brillouin) the scattered photon loses and gains energy which corresponds to that of the excitation energy (e.g., vibron), for Stokes and Anti-Stokes processes, respectively.

A general wave formulation of the interaction of matter with the external electrical field reads [20]:

$$P = \chi^{(1)} E_1 + \chi^{(2)} E_1 E_2 + \chi^{(3)} E_1 E_2 E_3 + \text{higher orders} \qquad (1)$$

Where P is the polarization induced, E_i are the electrical field components and $\chi^{(i)}$ are susceptibility tensors. It is the coefficient in the linear term, $\chi^{(1)}$, which corresponds to the dielectric constant that describes all aspects of linear optics like refraction, absorption, Rayleigh, Brillouin, and Raman scattering. For example, in the case of Raman scattering, phonons (or other excitations)

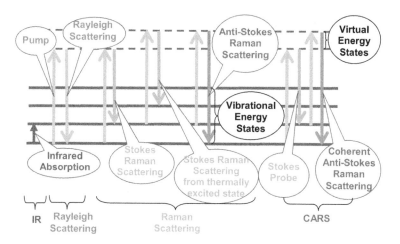

Figure 3. *Energy level diagram for the elementary processes which involve interaction of light with excitations in condensed matter.*

modulate $\chi^{(1)}$ causing light scattering on modified frequencies (beats). The corresponding selection rules are derived based on Equation (1). Briefly: IR modes have symmetry of polarization or linear coordinates, and Raman modes have symmetry of $\chi^{(1)}$ or squares of coordinates (xx, xy, etc). Higher-order terms in (1) are responsible for nonlinear phenomena. Since the values of higher-order $\chi^{(i)}$ are small, intense pulsed lasers are normally needed to observe these higher-orders effects. The second term in (1) accounts for such phenomena as hyper-Raman scattering, Pockels (linear electro-optic) effects, sum, difference and second harmonic generation. These effects require some intrinsic material anisotropy. In contrast, the third term can in principle be present in any material. This term is responsible for three wave-mixing phenomena, such as for example Coherent Anti-Stokes Raman Scattering (CARS) (Figure 3). CARS represents an example of a widely used technique (but not in high-pressure research yet), which is called a pump-probe technique (it can be also used in spectroscopic environment) (e.g., [21]). The (relatively) strong pump beam initiates some process of interest, e.g. an electronic transition. The probe beam, entering the sample later, will be amplified, attenuated or refracted because of the changes taking place in the sample. These changes are studied as a function of the delay between the pump and the probe beam, thereby allowing studying the dynamics of the processes in a very short time scale (e.g., tens of fs). The great advantage of time-domain spectroscopy is the ability to study fast transient phenomena, such as photoinduced bond breaking events and lattice vibrations. Since such events are relatively infrequent with continuous illumination, incoherent, and very short lived, they do not generate strong signals in comparison to static absorption in continuous-wave absorption experiments. Essentially, ultrafast pump-probe spectroscopy synchronizes many identical transient events to a common start time, leading to a large (i.e., detectable) macroscopic absorption change for many identical optical systems.

4 Techniques

Optical measurements in the DAC at high pressures in many cases pose a challenge for researchers because of a number of inherent problems connected to the design of the diamond anvil press. One of the main problems is the very small volume of the sample (of the order of pL) and its very small thickness (2-10 μm above 100 GPa), which requires very efficient ways of collecting the spectra. This problem becomes more severe for larger wavelengths (e.g., in the IR), where diffraction becomes the issue (see above). The use of optics with a large numerical aperture (NA>0.5) has become feasible recently (e.g., by using Boehler-Almax seats [22]), and it has been demonstrated that the requirements of stability of diamond anvils and supporting backing plates can be satisfied. These developments and also the availability of commercial lenses with a long focal length and fairly large numerical apertures (e.g., from Mitutoyo) made several of the problems manageable. Thus, the pressure limits of Raman [2], Brillouin [4], IR [6] measurements have been pushed to well above 1 megabar. The efficiency of the Raman technique can be substantially improved by using a high-throughput Raman system with holographic transmission entrance optics [23] and a single-stage fast imaging spectrograph with a multi-channel detector based on a charge-couple device (CCD) [24]. Use of these techniques for high-pressure research improved the capabilities of Raman systems very substantially [26, 27]. A second problem is related to the choice of diamonds used as high-pressure windows. Raman measurements require specially selected low-fluorescence diamonds (e.g., synthetic type II diamonds [29]) to suppress background signal and allow weak Raman bands (e.g., in metals) to be detectable [28, 30, 31]. Under very high pressure (depending on the diamond and excitation wavelengths, but sometimes > 130 GPa) a strong stress-induced red luminescence in diamond appears [18, 33], making even ruby fluorescence measurements challenging. In the case of IR measurements, nitrogen-impurity-free type-IIa diamonds should be used, but intrinsic second and third-order absorption in the region between 2.5 and 6 μm still blocks substantial amounts of radiation, thus reducing the signal-to-noise ratio. This signal reduction problems can be greatly corrected by using very thin diamonds (<1.2 mm height [27]), perforated diamonds [34], or the double-diamond anvil cell [35]. Infrared spectroscopy at ultrahigh pressures has been dramatically improved with synchrotron radiation as a light source [36, 37]. The synchrotron beam is essentially a point source and 2-4 orders of magnitude (depending on the wavelength) brighter than a blackbody source (Figure 4). These properties open possibilities to work beyond the classical

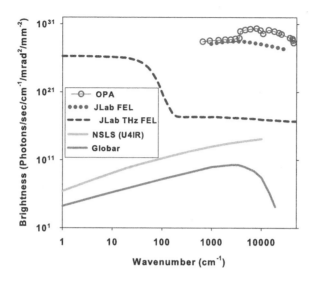

Figure 4. *Comparison of instantaneous brightness of the free electron laser of JLab (THz and IR) with that of a fs optical parametric amplifier (OPA) pumped by a Ti:Sapphire laser amplifier and time-averaged brightness of a synchrotron IR and conventional sources.*

diffraction limit and to study mid-IR absorption and reflectivity spectra of 5–25 microns samples [6, 38, 39, 40, 41]. The upgraded IR synchrotron facility at the NSLS has extended the spectral range for the DAC operation down to the far IR (20 cm^{-1}) [42, 43]. Rapid IR measurements (e.g., combined with shock compression or laser heating) require even higher system thoughput. This can be achieved by using pulsed IR lasers as the light sources. Current projects includes the use of a free-electron laser [44, 45] and a fs optical parametric amplifier (e.g., [46] (Figure 4).

Experimental methods based on the interaction of matter with continuous-wave visible or near-visible light (e.g., Raman, Brillouin, IR absorption and others) are well-established and versatile. They provide reliable information about equilibrium properties of materials under high pressure. Knowledge of materials dynamics such as thermal transport and acoustic wave propagation requires time-resolved measurements and pump-probe techniques mentioned above. The technique of impulsive stimulated light scattering (ISLS) [47] permits investigation of materials elasticity and thermal transport by generating and probing of transitory material acoustic and thermal gratings using laser pulses. Compared to the more familiar technique of Brillouin scattering (which in principle provides a similar information), ISLS has two important advantages. ISLS is naturally pulsed and gated and thus should be largely immune to the thermal emission that is associated with the high temperatures (>1000 K) of laser heating (as in case of pulsed Raman). Another advantage is the possibility to study the elasticity of opaque samples, for example metals [48, 49, 51, 52]. ISLS and Brillouin spectroscopy are often complementary to each other because the conditions for observation of different waves (e.g., transverse or longitudinal) are different in these techniques. For example, in Brillouin spectra, the transverse mode of diamond often interferes with the signal from the sample, (e.g., MgO, [53]), while it is readily detectable with ISLS [54].

5   Probing of intra- and inter-molecular interactions under pressure – the example of hydrogen

The atoms in simple molecular materials are strongly bonded by covalent intramolecular interactions, with much weaker intermolecular interactions of van der Waals type or hydrogen bond type (in hydrogen-containing materials). The closer association of the molecules that accompanies compression in general enhances intermolecular interactions. Under moderate compression, the molecules preserve their identity and materials can undergo a sequence of phase transformations (e.g., molecular orientational ordering) driven by various types of anisotropic forces, electronic charge redistribution due to the increasing overlap of the molecular charge density, charge transfer, as well as changes in magnetic and nuclear spin-dependent interactions. These changes are often described in terms of an effective intermolecular potential, which can be a function of the physical state of the material as well. At higher compression, there can be substantial modification of the molecules themselves. Indeed, an ensemble of molecules subjected to sufficiently high pressure ultimately undergoes molecular dissociation and transformation to a non-molecular state. An energy change arises from the reduction in kinetic energy of electrons that move from being localized in bonds to itinerant in the denser phase. The state and stability of molecules, inter- and intra-molecular interactions as well as molecular orientational ordering can be effectively probed by vibrational spectroscopy. Indeed, inter-particle potentials and ordering in molecular crystals can be probed through observations of elementary excitations, which give rise to Raman- and IR-active transitions (Figure 5). In the weak-coupling regime, one can expect the following optical modes to appear in the Raman and/or IR spectra of a molecular crystal (with m molecules in a unit cell, each molecule containing n atoms, and the total number of atoms in a unit cell $N = nm$):

- Intermolecular translations (phonons), N_{tr}=3m-3.

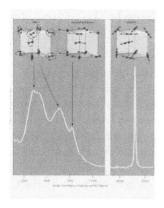 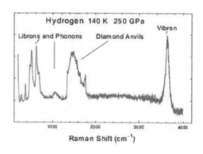

Figure 5. *Left: principal excitations in solid hydrogen phase I orientationally disordered hcp structure and the corresponding Raman spectrum ([56]). Right panel: Raman spectrum at 250 GPa and 140 K [25, 57] note the free rotations have become restricted to give rise to complex librons at low frequency.*

- Free and restricted rotations (rotons or librons)

$$N_{rot} = mN_{freedom}$$

where $N_{freedom}$ is the number of rotational degrees of freedom of the molecule.

- Intra-molecular modes (fundamentals or vibrons, the latter term is traditionally used for hydrogen and less often for other diatomic molecules),

$$N_{fundamental} = 3N - 3m - mN_{freedom}$$

The latter modes include vibrations of all kinds which include stretching, bending, wagging, rocking, twisting, and scissoring. Only Brillouin-zone-center optical modes can be probed because of the wave vector selection rule. Acoustic modes can be probed by Brillouin spectroscopy and ISLS (see above). Raman and IR activity are determined by selections rules according to the point symmetry of the crystal and symmetry of the mode examined. Here we give an example of hydrogen as the simplest diatomic molecule and also because of a number of fascinating properties which have been discovered under pressure (Figure 5). A full account of physical properties and phase diagram of hydrogen determined by optical spectroscopy (and other techniques) can be found in review articles [55, 56, 57, 66].

6 Optical properties of minerals in the deep Earth interior

The Earth's lower mantle contains vast amounts of rock, extending from just beneath the 660-km seismic discontinuity all the way to the core-mantle boundary at 2900 km depth. Lower-mantle materials are essentially semiconductors, insulating enough to promote convective heat transfer from the core, but heat conduction is gaining recognition as an important effect on geodynamic processes of the mantle at high pressure and temperature conditions. The optical properties of minerals, determined by optical absorption experiments, provide information on the radiative component thermal conductivity (κ_{rad}). The radiative component heat transfer is generally dependent on temperature [59], pressure [60, 61] and on the electronic structure of iron doping [62]. The presence of Fe in perovskite and ferropericlase strongly affects their physical properties compared with iron-free end-members enstatite (MgSiO3) and periclase (MgO). In addition to density, sound velocities, and rheology, iron can influence transport properties such as diffusion and conductivity. Optical properties and electrical and thermal conductivity depend considerably on Fe composition because iron belongs to d-block elements, called transition metals, which have d-shell electrons in their valence orbitals. Iron in compounds can exist in a number of oxidation states. Ferric iron is stable in most surface environments, while the Earth's mantle contains both ferrous and ferric iron, depending on the activity of oxygen in the mantle (i.e. oxidation fugacity) and on bulk minerals composition (such as Al-content) [63]. Since processes in the bio- and geosphere involve changes in redox state (Fe^{3+}/Fe^{2+} ratio), there is a number of broad consequences ranging from biogeochemical control to oxygenation of the atmosphere and oxidation state of the mantle. If both ferric and ferrous irons are present in the deep mantle as suspected, electrical conductivity can increase substantially because of electron hopping between iron ions with different oxidation states. The spin state of iron is another important factor affecting the physical properties of iron-containing minerals. Depending on the total spin of a system (high = spins unpaired or low = spins paired), the electronic structure of minerals changes, causing modifications of thermoelastic, magnetic and optical properties. The spin state depends upon the energy balance between the crystal field splitting and spin pairing energy (Figure 6), and this balance depends on thermodynamic conditions (e.g., pressure). Electronic spin transitions, predicted theoretically almost twenty years ago [64] have been recently discovered experimentally under high static pressures of 40–120 GPa (e.g., [65]) in both silicate perovskite and ferropericlase.

Optical properties of lower-mantle minerals, synthesized at various conditions, have now been studied over a wide spectral range including mid- and near-infrared, visible and ultraviolet (2000–35000 cm^{-1}) and at high-pressures and temperatures *in-situ*. Both iron free materials, enstatite and periclase, are

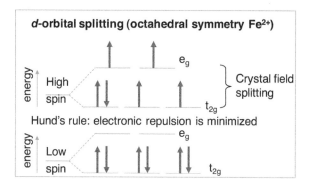

Figure 6. *Electronic configuration of the ground states in high- and low-spin configurations of $Fe^{2+}(3d^6)$.*

insulators with a wide band gap, so they are transparent in this spectral range. In Fe-bearing minerals, the major absorption band is believed to be caused by a crystal field transition of the high-spin Fe^{2+} ion. This transition is split into several components by the Jahn-Teller effect; the number of components and the symmetry of the ground and excited states are determined by a local symmetry of the iron ion environment. In ferropericlase, the $^5T_{2g} \rightarrow {}^5E_g$ crystal-field transition of the Fe^{2+} ion in an octahedral site is split into two components; in silicate perovskite, the $^5E_g \rightarrow {}^5T_{2g}$ crystal-field transition of the Fe^{2+} ion in a dodecahedral site is split into three components. These transitions are in the near-infrared spectral range ($7000–12500$ cm^{-1}), and they are expected to increase gradually in frequency with pressure. An abrupt change of absorption is expected at the spin-transition; the crystal field transitions were predicted to move to visible and ultraviolet, leaving the near-IR transparent [65]. Crystal-field bands of Fe^{3+} are of low intensity in the spectral range of interest because they are spin-forbidden. Besides the crystal-field transitions, other contributions include Fe^{2+}–Fe^{3+} intervalence transitions (near 15000 cm^{-1} in near-infrared to visible range) and Fe–O charge transfer (in the ultraviolet). Until quite recently, optical properties of mantle minerals at pressures comparable to the lower mantle ($24–133$ GPa) were unknown. The effects of temperature were estimated using corresponding measurements at ambient pressure. The effects of pressure were assumed to be small and were mainly attributed to those related to the crystal field band absorption. This approach becomes problematic if the absorption mechanism alters because of pressure-induced changes in electronic structure or if other absorption mechanisms than observed at ambient pressure come to play under pressure. These measurements deal with very small samples (typically $40 \times 40 \times 15$ μm^3) of laboratory-grown single-crystals (Figure 7) immersed into a transparent soft material serving as a pressure medium. The results of the experiments

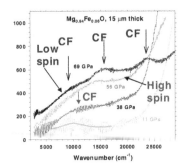

Figure 7. *Left: A microphotograph of a ferropericlase ($Mg_{0.8}Fe_{0.2}O$) single crystal of approximately $50 \times 50 \times 8$ micrometers dimensions on a diamond anvil at ambient conditions (top) and in the DAC under a 60 GPa pressure (bottom). Right: optical absorption spectra of $Mg_{0.94}Fe_{0.06}O$ ferropericlase through the spin transition[62].*

provide information about the wavelength-dependent absorption coefficient, $\alpha(cm^{-1}) = ln(10)/l_{10}$, where l_{10} is the path length in the material at which the incident radiation is attenuated by a factor 10.

Pressure-dependent optical absorption in ferropericlase [62] and silicate perovskite [66, 67] reveals a variety of unexpected phenomena which include substantial changes in absorption in the spectral range of interest (Figure 7). Contrary to the predictions, these changes are mainly related to an increase of absorption of Fe–O (in perovskite and ferropericlase) and Fe–Fe (in ferropericlase) charge transfer bands - these effects were largely overlooked in the past. The changes in absorption related to the spin transitions are notable, but the effects are much smaller. Moreover, these changes become even less pronounced at high temperature because of a smearing out of the spin transition. The results suggest that the absorption of mantle rocks is mainly governed by the iron concentration: Fe^{2+} in ferropericlase and Fe^{3+} in silicate perovskite. In case of ferropericlase, the effect is critically dependent on Fe concentration: the absorption increases drastically near the percolation limit ($\sim 12\%$ Fe). In case of silicate perovskite, the presence of Fe^{3+} increases intervalence Fe^{2+}–Fe^{3+} and Fe^{3+}–O charge transfer bands.

7   Prospects

Optical spectroscopy continues to be a very useful tool for studies of materials at ultrahigh pressures. With the development of new techniques, the abilities of Raman and IR methods can be extended to higher pressures and over a wider spectral range, for example, by using newly designed holographic optics and synchrotron light sources. Future progress in the field may be expected with the development of time-resolved Raman and IR spectroscopy and free-electron-laser-based IR techniques.

8   Acknowledgements

I acknowledge support from NSF/EAR, DOE/BES, DOE/ NNSA (CDAC) and the W. M. Keck Foundation. I thank V.V. Struzhkin, (Geophysical Lab), R.J.Hemley (Geophysical Lab), G.P.Williams (Jefferson Lab), J.C. Crowhurst (LLNL), J. M. Brown (U. Washington), S. D. Jacobsen (Northwestern U.), L. Carr (BNL) and Z. Liu (Geophysical Lab and BNL) for important information and useful discussions.

References

[1] Hemley R.J., Bell P.M., and Mao H.K., Laser techniques in high-pressure geophysics, Science, **237**, 605 (1987).

[2] Loubeyre P., Occelli F. and LeToullec R., Optical studies of solid hydrogen to 320 GPa and evidence for black hydrogen, Nature, **416**, 613 (2002).

[3] Shimizu H., Brody E.M., Mao H.K., and Bell P.M., Brillouin measurements of solid n-H2 and n-D2 to 200 kbar at room temperature, Phys. Rev. Lett., **47**, 128 (1981).

[4] Murakami M., Sinogeikin S.V., Hellwig H., Bass J.D., Li J., Earth Planet. Sci. Lett., **256**, 47 (2007).

[5] Hanfland M., Hemley R.J., Mao H.K., and Williams G.P., Synchrotron infrared spectroscopy at megabar pressures: vibrational dynamics of hydrogen to 180 GPa, Phys. Rev. Lett., **69**, 1129 (1992).

[6] Goncharov A.F., Struzhkin V.V., Somayazulu M., Hemley R.J., and Mao H.K., Compression of ice to 210 GPa: Evidence for a symmetric hydrogen bonded phase, Science, **273**, 218 (1996).

[7] Mao H.K., Hemley R.J., Wu Y., Jephcoat A.P., Finger L.W., Zha C.S., and Bassett W.A., High-pressure phase diagram and equation of state of solid helium from single crystal x-ray diffraction to 23.3 GPa, Phys. Rev. Lett., **60**, 2649 (1988).

[8] Dubrovinsky L., Dubrovinskaia N., Narygina O., Kantor I., Kuznetzov A.,Prakapenka V.B., Vitos L., Johansson B., Mikhaylushkin A.S., Simak S.I., and Abrikosov I.A., Body-Centered Cubic Iron-Nickel Alloy in Earth's Core, Science, **316** 1880 (2007).

[9] Hemley R.J., Mao H.K., Shen G., Badro J., Gillet P., Hanfland M., and Husermann D., X-ray imaging of stress and strain of diamond, iron, and tungsten at megabar pressures, Science, **276**, 1242 (1997).

[10] Mao, H. K., J. Xu, V.V. Struzhkin, J. Shu, R. J. Hemley, W. Sturhahn, M.Y. Hu, E. E. Alp, L. Vocadlo, D. Alf, G. D. Price, M. J. Gillan, M. Schwoerer-Bhning, D. Husermann, P. Eng, G. Shen, H. Gieffer , R. Lbbers, and G. Wortmann, Phonon density of states of iron to 153 GPa, Science, 292, 914-916 (2001).

[11] Struzhkin V.V., Timofeev Y.A., Hemley R.J., and Mao H.K., Superconducting Tc and electron-phonon coupling in Nb to 132 GPa: magnetic susceptibility at megabar pressures, Phys. Rev. Lett., **79**, 4262 (1997).

[12] Eremets M.I., Hemley R.J., Mao H.K., and Gregoryanz E., Semiconducting non-molecular nitrogen up to 240 GPa and its low-pressure stability, Nature, **411**, 170 (2001).

[13] Fei Y. and Mao H.K., In situ determination of the NiAs phase of FeO at high pressure and high temperature, Science, **266**, 1678 (1994).

[14] Gregoryanz E., Sanloup C., Somayazulu M., Badro J., Fiquet G., Hemley R.J., and Mao H.K., Synthesis and characterization of a binary noble metal nitride, Nature Mat., **3**, 294 (2004).

[15] Goncharov A.F., Gregoryanz E., Mao H.K., Liu Z., and Hemley R.J., Optical evidence for a nonmolecular phase of nitrogen above 150 GPa, Phys. Rev. Lett., **85**, 1262 (2000).

[16] Gregoryanz E., Degtyareva O., Somayazulu M., Hemley R.J., and Mao H.K., Melting of dense sodium, Phys. Rev. Lett., **94**, 185502 (2005).

[17] Struzhkin V.V., Hemley R.J., Mao H.K., and Timofeev Y.A., Superconductivity at 10 to 17 K in compressed sulfur, Nature, **390**, 382 (1997).

[18] Mao H.K., and Hemley R.J., Experimental study of the Earth's deep interior: accuracy and versatility of diamond cells, Proc. Roy. Soc. London A, **354**, 1315 (1996).

[19] Wilks J. and Wilks E., Properties and Applications of Diamond, Butterworth-Heinemann Ltd., Oxford, (1991).

[20] Bloembergen N., "Conservation laws in nonlinear optics," J. Opt. Soc. Am., **70**, 1429 (1980).

[21] Reid G.D. and Wynne K. Ultrafast laser technology and spectroscopy. In Encyclopedia of Analytical Chemistry (ed. Meyers, R. A.) 13644-13670 (John Wiley & Sons, Ltd., Chichester, 2000).

[22] Boehler R. and De Hantsetters K. New anvil designs in diamond-cells. High Pressure Res., **24**, 391 (2004).

[23] Tedesco J.M., Owen H., Pallister D., Morris M.D., Anal. Chem., **65**, 441A (1993).

[24] Delhaye M., Barbillat J., Aubard J., Bridoux M., Da Silva E. In Raman Microscopy, Turrell G., Corset J. (eds). Academic Press: London, **51**, (1996).

[25] Williams K.P.J., Pitt G.D., Smith B.J.E., Whitley A., Batchelder D.N., Hayward I.P., J. Raman Spectrosc., **25**, 131, (1996).

[26] Goncharov A.F., Struzhkin V.V., Hemley R.J., Mao H.K., Raman Spectroscopy of Dense H2O and the Transition to Symmetric Hydrogen Bonds, Phys. Rev. Lett, **83**, 1998, (1999).

[27] Goncharov A.F., Struzhkin V.V., Hemley R.J., Mao H.K., and Liu Z., New techniques for optical spectroscopy at ultrahigh pressures. Science and Technology of High Pressure, Honolulu, Hawaii, Universities Press, Hyderabad, India: 9095 (2000).

[28] Goncharov A.F. and Struzhkin V.V., Raman spectroscopy of metals, high temperature superconductors and related materials under high pressure, J. Raman Spectrosc., **34**, 532 (2003).

[29] Goncharov A.F., Hemley R.J., Mao H.K., J. Shu, New High-Pressure Excitations in Parahydrogen, Phys. Rev. Lett., **80**, 101, (1998).

[30] Merkel S., Goncharov A.F., Mao H.K., Gillet P., Hemley R.J. Raman spectroscopy of iron to 152 gigapascals: implications for Earth's inner core, Science, **288**, 1626, (2000).

[31] Goncharov A.F., Gregoryanz E., Struzhkin V.V., Hemley R.J., Mao H.K., Boctor N., Huang E., Raman scattering of metals to very high pressures, In High Pressure Phenomena, Hemley RJ, et al. (eds). Proceedings of the International School of Physics Enrico Fermi, Course CXLVII. Societ'a Italiana di Fisica, p. 297 (2002).

[32] Mao H.K., Hemley R.J., Optical transitions in diamond at ultrahigh pressures, Nature, **351**, 721, (1991).

[33] Vohra Y.K., in: Recent Trends in High Pressure Research: Proceedings of the XIII AIRAPT International Conference on High Pressure Science and Technology, Bangalore, India, 7-11 October 1991, ed. by A. K. Singh (Oxford & IBH Publishing Co., New Delhi, 1992).

[34] Dadashev A., Pasternak M.P., Rozenberg G.K. and Taylor R.D. Applications of perforated diamond anvils for very high-pressure research, Rev. Sci. Instrum., **72**, 2633, (2001).

[35] Silvera I.F., The double-diamond anvil cell, the poor-man's megabar pressure cell, Rev. Sci. Instrum., **70**, 4609, (1999).

[36] Carr G.L., Hanfland M., Williams G., Rev. Sci. Instrum., **66**, 1643, (1995).

[37] Hemley R.J., Goncharov A.F., Lu R., Li M., Struzhkin V.V. and Mao H.K., High-pressure synchrotron infrared spectroscopy at the National Synchrotron Light Source, Il Nuovo Cimento, **20**, 539, (1998).

[38] Hemley R.J., Mao H.K., Goncharov A.F., Hanfland M., Struzhkin V.V., Phys. Rev. Lett., **76**, 1667, (1996).

[39] Struzhkin V.V., Goncharov A.F., Hemley R.J., Mao H.K., Phys. Rev. Lett., **78**, 4446, (1997).

[40] Hemley R.J., Mazin I.I., Goncharov A.F. and Mao H.K., Europhys. Lett., **37** 6, 403, (1997).

[41] Hemley R.J., Goncharov A.F., Mao H.K., Karmon E. and Eggert J.H., Spectroscopic studies of p-H2 to above 200 GPa, J. Low. Temp. Phys., **110**, 75, (1998).

[42] Klug D.D., Zgierski M.Z., Tse J.S., Liu Z., Kincaid J.R., Czarnecki K. and Hemley R.J., "Doming modes and dynamics of model heme compounds", PNAS, **99**, 1256, (2002).

[43] Klug D.D., Tse J.S., Liu Z., Gonze X., and Hemley R.J., Anomalous transformations in Ice VII, Phys. Rev. B, **70**, 144113, (2004).

[44] Neil G.R., Bohn C.L., Benson S.V., Biallas G., Douglas D., Dylla H.F., Evans R., Fugitt J., Grippo A., Gubeli J., Hill R., Jordan K., Li R., Merminga L., Piot P., Preble J., Shinn M., Siggins T., Walker R., Yunn B., Sustained Kilowatt Lasing in a Free-Electron Laser with Same-Cell Energy Recovery, Phys. Rev. Lett., **84**, 662, (2000).

[45] Neil G.R., Behre C., Benson S.V., Bevins M., Biallas G., Boyce J., Coleman J., Dillon-Townes L.A., Douglas D., Dylla H.F., Evans R., Grippo A., Gruber D.,

Gubeli J., Hardy D., Hernandez-Garcia C., Jordan K., Kelley M.J., Merminga L., Mammosser J., Moore W., Nishimori N., Pozdeyev E., Preble J., Rimmer R., Shinn M., Siggins T., Tennant C., Walker R., Williams G.P., Zhang S., The JLab High Power ERL Light Source, Nucl. Instrum. Methods, **A557**, 9, (2006).

[46] Moore D.S., Funk D.J. and McGrain D.S., At the Confluence of Experiment and Simulation: Ultrafast Laser Spectroscopic Studies of Shock Compressed Energetic Materials, in Chemistry Under Extreme Conditions (ed. Manaa, R.) pp. 360–398 (Elsevier Science, 2005).

[47] Yan Y. and Nelson K.A., Impulsive stimulated scattering. I. general theory. J. Chem. Phys., **87**, 6240, (1987).

[48] Crowhurst J.C., Abramson E.H., Slutsky L.J., Brown, J. M., Zaug, J. M. and Harrell, M. D. Surface acoustic waves in the diamond anvil cell: An application of impulsive simulated light scattering. Phys. Rev. B, **64**, 100103, (2002).

[49] Goncharov A.F., Crowhurst J.C., Zaug J.M., Elastic and Vibrational Properties of Cobalt to 120 GPa, Phys. Rev. Lett., **92**, 115502, (2004).

[50] Crowhurst J.C., Goncharov A.F., Zaug J.M. Impulsive stimulated light scattering from opaque materials at high pressure. J. Phys.: Condens. Matter, **16**, S1137, (2004).

[51] Crowhurst J.C., Goncharov A.F., and Zaug J.M. (2005). Direct measurements of the elastic properties and of Fe and Co to 120 GPa implications for the composition of Earths core. In Advances in High-Pressure Technology for Geophysical Applications (eds. J. Chen, Wang, Y., Duffy, T., Shen, G. and Dobrzhinetskaya, L.) 3-19 (Elsevier, Amsterdam, 2005) (AGU).

[52] Crowhurst J.C., Antonangeli D., Brown J.M., Goncharov A.F., Farber D.L., Aracne C.M., Determination of the high pressure elasticity of cobalt from measured interfacial acoustic wave velocities, Appl. Phys. Lett., **89**, 111920, (2006).

[53] Zha C.S., Mao H.K. and Hemley R.J., Elasticity of MgO and a primary pressure scale to 55 GPa, PNAS, **97**, 13494, (2000).

[54] Crowhurst J.C., Brown J.M., Goncharov A.F. and Jacobsen S.D., Elasticity of (Mg,Fe)O through the spin transition of iron in the lower mantle, Science, **319**, 451, (2008).

[55] Silvera I.F., Rev. Mod. Phys., **52**, 393, (1980).

[56] Mao H.K. and Hemley R.J., Ultrahigh-pressure transitions in solid hydrogen, Rev. Mod. Phys., **66**, 671, (1994).

[57] Goncharov A.F. and Hemley R.J., Probing hydrogen-rich molecular systems at high pressures and temperatures, Chem. Soc. Rev., **35**, 899, (2006).

[58] Goncharov A.F., Crowhurst J.C., Proton delocalization under extreme conditions of high pressure and temperature. Phase Transitions, **80**, 1051, (2007).

[59] Hofmeister A.M., Mantle Values of Thermal Conductivity and the Geotherm from Phonon Lifetimes, Science, **283**, 1699, (1999).

[60] Mao H.K. and Bell P.M., Electrical Conductivity and the Red Shift of Absorption in Olivine and Spinel at High Pressure, Science, **176**, 403, (1972).

[61] Keppler H., Smyth J.R., Optical and near infrared spectra of ringwoodite to 21.5 GPa: Implications for radiative heat transport in the mantle. Am. Miner., **90**, 1209, (2005).

[62] Goncharov A.F., Struzhkin V.V., Jacobsen S.D., Reduced Radiative Conductivity of Low-Spin (Mg,Fe)O in the Lower Mantle, Science, **312**, 1205, (2006).

[63] McCammon C., Perovskite as a possible sink for ferric iron in the lower mantle, Nature, **387**, 694, (1997).

[64] Sherman D.M. The high-pressure electronic structure of magnesiowustite (Mg,Fe)O: applications to the physics and chemistry of the lower mantle. J. Geophys. Res., **96**, 14299, (1991).

[65] Badro J., Rueff J.P., Vanko G., Monaco G., Fiquet G., Guyot F., Transitions in Perovskite: Possible Nonconvecting Layers in the Lower Mantle, Science, **305**, 383, (2004).

[66] Goncharov A.F., Haugen B.D., Struzhkin V.V., Beck P. and Jacobsen S.D., Radiative conductivity in the Earths lower mantle, Nature, **456**, 231, (2008).

[67] Keppler H., Dubrovinsky L.S., Narygina O., Kantor I., Optical Absorption and Radiative Thermal Conductivity of Silicate Perovskite to 125 Gigapascals, Science, **322**, 1529 (2008).

[68] http://www.ioffe.rssi.ru/SVA/NSM/Semicond/Diamond/optic.html

Chapter 10
Magnetism and High Pressure

F. Malte Grosche

The University of Cambridge, United Kingdom

Among the various forms of electronic order in condensed matter, magnetism takes a special place. Ferrromagnetism has been known for several thousand years, but new forms of magnetism are being discovered at an increasing rate. The theoretical understanding of magnetism, similarly, has a long history and continues to present interesting challenges. Moreover, it appears increasingly plausible that further to the magnetic structure itself, deviations away from the ordered structure, magnetic *fluctuations*, can couple to the electrons from which the magnetic order arises in the first place, and lead to novel electronic states, such as unconventional superconductivity. This section aims to give a brief introduction to elements of the theory of magnetism in solids.

1 Magnetic equation of state, feedback instability

If we want to consider the magnetic properties of a material which may become ferromagnetic at sufficiently low temperature T, then it will be necessary to introduce the magnetisation, M, as the response of the material to an externally applied field, H. At small applied fields, at least, we may expect the magnetisation to follow the field, at constant temperature, according to an *equation of state* of the form

$$H = aM + bM^3, \tag{1}$$

where, to keep things simple, we neglect the vector nature of H and M. In this equation of state, a takes the role of the inverse susceptibility $\chi^{-1} = dH/dM$ and b ensures that the magnetisation bends over towards saturation for high fields.

For a material to be ferromagnetic, we require a finite M, a *remanent magnetisation*, even for zero H. This would appear to be possible only, if the parameter a in the equation of state is negative. Systems of non-interacting electrons, however, do not exhibit a negative susceptibility: the Curie law for isolated local moments gives $a \propto T$, whereas in metals, the Pauli susceptibility (to which we return briefly below) is positive and weakly temperature dependent. The transition to magnetism must, therefore, be caused by interactions between the electrons.

The simplest way to incorporate these interactions is to introduce an *exchange molecular field*, h, into the equation of state: $H + h = aM + bM^3$. The exchange molecular field is not a real magnetic field, which could deflect a compass needle or induce voltages in a pick-up coil. It is a way to represent, in a mean field sense, the influence of other electrons on a test electron. If we assume that the exchange field is simply proportional to the overall magnetisation (with constant of proportionality λ, this is the Weiss molecular field concept), then we arrive at a feed-back equation:

$$H + \lambda M = aM + bM^3,$$

which can be recast in the form of the original equation of state, with a modified linear coefficient $a^* = a - \lambda$:

$$H = (a - \lambda)M + bM^3 = a^*M + bM^3.$$

We see, then, that although the noninteracting susceptibility is finite, interactions between the electrons give rise to a feed-back effect, which boosts the magnetic susceptibility $\chi = 1/a^* = \chi_0/(1 - \lambda)\chi_0$, where $\chi_0 = 1/a$ is the noninteracting susceptibility. This leads to a magnetic instability, if

$$\lambda\chi_0 > 1 \qquad\qquad (2)$$

The above equation represents a simple form of the *Stoner criterion*, which can be generalised to investigate many forms of electronic order.

As we have seen, magnetism is produced by interactions between electrons, but in the Weiss field concept, the interaction field couples to the magnetisation produced by the electrons. This is surprising, because the large Coulomb interaction between the electrons couples only to the charge, not to the spin of the electrons. The first idea might just be that the moments could couple through the dipole magnetic fields they generate. However, this is very small: the energy of interaction of two magnetic dipoles of strength m at a distance r is of order $\mu_o m^2/4\pi r^3$. Putting in a magnetic moment of order a Bohr magneton, we get

$$U_{dipolar} \approx \frac{\mu_o}{4\pi}\left(\frac{e\hbar}{2m}\right)^2 \frac{1}{r^3} \approx \pi\alpha^2\left(\frac{a_{Bohr}}{r}\right)^3 \text{ Ryd.} \qquad (3)$$

where $\alpha \approx 1/137$ is the fine structure constant. At typical atomic separations of 2 nm, this is about 4×10^{-5} eV, or less than a degree Kelvin.

If the dipolar interactions are too weak to explain the robust magnetic order observed in real material, then how can a spin-dependent interaction arise from the starting Hamiltonian governing the electrons and nuclei in the solid, in which only Coulomb interactions appear to be relevant? A number of distinct mechanisms have been identified, all of which start with a reduction in the energy scale over which the model claims to apply (*"integrating out high-energy degrees of freedom"*), which then results in a renormalisation of the effective interaction. In other words, once the electrons have been confined to a particular sub-set of low energy states, the effects of Pauli exclusion are such that the total energy depends on the spin configuration. There are complementary views of magnetism as originating either from the alignment of *local moments* or from a spontaneous spin polarisation of itinerant electrons. We begin with the former.

2 Types of magnetic interactions

Direct exchange

As a model system, let us consider two electrons in two orbitals, $|a>, |b>$, which are orthogonal to each other. Because the electrons are indistinguishable, the two-body wavefunction has to be antisymmetric under particle exchange:

$$\Psi(\vec{r}_1, \vec{r}_2) = -\Psi(\vec{r}_2, \vec{r}_1).$$

A simple approximation to the full two-body wavefunction can be formed from antisymmetrised product wavefunctions:

$$\Psi(\vec{r}_1, \vec{r}_2) = \frac{1}{\sqrt{2}} \left(|a\rangle|b) - |b\rangle|a) \right),$$

where – just for the moment – we use round brackets to denote the state of the second electron and angled brackets for the state of the first electron. For many-electron systems, this expression generalises to the Slater determinant.

If we now consider spin, as well, then we find four possible antisymmetrised two-particle states, which can be combined grouped into one state with a singlet spin wavefunction, for which the spatial state is symmetric under particle exchange

$$\frac{1}{2}(|ab\rangle + |ba\rangle)(|\uparrow\downarrow\rangle - |\downarrow\uparrow\rangle)$$

and three states with triplet spin wavefunctions, for which the spatial state is antisymmetric under particle exchange

$$\frac{1}{2}(|ab\rangle - |ba\rangle) \begin{pmatrix} |\uparrow\uparrow\rangle \\ |\uparrow\downarrow\rangle + |\downarrow\uparrow\rangle \\ |\downarrow\downarrow\rangle \end{pmatrix}.$$

We will now find that the singlet state has a higher energy than the triplet state in the presence of Coulomb repulsion between the electrons:

We introduce some shorthand notation:

$$|ab\rangle \equiv |a\rangle|b\rangle, |ba\rangle \equiv |b\rangle|a\rangle \tag{4}$$

$$E_0 = \langle ab|\hat{H}|ab\rangle - E_a + E_b + E_{Coul} \tag{5}$$

$$E_{Coul} \equiv \langle ab|\hat{H}_{1,2}|ab\rangle \tag{6}$$

$$= \int d^3\vec{r}_1 d^3\vec{r}_2 |\psi_a(\vec{r}_1)|^2 |\psi_b(\vec{r}_2)|^2 V(\vec{r}_1 - \vec{r}_2) \tag{7}$$

$$E_{ex} \equiv \langle ba|\hat{H}_{1,2}|ab\rangle \tag{8}$$

$$= \int d^3\vec{r}_1 d^3\vec{r}_2 \psi_b^*(\vec{r}_1)\psi_a(\vec{r}_1)\psi_a^*(\vec{r}_2)\psi_b(\vec{r}_2) V(\vec{r}_1 - \vec{r}_2). \tag{9}$$

Here, E_{Coul} looks like Coulomb repulsion between charge densities, and E_{ex} resembles E_{Coul}, but the electrons have traded places ($\Rightarrow$ exchange term). For short range interactions, such as $V = \delta(\vec{r}_1 - \vec{r}_2)$, $E_{Coul} \to E_{ex}$.

We now find that the energy of the singlet state is

$$E_{singlet} = \frac{1}{2}\left(\langle ab + ba|\hat{H}|ab + ba\rangle\right) \tag{10}$$

$$= E_0 + E_{ex}. \tag{11}$$

The energy of the triplet state, however, is lower:

$$E_{triplet} = \frac{1}{2}\left(\langle ab - ba|\hat{H}|ab - ba\rangle\right) \tag{12}$$

$$= E_0 - E_{ex} \tag{13}$$

There is, therefore, a spin dependent effective interaction in this simple model system. Note that this interaction arises, because the electrons have been constrained to single occupancy of the two orbitals, leaving only spin flips as the remaining degrees of freedom.

This simple example reflects a general phenomenon: the spin triplet state is symmetric under particle exchange and must therefore be multiplied by an antisymmetric spatial wavefunction. An antisymmetric spatial wavefunction must have nodes whenever two spatial coordinates are equal: $\psi(...., r_i = r, ...r_j = r, ...) = 0$. So it is then clear that the particles stay farther apart in an antisymmetrised state than in a symmetric state, and because of the Pauli principle an antisymmetric wavefunction (which will generally have high spin) has lower energy.

The physical reason for the existence of local moments on atoms is then a combination of the Pauli principle together with repulsive interactions between electrons. If we consider, say, d-levels in an ion, since the d-states are degenerate, we shall always get high spin configurations. However, in the environment of a solid, the d-levels are split because the atom is no longer in a potential of spherical symmetry. If this *crystal field splitting* is large enough, then the orbitals will be filled one after another - and generally the atom will have low spin.

When the orbitals concerned are orthogonal, E_{ex} is *positive* in sign, i.e. the lowest energy state is a triplet. However, if the overlapping orbitals are not orthogonal – as will happen between two orbitals between neighbouring atoms – the interaction may be of a *negative* sign, so the lowest energy is a singlet.

Heisenberg Hamiltonian

We can express the spin-dependent interaction between the electrons, which has arisen from the direct exchange term E_{ex}, in terms of the spin states of the two electrons, which are probed by the spin operators $\hat{S}_1$ for electron 1 and $\hat{S}_2$ for electron 2. Because triplet and singlet states differ in the expectation value of the magnitude of the total spin $\hat{S} = \hat{S}_1 + \hat{S}_2$, we can use this to distinguish between the singlet and triplet states:

$$\hat{S}^2 = (\hat{S}_1 + \hat{S}_2)^2 = \frac{3}{2} + 2\hat{S}_1 \cdot \hat{S}_2.$$

This leads to the definition of a new operator $\hat{H}_{spin}$

$$\hat{H}_{spin} = \frac{1}{4}(E_{singlet} + 3E_{triplet}) - (E_{singlet} - E_{triplet})\hat{S}_1 \cdot \hat{S}_2.$$

The eigenvalues of this operator are $E_{singlet}$ for the singlet state, and $E_{triplet}$ for the triplet state. They therefore reproduce the spectrum of the full hamiltonian, provided that only spin state changes are allowed, i.e. we restrict ourselves to low energy excitations.

By defining $J = (E_{singlet} - E_{triplet})/2$ and shifting the zero in energy, we then obtain the *Heisenberg Hamiltonian* for two electrons

$$\hat{H}_2 = -2J\hat{S}_1 \cdot \hat{S}_2,$$

which can be extended naturally to a collection of spins

$$\hat{H}_{Heisenberg} = -\sum_{ij} J_{ij}\hat{S}_i\hat{S}_j.$$

Superexchange and insulating antiferromagnets

When there is strong overlap between orbitals, as in a typical covalent bond, then it is advantageous for the system to form hybridised molecular orbitals and to occupy them fully. In this case, the singlet state has far lower energy than the triplet state, and the system has no magnetic character. However, a

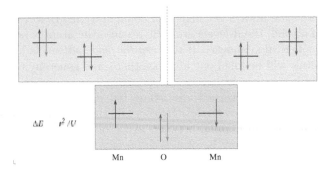

Figure 1. *An illustration of the superexchange mechanism in antiferromagnetic insulators. Two magnetic moments are separated by a non-magnetic ion. The large separation rules out a direct exchange interaction between the magnetic moments. The excited states (top left and top right), which involve double occupancy of an atomic orbital on the magnetic ions, are only possible, if the original spin state of the two magnetic ions was anti-aligned. Only for anti-aligned magnetic moments can the system therefore benefit from the lower energy, which admixing an excited state brings in second order perturbation theory. This results in a ground state energy, which depends on the mutual spin orientation of the two magnetic moments.*

special class of much weaker interactions can be important when two magnetic moments are separated by a non-magnetic ion (often O^{2-}) in an insulator (Fig. 2). Direct exchange between the two local moments is unimportant, because they are too far apart. We consider a ground state in which the relevant valence state of each magnetic ion is singly occupied and that of the non-magnetic ion is doubly occupied. The spectrum of excitations from this ground state is now dependent on the spin orientation of the electrons on the magnetic moments: if the two spins are antiparallel, then it is possible for an electron from the non-magnetic ion to hop onto one of the magnetic ions, and be replaced by an electron from the other magnetic ion. Although the state created in this way has a higher energy than the ground state, it can be admixed to the initial ground state and will – in second order perturbation theory – always cause the new, perturbed, ground state energy to be lowered. This admixture is not possible, if the two magnetic moments were aligned. We arrive, therefore at a total energy for the system, which depends on the mutual orientation of the two magnetic moments.

    This effective *superexchange interaction* is of order $J \sim -t^2/U < 0$, where t is the matrix element governing hopping between the magnetic moment and the non-magnetic ion, and U is the Coulomb repulsion energy on the magnetic moment. When extended to a lattice, it favours an antiferromagnetic ground state, in which alternate sites have antiparallel spins. On complicated lattices, very complex arrangements of spins can result.

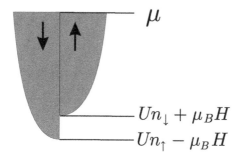

$$\mu$$

$$Un_\downarrow + \mu_B H$$
$$Un_\uparrow - \mu_B H$$

Figure 2. *Spin-split bands in the Stoner approximation.*

The magnitude of this interaction is often quite small, in the range of a few to a few hundred degrees Kelvin. Consequently, these systems will often exhibit phase transitions from a magnetically ordered to a disordered paramagnetic state at room temperature or below.

While it is straightforward to measure the magnetisation in a ferromagnet, measuring the order parameter of an antiferromagnet is more tricky because it corresponds to spins ordering with a finite wavevector. Such order can be observed by elastic neutron scattering, since the neutron has a magnetic moment.

Band magnetism in metals

Let us start with Pauli paramagnetism. We consider a Fermi gas with energy dispersion $\epsilon_\mathbf{k}$ in a magnetic field H. In a magnetic field, the spin-up and spin-down bands will be Zeeman-split (see Fig. 2):

$$
\begin{aligned}
\epsilon_{\mathbf{k}\uparrow} &= \epsilon_\mathbf{k} + \mu_B H, \\
\epsilon_{\mathbf{k}\downarrow} &= \epsilon_\mathbf{k} - \mu_B H.
\end{aligned}
\tag{14}
$$

Since the chemical potential must be the same for both spins, there must be a transfer of carriers from the minority spin band to the majority spin band:

$$
n_\uparrow - n_\downarrow = \mu_B H g(\mu)
\tag{15}
$$

where $g(\mu)$ is the density of states at the Fermi level. The magnetisation is $M = \mu_B(n_\uparrow - n_\downarrow)$ which gives the static spin susceptibility

$$
\frac{M}{H} = \chi_\sigma = \mu_B^2 g(\mu).
\tag{16}
$$

Now let us include in a very simple fashion the effect of interactions. The Stoner-Hubbard model, which provides arguably the simplest way forward, includes an effective interaction U between up and down spin densities on each lattice site:

$$\hat{H}_{int} = U n_\uparrow n_\downarrow. \tag{17}$$

If we treat this interaction in a mean-field approximation, it leads to a shift of the energies of the two spin bands (see Fig. 2)

$$
\begin{aligned}
\epsilon_{k\uparrow} &= \epsilon_k + U n_\downarrow + \mu_B H \\
\epsilon_{k\downarrow} &= \epsilon_k + U n_\uparrow - \mu_B H.
\end{aligned} \tag{18}
$$

With the same approximation as before - that the density of states can be taken to be a constant, we can then self-consistently determine the average spin density

$$n_\uparrow - n_\downarrow = [U(n_\uparrow - n_\downarrow) + 2\mu_B H]\frac{1}{2}g(\mu). \tag{19}$$

The magnetisation is $M = \mu_B(n_\uparrow - n_\downarrow)$ which then gives us the static spin susceptibility

$$\chi_\sigma = \frac{\mu_B^2 g(\mu)}{1 - \frac{U g(\mu)}{2}}. \tag{20}$$

In comparison to the non-interacting case, the magnetic susceptibility is enhanced, and will diverge if U is large enough that the Stoner criterion is satisfied

$$\frac{U g(\mu)}{2} > 1, \tag{21}$$

which marks the onset of ferromagnetism in this model (note the equivalence with the earlier form of the Stoner criterion, Eq. 2).

Local moment magnetism in metals – indirect exchange

In a d-band metal, such as iron, or in f-band metals, such as gadolinium or erbium there are both localised electrons with a moment derived from the tightly bound orbitals, and *itinerant* electrons derived from the $s - p$ bands. The itinerant bands are weakly, if at all, spin-polarised by themselves because the exchange interactions are small and the kinetic energy large. However, the itinerant carrier acquires an *induced* spin polarisation due to its interaction with the core spin on one atom. This spin polarisation can then be transmitted to a neigbouring ion, where it attempts to align the neighbouring spin. There is then an interaction between the localised electron spins, which is mediated by the itinerant electrons, often called RKKY (for Ruderman-Kittel-Kasuya-Yoshida).

A more detailed view of this process can be given. If we have an ion of spin $\mathbf{S}$ embedded in the conduction electrons, one would expect that the local direct exchange will give rise to a contact interaction of the form

$$H_{int} = -J\mathbf{S} \cdot \mathbf{s}\delta(\mathbf{r}), \tag{22}$$

with **s** the conduction electron spin density, and J a direct exchange interaction. The spin density is not otherwise polarised, but the perturbation will induce a weak spin density modulation in the conduction cloud, which will of course decay away to zero at large distance from the ion. The induced spin density is just

$$s(\mathbf{r}) = J\chi_\sigma(\mathbf{r})S \qquad (23)$$

where we have introduced the spin susceptibility χ_σ. (Above we considered the average spin susceptibility to a uniform field, this is a generalisation to non-uniform fields).

At a nearby lattice site (say **r**), the induced spin density caused by the polarisation of one atom interacts with the spin of another, and the energy is then

$$-J\mathbf{S}(\mathbf{r}) \cdot \mathbf{s}(\mathbf{r}) = J^2\chi_\sigma(\mathbf{r})\mathbf{S}(\mathbf{r}) \cdot \mathbf{S}(0). \qquad (24)$$

Summing over all pairs of sites in the crystal we obtain

$$H_{RKKY} = -\sum_{ij} J^2\chi_\sigma(\mathbf{r}_{ij})\mathbf{S}(\mathbf{r}_i) \cdot \mathbf{S}(\mathbf{r}_j). \qquad (25)$$

If we could replace $\chi_\sigma(\mathbf{r}_{ij})$ by its average (say Eq. (16)) then one would predict a long range ferromagnetic interaction, which is not far from the truth for many materials. Of course, in a more accurate theory, χ decays as a function of distance. A careful analysis shows in fact that χ *oscillates*, changing sign as it decays, with a wavelength π/k_F. The origin of these *Friedel oscillations* is the fermi sea itself. Since the electron gas occupies states of momenta smaller than k_F, it is not possible for it to respond to a spatial frequency faster than $2k_F$, and there is a sharp kink in $\tilde{\chi}(q)$ [1] at the momentum $q = 2k_F$. Sharp features in momentum space give rise to oscillations in real space for the fourier-transformed $\chi(r)$, appearing as oscillations visualised in 3.

We saw that χ_o is of order $g(\mu)$, the density of states per unit energy, so the the Heisenberg interatomic exchange parameter is of order $J^2N(\mu)$ which can be large - up to fractions of an eV .

3   Magnetic phase transitions

The Heisenberg model, however complicated the mechanisms that generate the interactions, provides a very good description of the low energy spin dynamics of many magnetic materials. For most purposes, and especially to describe phenomena at finite temperatures, it turns out that the spins can be treated classically and so the analysis of magnetic ground states and magnetic ordering becomes a topic in classical statistical physics, that is somewhat removed from the agenda of this course. Because the interaction J is usually small in comparison to other electronic energies in the problem, we need to

---

[1] Here $\tilde{\chi}(q) = \int dr \exp(iq \cdot r)\chi(r)$ is defined in Fourier space

Figure 3. *In metals, a local moment will polarise the conduction electron spins, producing a spin density that decays away and oscillates in sign with period $1/2k_F$. The interaction of the induced spin density with a neighbouring local moment produces the RKKY interaction.*

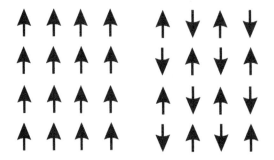

Figure 4. *Schematic picture of the ground state of a ferromagnet and an antiferromagnet. The order parameter for the ferromagnet is the uniform magnetisation, and for an antiferromagnet it is the staggered magnetisation $< S(Q) >$, where Q is the wavevector corresponding to the period of the order.*

include the thermal fluctuations only of the spins at low temperatures, because other degrees of freedom are comparatively stiff, so produce only small changes to the free energy at the temperatures where macroscopic magnetic phenomena are seen. The transition temperature of a magnet is determined by a competition between the energetics of the interaction between spins – favouring ordering – and the entropy, which is larger in a disordered state. Only in rare cases do we need to go beyond simple classical models of interacting moments to understand the magnetic behaviour of real materials.

Depending on the sign of J, the ground state will be ferromagnetic (aligned spins) or anti-ferromagnetic (anti-aligned spins on neighbouring sites); more complicated magnetic states can arise if we have different magnetic ions in the unit cell, and also on taking account of magnetic anisotropy.

While it is straightforward to measure the magnetisation in a ferromagnet, measuring the order parameter of an antiferromagnet is more tricky because it corresponds to spins ordering with a finite wavevector. Such order can, however, be cleanly observed by elastic neutron scattering.

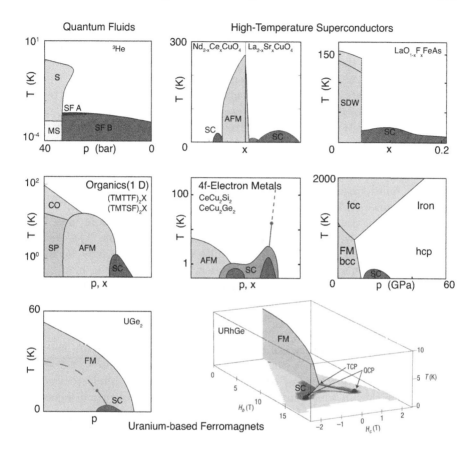

Figure 5. Phase diagrams of quantum matter, *illustrating the chance of discovery on the border of known ordered phases. Further examples could be* ^4He, *the ruthenates and the fractional quantum Hall effect.*

Discovery at quantum phase transitions

Depending on the details of the effective quasiparticle interaction, the electron liquid can be unstable to a variety of low temperature structures, such as magnetism and superconductivity, but also charge density wave order, orbital order and a variety of more exotic forms of order, such as the nematic high field state found in $Sr_3Ru_2O_7$ and the Skyrmion lattice discovered recently in MnSi. Could we turn this around and tip the electron liquid into different low temperature states by adjusting the effective quasiparticle interaction? Indeed, the effective interaction can be tuned indirectly by varying material properties such as band structure, density or magnetisation. This interaction-tuning is most effective near a *quantum phase transition*, the threshold of order, where a known ordered state melts at low temperature as a function of quantum control parameters, such as pressure or magnetic field. Close to quantum phase transitions that channel in the quasiparticle interaction which is responsible for the eventual ordering phenomenon is expected to outgrow all other interaction channels at low temperature.

The fundamental tool by which control over quantum matter is attained is the tunability of the quasiparticle interaction near quantum phase transitions. This recognition motivates a search strategy for novel low temperature states: as the electron liquid is pushed to a quantum phase transition, the opportunity for new forms of electronic self-organisation arises, either through a discontinuous, first-order-like jump into a novel phase or as a result of the divergent quasiparticle interaction. In a nutshell, the strategy urges to look for the unknown on the border of the known.

The periodic table of the elements supplies us with about one hundred materials in which to explore electronic self-organisation. The number of *compounds*, however, is practically limitless: while there may be about $100 \cdot 100 = 10^4$ binary compounds, the present state of the art in high purity materials synthesis combines four elements, giving rise to roughly 10^8 possible materials. How can we search out novel phenomena in the vast space of available materials, which is set to grow exponentially as our control over materials synthesis improves?

Examining past discoveries of quantum order (some of which are listed in Fig. 3) brings out a pattern which we may use to guide future exploration: The search for novel types of order in condensed matter tends to lead us towards materials in which numerous electronic states are nearly degenerate, so that small adjustments of external *quantum control parameters* – pressure, magnetic field or composition – can tip the system into a variety of ordered phases.

In order to explore quantum matter phase diagrams, the state of a particular system has to be varied continuously and cleanly. Composition is the most commonly used control parameter, but it has disadvantages: alloying introduces disorder, whereas adjusting the composition by changing from one compound to another introduces discretisation. Pressure, which provides precise control over the lattice density, and magnetic field emerge, consequently, as the most suitable vehicles for surveying phase diagrams. By complementing magnetic field with a second tuning parameter, lattice density, a huge volume of parameter space opens up for investigation, exploration and discovery. This requires, however, high resolution experimental probes, especially for investigating magnetic properties, at high pressures.

4   Examples of high pressure magnetic measurement methods

A variety of techniques have been developed to measure magnetic properties at high pressure. The following sections may serve to give some examples of

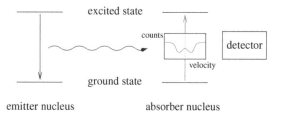

excited state

counts

detector

velocity

ground state

emitter nucleus   absorber nucleus

Figure 6. *Principle of operation of Mössbauer spectroscopy (see, e.g. [1]).* *The hyperfine shifts induced at the emitter nucleus sites by local magnetic* *fields can be determined by measuring the gamma-ray absorption in a* moving *target.*

what has been tried without, however, pretending to be an exhaustive survey.

4.1 Spectroscopic high pressure measurement methods

In high pressure measurements, spectroscopic techniques have the clear advantage that they usually do not require any electrical connections into the sample volume. It is possible to access the magnetic moments in the sample locally by probing the nuclear magnetisation, and it has more recently also become possible to determine the collective spin polarisation by electronic spectroscopy.

Magnetic measurement methods involving nuclear spectroscopy

It is possible to access nuclear moments in a pressure cell by using *nuclear magnetic resonance* (NMR) techniques. This approach usually requires delicate coil-sets inside the high pressure volume. It has been very successful in piston-cylinder cells (see, e.g. [3, 4]), and it may be possible to carry out NMR in anvil cells, using a miniature coil inside the sample volume, as shown below for susceptibility measurements. Fig. 4.1 shows an example of a high pressure NMR experiment on the hidden order state of URu_2Si_2.

Mössbauer spectroscopy, on the other hand, avoids some of the complications associated with NMR. Fig. 4.1 sketches out the principle of this technique. Mössbauer-active nuclei in the sample emit γ-radiation, which can cause the reverse transition in an identical absorber nucleus outside the sample. By detecting the absorption in the absorber while the absorber is being moved, tiny shifts in the emitter's line spacing can be detected. These frequency shifts – like in NMR – serve as a probe for the local environment of the emitter nucleus.

Synchrotron-based electronic spectroscopy

With the availability of high-intensity x-rays at synchrotron facilities, it has become possible to use electronic spectroscopy to investigate magnetic order

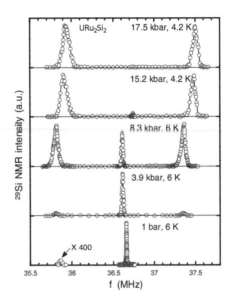

Figure 7. *An example of the power of high pressure NMR measurements in a piston-cylinder cell [2]. Here, the enigmatic hidden order state of URu_2Si_2, which causes no splitting or broadening of the NMR line at ambient pressure, is quickly suppressed with increasing pressure, and replaced by uniform antiferromagnetic order.*

at high pressure. *X-ray magnetic circular dichroism* (XMCD) measurements, for instance, detect the difference in absorption of right hand versus left hand circularly polarised x-rays (Fig. 4.1).

In its simplest form, this technique probes transitions from core p levels to empty band states near the Fermi level, which would have largely d character (L edge). A uniform magnetisation splits the density of states curve near the Fermi level into majority and minority bands. Circularly polarised x-rays then selectively induce transitions into either the majority or the minority band, causing a difference between the absorption rates for the two polarisation states.

Because p to d transitions typically involve energies on the scale of soft x-rays, which are absorbed strongly in the pressure cell diamonds, real high pressure experiments involving this technique probe transitions from s core levels into the conduction band (K edge). These transitions are on the energy scale of hard x-rays, to which diamonds are more transparent. Selection rules forbid transitions from s to d orbital angular momentum states, so in this case, the electrons are promoted into those p states which contribute, by hybridisation with d orbitals, to the conduction band close to the Fermi level. Because this less direct method depends on the $d - p$ hybridisation, the magnetic contrast can be weaker. It has been demonstrated to work beauti-

Spin and Orbital Moments: X-Ray Magnetic Circular Dichroism

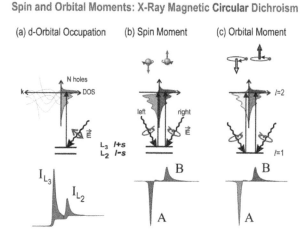

Figure 8. *An illustration of the principle underlying x-ray magnetic circular dichroism measurements as a tool for exploring electronic magnetic order, from [6]. Left- and right-hand circularly polarised light will be absorbed differently in magnetically polarised metals, because transitions from core states into the conduction band will occur selectively into the majority or minority carrier band, depending on the polarisation state of the incoming x-rays.*

fully, however, in some cases, such as in following iron magnetism up to high pressure [5] (Fig. 4.1).

4.2 Inductive or SQUID-based methods

SQUID magnetometry under pressure

A variety of methods have been developed for measuring magnetic properties directly. A relatively straightforward approach consists in adapting existing methods for ambient pressure measurements to accept an entire pressure cell. For example, a number of groups have developed miniature piston-cylinder cells (e.g. [7], Fig. 4.1) or even diamond anvil cells [8] (Fig. 4.1), which fit into commercial SQUID magnetometers. Similar cells, but with larger sample volumes, can also be used for neutron scattering studies [9]. Because of the extremely high sensitivity of a SQUID magnetometer, the resolution of the instrument is easily sufficient for detecting superconducting or ferromagnetic transitions even in samples small enough for a diamond anvil cell. The background from the cell body presents the real, practical limitation of this method.

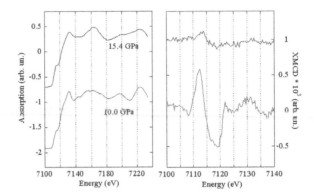

Figure 9. *An example of a high pressure XMCD measurement on iron, which demonstrates the suppression of magnetism in iron above 10 GPa [5]. The left panel shows total x-ray absorption, which is not affected very strongly under pressure. The right panel shows the difference between the absorption coefficients of the two circular polarisation states of the incoming x-rays. Here, a clear reduction in difference signal can be observed, when the pressure is increased from 10 GPa to 15 GPa.*

Vibrating coil magnetometry in diamond anvil cells

Vibrating sample magnetometry is a very convenient and fast technique, which can typically be used to higher magnetic fields than SQUID magnetometry. The fundamental idea, to employ the relative motion between a sample and a coil-set to detect the magnetic moment of a sample, can be put to use also by moving the coil-set close to – or even inside – the pressure cell body.

A beautiful demonstration of this *vibrating coil magnetometry* under high pressure has been presented by Ishizuka et al. [10, 11]. The voltage picked up by the vibrating coil is proportional to the variation of the magnetic flux density along the direction of motion of the coil, at the coil position. Because the magnetic field caused by the sample, $B_{induced}$, varies differently with position than the field caused by the gasket and the cell, the coil-set can be positioned for optimal background cancellation (Fig. 4.2).

Susceptibility measurements in piston-cylinder cells

Methods for detecting the magnetisation directly become increasingly cumbersome at very low temperatures and at high magnetic fields. In many cases, it may be easier to detect the magnetic susceptibility, dM/dH, usually by the *mutual inductance* technique. This consists in picking up the voltage which a time-varying sample moment induces in a coil.

Piston-cylinder cells, with their large sample volumes of typically 6mm diameter and 10-20mm length, provide ample space for an entire mutual induc-

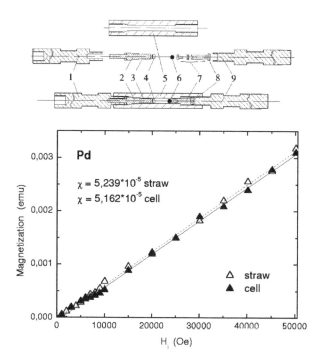

Figure 10. *Top: Miniature CuBe pressure cell with inner diameter 2.5 mm and outer diameter 8.6 mm, max. pressure ~ 20 kbar, developed by Kamarad et al. [7]. 1, 9: upper and lower pressure clamping bolts, 2: plug with 3: sealing, 4: sample on a holder, 5: pressure cell, 6: lead pressure sensor, 7: piston with Bridgman mushroom-type seal, 8: piston backup. Bottom: results of a test measurement on palladium.*

tance susceptometer *within* the high pressure volume. Typically, this means having a modulation coil of about 400-600 turns and two counterwound pick-coils, each with a similar number of turns, inside the pressure cell.

Because the coil is now tightly packed around the sample, filling factors close to 1 can be achieved and background subtraction becomes straightforward. Fig. 4.2 shows a comparatively elaborate setup, which was developed by C. Pfleiderer in the early 1990's and was used to follow the magnetic susceptibility of $ZrZn_2$ under pressure [12]. The modulation field was of the order of 10^{-4} T (1 G) and the measurement resolution was $\Delta\chi \sim 10^{-6}$ (SI). Background cancellation to $\sim 10^{-3}$ could be achieved by mechanically adjusting the pick-up coil.

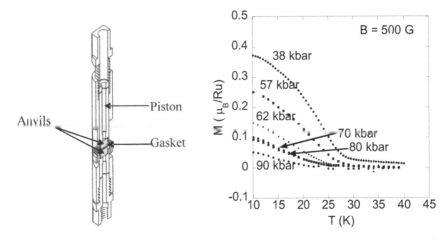

Figure 11. Left: *A diamond anvil cell for SQUID magnetometry, designed by Patricia Alireza [8]. Right: High pressure magnetisation of Ca_2RuO_4 measured in a SQUID magnetometer, using the cell design shown in the left panel.*

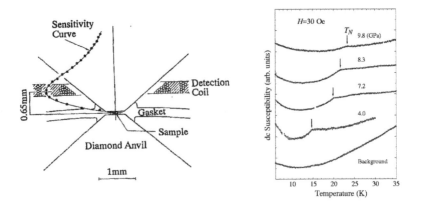

Figure 12. *Magnetisation of $PrSn_3$, measured by the vibrating coil method. The diamond anvil sample contains a $50\mu m$ sample surrounded by 5 mm gasket, which corresponds to a coil-set filling factor of $\sim 10^{-6}$. The coil is vibrating at a frequency of 155 Hz in an applied field of 30 Oe; the flux through the detection is detected by means of a DC-SQUID [10, 11].*

Susceptibility measurements in diamond anvil cells

In order to overcome the pressure limitations of piston-cylinder cells for susceptibility measurements, the mutual inductance technique has to be adapted and miniaturised for use in diamond anvil cells.

If the pick-up coil is placed outside the sample volume, then the filling factor is very low ($\sim 10^{-6}$), but the sensitivity can still be more than sufficient

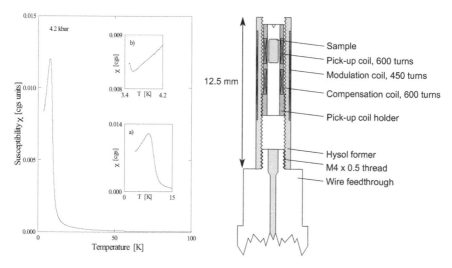

Figure 13. *Magnetic susceptibility of the low temperature ferromagnet* $ZrZn_2$ *under pressure* ($T_c \simeq 20$ *K at* $p = 0$). *The small anomaly at* 3.4 *K in inset (a) is due to the superconducting transition of a tiny tin sample,* outside *the coil-set, which was used as a manometer.*

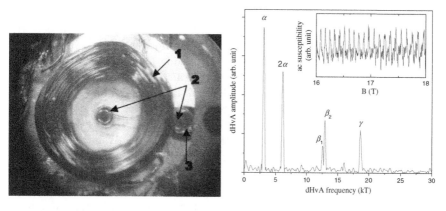

Figure 14. *Quantum oscillation measurements in a high pressure anvil cell. Left: photograph of the susceptometer [13]. The anvil culet diameter is* ~ 1 *mm. (1) modulation coil, (2) pickup-coil pair, (3) additional compensation coil. The gasket has been insulated with* Al_2O_3-*loaded epoxy. Right: de Haas van Alphen signal in* Sr_2RuO_4 *at 0.55 GPa [14].*

to detect superconducting or ferromagnetic transitions. Because the sensitivity of a flat coil scales with 1/diameter, it would appear that a coil wound around the diamonds rather than directly around the sample is very disadvantageous. However, the space available for the coil grows as well, and if the number of

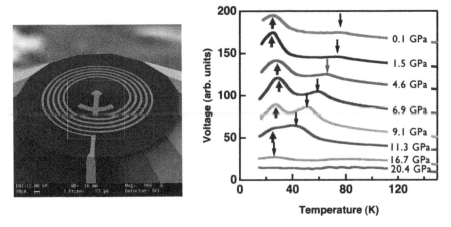

Figure 15. *Left panel: scanning electron microscopy image of a multi-turn microcoil embedded into the culet of a high pressure diamond. Leads have been encapsulated by using chemical vapour deposition in a microwave-heated hydrogen plasma. Right panel: the resulting superanvils have been demonstrated already in high pressure experiments into the megabar range, e.g. by susceptibility measurements on rare earth magnets such as erbium [15].*

turns grows ∝ diameter, then a large coil around the diamond can perform just as well as a small coil inside the sample volume, at least on paper.

Background is the main problem with placing a pick-up coil outside the sample volume in anvil cell susceptibility measurements. This problem can largely be eliminated by placing a smaller coil inside the sample volume. In a nice set of experiments, P. Alireza and S. Goh [14] have recently demonstrated the potential of this approach (Fig. 4.2). In particular, it is possible to detect the quantum oscillations associated with the de Haas van Alphen effect, which enables the determination of the geometry of the Fermi surface of metals under anvil-cell pressures.

Pattern leads into the anvil: transport and susceptibility measurements using superanvils

The quantum oscillation measurements shown above require the operator to pass fragile and unreliable fine wire leads into the high pressure volume. The difficulties associated with this procedure may in the long term prove a serious obstacle to the further development of magnetic measurement techniques under high pressure. It may be fruitful to consider a drastic departure from traditional pressure methods.

Modern microfabrication facilities can easily lay down evaporated or sputtered leads with submicron dimensions and micromanipulators for handling sub-cellular objects of far smaller dimensions are commonplace in biotechnology. In contrast to traditional anvil cell methods, patterning onto the anvil benefits from the hardness of the anvil material, which deforms much less than the gasket under pressure.

Microlithography and superanvil approaches in diamond anvil cells have been pioneered at large facilities such as the Carnegie Institution and the Lawrence Livermore National Laboratory in the U.S.A and at the Center for Science at Extreme Conditions in Edinburgh. Initially, patterned anvils were used to measure electrical transport into the megabar range [16]. By patterning a coil onto the surface of a diamond anvil [15] (Fig. 4.2), high-quality susceptibility measurements have been demonstrated into an unprecedented pressure range.

References

[1] Dominic P. E. Dickson and Frank J. Berry, editors. *Mössbauer Spectroscopy*. Cambridge University Press, 2005.

[2] K Matsuda, Y Kohori, T Kohara, H Amitsuka, K Kuwahara, and T Matsumoto. The appearance of homogeneous antiferromagnetism in URu_2Si_2 under high pressure: a Si-29 nuclear magnetic resonance study. *Journal of Physics-Condensed Matter*, 15(14):2363–2373, Apr 16 2003.

[3] Y. Yamada, J. G. M. Armitage, R. G. Graham, and P. C. Riedi. Pressure-dependence of magnetic-properties of $Nb_{1-y}Fe_{2+y}$. *Journal Of Magnetism And Magnetic Materials*, 104:1317–1318, February 1992.

[4] C. Thessieu, K. Ishida, Y. Kitaoka, K. Asayama, and G. Lapertot. Pressure effect on MnSi: An nmr study, Jan 1998.

[5] O. Mathon, F. Baudelet, J. P. Itie, S. Pasternak, A. Polian, and S. Pascarelli. XMCD under pressure at the Fe-K edge on the energy-dispersive beamline of the ESRF. *Journal of Synchrotron Radiation*, 11:423–427, September 2004.

[6] J. Stöhr. Exploring the microscopic origin of magnetic anisotropies with x-ray magnetic circular dichroism (XMCD) spectroscopy. *Journal of Magnetism and Magnetic Materials*, 200(1-3):470 – 497, 1999.

[7] J Kamarad, Z Machatova, and Z Arnold. High pressure cells for magnetic measurements - Destruction and functional tests. *Review of Scientific Instruments*, 75(11):5022–5025, Nov 2004.

[8] Patricia Lebre Alireza, Samira Barakat, Anne-Marie Cumberlidge, Gil Lonzarich, Fumihiko Nakamura, and Yoshiteru Maeno. Developments on susceptibility and magnetization measurements under high hydrostatic pressure. *Journal of the Physical Society of Japan*, 76SA(Supplement A):216–218, 2007.

[9] C Pfleiderer, AD Huxley, and SM Hayden. On the use of Cu : Be clamp cells in magnetization and neutron scattering studies. *Journal of Physics-Condensed Matter*, 17(40, Sp. Iss. SI):S3111–S3120, Oct 12 2005. International Workshop on Medium Pressure Advances for Neutron Scattering, Grenoble, France, Oct 20-23, 2004.

[10] M. Ishizuka, K. Amaya, and S. Endo. Precise magnetization measurements under high pressure in a diamond-anvil cell. *Review of Scientific Instruments*, 66(5):3307–3310, 1995.

[11] M. Ishizuka and S. Endo. Detection of antiferromagnetic signals in a diamond-anvil cell using a squid vibrating coil magnetometer. *Journal of Physics-Condensed Matter*, 14(44):10719–10722, 2002.

[12] F. M. Grosche. *Pressure studies on strongly correlated electron systems.* PhD thesis, University of Cambridge, 1995.

[13] Patricia Lebre Alireza and Stephen R. Julian. Susceptibility measurements at high pressures using a microcoil system in an anvil cell. *Review of Scientific Instruments*, 74(11):1728–1731, 2003.

[14] S. K. Goh, P. L. Allreza, P. D. A. Mann, A. M. Curnberlidge, C. Bergernann, M. Sutherland, and Y. Maeno. High pressure de Haas-van Alphen studies of Sr_2RuO_4 using an anvil cell. *Current Applied Physics*, 8(3-4):304–307, May 2008. 3rd International Conference on Advanced Materials and Nanotechnology, Wellington, New Zealand, Feb 11-16, 2007.

[15] DD Jackson, V Malba, ST Weir, PA Baker, and YK Vohra. High-pressure magnetic susceptibility experiments on the heavy lanthanides Gd, Tb, Dy, Ho, Er, and Tm. *Phys. Rev. B*, 71(18), May 2005.

[16] J. R. Patterson, C. M. Aracne, D. D. Jackson, V. Malba, S. T. Weir, P. A. Baker, and Y. K. Vohra. Pressure-induced metallization of the Mott insulator MnO. *Phys. Rev. B*, 69(22):220101, Jun 2004.

Chapter 11
The Deep Earth

Chrystéle Sanloup

Université Pierre et Marie Curie, Paris, France and The University of Edinburgh, United Kingdom

1 Introduction

By the end of this chapter, we will have described the Earth's interior in terms of structure and composition as presently known and shown on figure 1. It can be briefly described as consisting of an outer silicate solid shell, the mantle, and an inner core mostly made of Fe. The core itself is divided into a liquid outer part and a solid inner part, its radius is about half the Earth's radius. The relative abundance of geophysical data (geodetic, seismic) gives us a much more precise picture of the Earth's interior compared to other planets. In addition to geophysical data, a wealth of information comes from the petrological and geochemical studies of rocks, especially regarding the time evolution of the planet. Knowledge of Earth's interior can indeed not be dissociated from that of the history of the Earth, including its accretion, its differentiation into the main reservoirs (atmosphere, crust, mantle and core), and up to its present state and dynamics. Table 1 summarizes what needs to be measured to solve those questions.

2 Geophysical constraints

The idea of a stratified Earth first came with the measurement of the density of the Earth by H. Cavendish in 1798. Cavendish measured the force of gravity f between masses m and M in a lab (figure 2, equation 1), and the value of the gravitational field g being known from Newton's work, he estimated the mass (M_E) and hence the density (d_E) of the Earth to 5.48 g·cm−3 (the current estimate is 5.52 g·cm−3).

$$\| \vec{f} \| = \frac{mMG}{r^2} = m\vec{g}, \quad \| \vec{g} \| = \frac{M_E G}{r_E^2} \qquad (1)$$

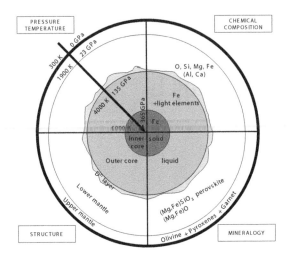

Figure 1. *A schematic of the Earth's interior.*

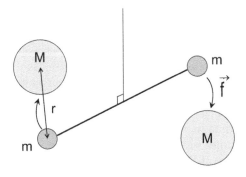

Figure 2. *Measurement of the force of gravity between masses set in a horizontal plane.*

This value is well above the density of rocks to be found on the Earth's surface, ranging from 1.5-2.5 g·cm^{-3} for sedimentary rocks to 3.2 g·cm−3 for mantle rocks. Denser materials must therefore be hidden at depth. This density stratification can also be deduced from the moment of inertia of the Earth, $0.33 \times M_E r_E^2$, as it departs from the value for a homogeneous sphere ($0.4 \times Mr^2$).

A much more precise description of the Earth's interior emerged with global seismological models, the most famous and still most widely used being PREM (Preliminary Reference Earth Model [1]). Briefly, PREM gives seismic velocity-depth profiles, $v_P(z)$ and $v_S(z)$ for the longitudinal and shear waves respectively (Figure 3), and derives from those density and pressure

Table 1. *The Earth: what we need to measure?*

Big questions	What we need to measure
Composition of the planet: $\rightarrow$ chemistry $\rightarrow$ mineralogy	Geochemistry: partition coefficients
Structure	Density of planetary materials: $\rightarrow$ silicates and Fe-alloys $\rightarrow$ in the solid AND liquid states
Dynamics $\rightarrow$ volcanology, crustal dynamics $\rightarrow$ mantle dynamics, heat transfer $\rightarrow$ geomagnetism	Transport properties: $\rightarrow$ viscosity, rheology $\rightarrow$ thermal conductivity $\rightarrow$ electrical resistivity

profiles (Figure 4). Seismic velocities are indeed a function of the density and elastic properties (isentropic bulk modulus, K_S, and shear modulus, μ_S) of the medium. A major discontinuity at 2900 km separates the mantle from the core; no shear wave can travel through the liquid outer core. A more precise description of the PREM model can be found in [2].

Pressure conditions inside the Earth are therefore precisely known and reach 365 GPa at the center of the planet. Such conditions can be reproduced in the lab using either diamond-anvil cell under static pressure, or shock-wave experiments under dynamic pressure. However, there is a compromise between the size of the sample and the pressure to be reached; measurements requiring relatively large amount of samples (*e.g. ex situ* chemical analyses or *in situ* diffraction data on liquids) might be better done using large volume presses instead (see S.Klotz chapter 1).

3 Phase transitions

Phase transitions in the mantle

A number of small discontinuities can be observed on the seismic profiles within the mantle (Figure 3), *i.e.* down to 2900 km. F. Birch soon identified phase transitions of major mantle minerals as the cause of these discontinu-

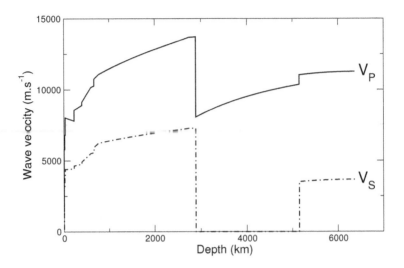

Figure 3. *Seismic velocity profiles (PREM model [1]).*

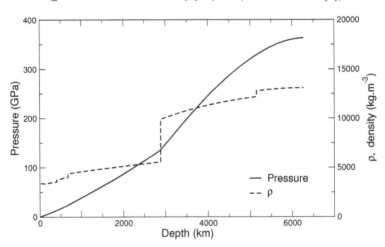

Figure 4. *Density and pressure profiles (PREM model [1]).*

ities [3]. The mineralogical composition of peridotites, the mantle rocks, is
dominated by olivine ($\sim$ 70%) plus minor amounts of pyroxene and garnet.
The phase transitions of the magnesian end-member of olivine (Figure 5) do
match the seismological continuities, the main one at 670 km corresponding to
a breakdown of the olivine into an oxide plus a silicate perovskite, and defines
the upper mantle to lower mantle transition. This transition is accompanied
by a coordination change of silicon, from 4 to 6. The basic unit of perovskite

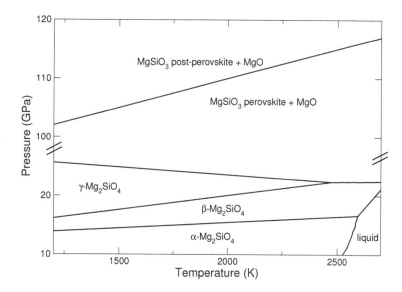

Figure 5. *The phase diagram of Mg_2SiO_4.*

is therefore an octahedron instead of the silica tetrahedron that is found in lower-pressure silicates.

More recently, the discovery of a perovskite to post-perovskite transition [4] above 120 GPa has stimulated a lot of experimental and theoretical work. Among the interesting properties of this newly discovered deep mantle mineral are 1) its anisotropic deformation [5] and 2) an increased electrical conductivity [6]. The first property could explain the strong seismic anisotropy of the D" layer where it is to be found (the D" layer corresponds to the few hundred kilometers of mantle on top of the core), and the second provides a mechanism for the observed length-of-day variations [7] by an electro-magnetic coupling between mantle and core flow.

The Fe phase diagram

The Fe phase diagram remains to be explored at the challenging P-T conditions of the solid inner core (Figure 6). The structure of Fe in the inner core remains an open question, although theoretical work and experiments carried on an $Fe_{0.9}Ni_{0.1}$ alloy favour a bcc phase [8]. However, the melting curve of Fe has been measured below 100 GPa by diamond-anvil cell techniques and up to core conditions by shock-wave techniques. Combining experimental error bars with the mismatch between both techniques gives an uncertainty of ± 1000 K on the melting point of iron at 330 GPa, *i.e* at the transition between the liquid outer core and the solid inner core.

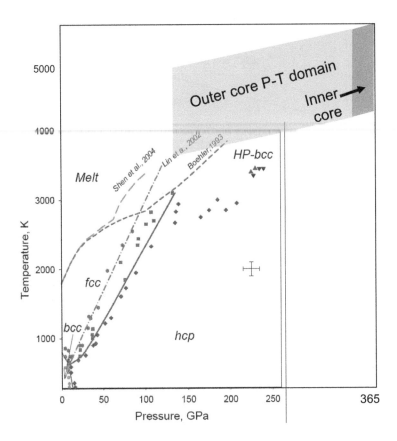

Figure 6. *Phase diagram of iron (adapted from [8]).*

Building the geotherm

Unlike the pressure profile, the temperature profile within the Earth, also called the geotherm, is not precisely known and cannot be derived directly from seismological profiles. It can be estimated (as plotted on Figure 7) using the following guidelines:

- phase transitions are used as anchor points: transitions of olivine in the mantle and the solid→liquid Fe transition in the core;

- the thermal gradient is adiabatic in the convective parts of the mantle and core;

- the thermal gradient is conductive in the parts of the mantle which are not convective (*i.e.* the lithosphere and the D" layer).

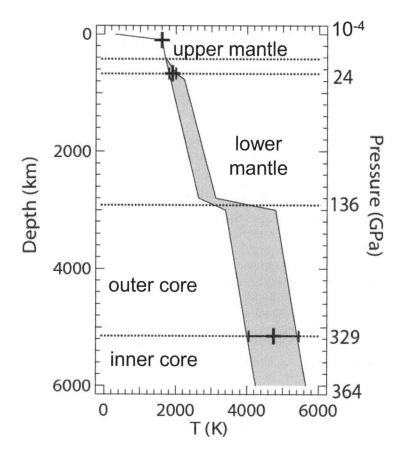

Figure 7. *Temperature profile in the deep Earth (adapted from [9]).*

4 Refining the chemical composition of the deep reservoirs

Deep mantle: radial models

Peridotites, as sampled by volcanic lavas upon their ascent through the mantle, are taken as rocks representative of the upper mantle. For the lower mantle, no natural sample is available at the exception of rare mineralogical assemblages trapped in diamonds. The mineralogical composition of the lower mantle is tested essentially against seismological profiles. Such is the method initiated by F. Birch [10] by comparing density-velocity relationships (the so called "Birch's law") between simple compounds and seismological data. Ideally, one should compare experimental data directly with seismological ve-

locity profiles. However, measuring accoustic velocities on minerals at high
P-T conditions, although feasible, is yet restricted to upper mantle conditions
[11, 12]. Instead, the comparison is more commonly done on density profiles
as the density of minerals can now be relatively easily measured at all mantle
conditions.

But, as seen in paragraph 3, the lower mantle temperature profile is not well
determined, with an error bar of 500 K. There is a consequent trade-off between the composition and the temperature. Seismologically satisfying assemblages are either Si- and Fe-enriched (relative to a peridotite) and on the high
T side, or chemically identical to a peridotite and on the low T side.

Deep mantle: Seismic tomography

PREM is by construction a radially homogeneous model. Seismic velocities
do vary radially though (rarely by more than 1%), and such variations are
mapped by seismic tomography methods. This powerfull technique has revealed amazing snap shots of mantle dynamics and fine structure. Variation of
seismic velocities can be due to either temperature or chemical heterogeneities.
Temperature heterogeneities are the most frequent and reveal convective flow
patterns in the mantle. Chemically distinct zones are also invoked, for instance
to explain low-shear velocity provinces [13] emanating from the base of the
lower mantle. Stronger heterogeneities in the D" layer (up to 2% variations for
shear wave velocities) are linked to the presence of the post-perovskite phase
or to melt pockets (ultra-low velocity zones).

Core

As, once more, first pointed out by F. Birch [3], the core is mostly made of Fe
plus some light elements. Those are needed to reduce the $\sim 10\%$ mismatch
between seismological and experimental accoustic velocity profiles for pure
liquid Fe [14] (Figure 8). The experimental velocity profile was not obtained
directly but calculated from shock-wave density measurements, as $v_P = \sqrt{\frac{\rho}{K}}$
in the liquid state.

The presence of light elements is also required in the solid inner core as the
seismological densities are smaller than those measured along the equation of
state of Fe [15], even accounting for large temperature corrections.

The nature and precise amount of light elements is a long standing issue, with
Si, O, S, C, and H as the most likely ones [16]. Several criteria are classically
used for or against the presence of a particular element X in the core:

- its chemical affinity: whether or not it is siderophile (*i.e.* it partitions
  preferentially into the Fe phase during core formation);

- its initial abundance after the accretion phase, which is basically controlled by its volatility;

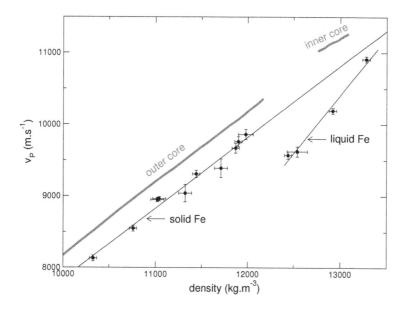

Figure 8. *Accoustic wave velocity in Fe as a function of density (adapted from [14]).*

- the physical properties of the Fe-X alloy, *i.e.* if they match or not those of the seismological profiles;

- its presence must depress the melting point of Fe as the geodynamo is at least partially fuelled by chemical convection.

All these criteria are pressure-dependent. That is obvious for the physical properties but is also true for the chemical affinity of elements. Oxygen and silicon, typical lithophile elements (*i.e.* major constituents of rocks) at ambient pressure conditions, have been shown to partition into molten Fe above 70 GPa [17] and even more in the P-stability field of the post-perovskite phase [18]. It follows that there are likely to be the most abundant light elements in the core, sulphur being a siderophile element at all investigated P-T conditions but too volatile to be present in more than a few % [19]. As for the melting behavior, Fe-S alloys have received the largest attention so far. A major effect of pressure is to shift the eutectic composition towards the Fe pole [20, 21], resulting in the Fe-S phase being present at the solidus for potential core compositions. That would obstruct chemical convection in the core and brings another argument against sulphur being a major light element.

5   Core dynamics

A quite precise picture of core dynamics has emerged from satellite mappings of the Earth's magnetic field. By comparing the magnetic field at a 20 years interval difference, geophysicists could reconstruct the flow at the surface of the core [22]. It can briefly be described as vortexes imbedded in a globally westward flow. Several of the vortexes can be paired between the northern and southern hemispheres, and are therefore interpreted as the surface expression of cylindrical flows. The viscosity of molten Fe at core conditions is a key parameter of core convection. In the case of mantle convection, the viscosity is estimated within a reasonable range from post-glacial rebond of the lithosphere during the quaternary era in Scandinavia or Canada. In the case of core convection, geophysical estimates of the viscosity vary over several order of magnitudes, from 10^{-2} to 10^{10} Pa·s depending on the method (geodesy, seismic wave attenuation, geomagnetism). On the lower end of this range, core flow would be turbulent and dominated by Coriolis forces, while it would be more regular and dominated by viscous forces for the highest values. Experimentally, viscosity can be measured by different methods under pressure. The sink-float method records the falling of a sphere in the liquid by x-ray radiography and deduces the viscosity from the Stokes' equation [24, 25]. In the case of Fe, the viscosity can also be extracted from ^{57}Fe diffusion profiles measured on the sample after quench, assuming the liquid can be described as a packing of hard-spheres. Both type of measurements have confirmed the lower values of 1-2×10^{-2} Pa·s [23], and light elements such as S and C seem to have little effect.

6   Differentiation of the Earth

Planets were born molten [26], this early stage being referred to as the magma ocean. During this period, elements were partitioned between the different main reservoirs (atmosphere, crust, mantle and core) according to their chemical affinities (volatiles in the atmosphere, lithophiles in the mantle and crust, siderophiles in the core), and the reservoirs then segregated according to the density of their materials. Our knowledge of Earth's differentiation is therefore intrinsically related to our knowledge of the present structure and composition of the Earth's reservoirs.

Tracing the formation of Earth's reservoirs

The chemical affinity of the elements is a key property to derive the composition of hidden reservoirs, such as the deep mantle and core. But it is also used to trace planetary differentiation in terms of processes and timing. In terms of timing, the principle is to consider a radioactivity for which 1) mother and daughter elements have opposite chemical affinities, and 2) mother's half-life

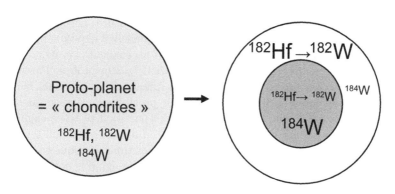

Figure 9. *Partitioning of W and Hf upon core formation.*

is comparable to the duration of the event of interest, core or atmosphere formation for instance. In the case of core formation, geochemists use the $^{182}\text{Hf}/^{182}\text{W}$ system. Hafnium is a lithophile element, tungsten is siderophile, and ^{182}Hf decays into ^{182}W with a half-life of 9 My. W isotopes are measured in chondrites (*i.e.* undifferentiated meteorites taken as a proxy for the bulk Earth) and in terrestrial samples. If core formation took place while ^{182}Hf had not completely decayed, that will result into an excess of ^{182}W in terrestrial rocks relative to chondrites (as illustrated on Figure 9). It is indeed what has been measured [27], therefore core must have formed within 30 My after solar system formation. In the case of atmosphere formation, the $^{129}\text{I} \rightarrow {}^{129}\text{Xe}$ radioactivity is used, with a half-life of 17 My. In that case iodine is a lithophile element and xenon, as a gas, a volatile. ^{129}Xe excesses measured in the cortex of pillow-lavas show that the mantle was largely degassed within 100 My. Argon and helium isotopes are also used to assess atmosphere formation, but on much larger timescales.

In these datation processes, the potential pressure dependence of the chemical affinity is not considered. However, heavy rare gases do change behaviour under high P-T conditions. Upon melting at room pressure, rare gases strongly partition into the melt which is an incompatible behaviour (*i.e.* incompatible with the cristalline network). Two independent groups have shown that such is not the case under pressure, as they observed a solubility drop of Ar at 5 GPa in olivine and silica melts [28, 29, 30], and higher in more complex melts [30]. Even more problematic is the chemistry of xenon at high P-T conditions. Although an archetypical inert element at ambient conditions, xenon can be covalently bonded to oxygen in silica at the conditions of the Earth's continental crust, through a Xe-Si substitution in the tetrahedra [31]. From atmophile at ambient conditions, Xe therefore becomes lithophile at depth. Those results question the use of rare gases isotopes as tracers of Earth's dynamics.

More generally, the chemical affinity of tracer elements should not be assumed constant and equal to that known at ambient pressure, neither should the partitioning of volatile elements be studied on quenched experiments. Ideally, one should measure the partition coefficient of an element, *i.e.* its concentration ratio between two co-existing phases, with *in situ* probes. Partitioning coefficients are traditionally measured from quenched experiments, with the associated draw-backs such as exsolutions or precipitations upon quenching. Pioneer studies have thus been carried on the partition of trace elements in fluids using X-ray fluorescence combined with diamond-anvil cell techniques [32, 33]. The presence of diamonds limits however this technique to the study of elements heavier than Fe, so that the fluorescence signal is not absorbed by the anvils. These studies are also for now limited to moderate pressures (<5 GPa).

Silicate melts at high pressure

Silicate liquids played a large part in the transport of volatiles to or from the deep interior which resulted in the formation of the atmosphere. Segregation of Fe towards the center of the planet was also facilitated through the molten silicate matrix. Despite their importance, very few studies have been dedicated to silicate melts at high pressure, and almost none *in situ*. Those are indeed very challenging experimentally as silicate melts are difficult to confine at the required high temperatures, and they give very weak signal in both X-ray diffraction and Raman scattering experiments. Few shock-wave studies are nonetheless available [34, 35]. Their principal result is that there is a density cross-over between melt and solid silicates, which has two main consequences: 1) the origin of magmas is limited in depth ($\sim$ 300 km), and 2) silicate melts can be segregated at depth. Similar conclusions have been reached from *ex situ* experiments using the sink-float method [36]. This method allows only few (1-3) data points to be collected along the equation of state. Future directions should include *in situ* density measurements of silicate melts under static pressures.

7   Conclusions

Earth's sciences have profoundly benefited from the development of high P-T experimentation, as Earth's interiors would have remained inaccessible otherwise. The confrontation of experimental and seismological data has unveiled the deep structure and composition of the planet, and a clear understanding of its formation and differentiation is underway.

Nonetheless, one should keep in mind the limitations of high pressure experimentation in Earth's sciences. The first and main one is certainly the extrapolation of scales in space (from the μm to mm lab scale to the 1000 km Earth's scale) and time (from the minute to maximum few days lab scale to the 10^9 y geological scale). The second limitation comes from the experimental chemical systems, which are necessarily less complex than real terrestrial rocks for the sake of proper data processing and understanding.

References

[1] Dziewonski, A. M. & Anderson, D. L. Preliminary reference Earth model. *Phys. Earth Planet. Int.* **25**, 297–356 (1981).

[2] Poirier, J. P. *Earth's Interior* (Cambridge University Press, Cambridge, UK, 2000).

[3] Birch, F. Elasticity and constitution of the Earth's interior. *J. Geophys. Res.* **57**, 227–286 (1952).

[4] Murakami, M., Hirose, K., Kawamura, K., Sata, N. & Ohishi, Y. Post-perovskite phase transition in MgSiO$_3$. *Science* **304**, 855–858 (2004).

[5] Merkel, S. *et al.* Deformation of (Mg,Fe)SiO$_3$ post-perovskite and D" anisotropy. *Science* **316**, 1729–1732 (2007).

[6] Ohta, K. *et al.* The electrical conductivity of post-perovskite in Earth's d" layer. *Science* **320**, 89–91 (2008).

[7] Courtillot, V., Mouel, J. L. L., Ducruix, J. & Cazenave, A. Geomagnetic secular variations as a precursor of climatic change. *Nature* **297**, 386–387 (1982).

[8] Dubrovinsky, L. *et al.* Body-centered cubic iron-nickel alloy in Earth's core. *Science* **316**, 1880–1883 (2007).

[9] Dewaele, A. & Sanloup, C. *L'intérieur de la Terre et des planètes* (Belin, Paris, 2005).

[10] Birch, F. Density and composition of mantle and core. *J. Geophys. Res.* **69**, 4377–4388 (1964).

[11] Li, B., Jackson, I., Gasparik, T. & Liebermann, R. Elastic wave velocity measurement in multi-anvil apparatus to 10 GPa using ultrasonic interferometry. *Phys. Earth Planet. Int.* **98**, 79–91 (1996).

[12] Sinogeikin, S. V. & Bass, J. D. Single-crystal elasticity of MgO at high pressure. *Phys. Rev. B* **59**, 14141 (1999).

[13] Garnero, E. J. & McNamara, A. K. Structure and dynamics of Earth's lower mantle. *Science* **320**, 626–628 (2008).

[14] Brown, J. M. & McQueen, R. J. Phase transitions, Grüneisen parameter, and elasticity for shocked iron between 77 GPa and 400 GPa. *J. Geophys. Res.* **91**, 7485–7494 (1986).

[15] Dewaele, A. *et al.* Quasihydrostatic equation of state of iron above 2 Mbar. *Phys. Rev. Lett.* **97**, 215504 (2006).

[16] Poirier, J.-P. Light elements in the Earth's outer core: a critical review. *Phys. Earth Planet. Int.* **85**, 319–337 (1994).

[17] Jeanloz, R. & Knittle, E. Earth's core-mantle boundary: results of experiments at high pressures and temperatures. *Science* **251**, 1438–1443 (1991).

[18] Sakai, T. *et al.* Interaction between iron and post-perovskite at core-mantle boundary and core signature in plume source region. *Geophys. Res. Lett.* **33**, L15317 (2006).

[19] Dreibus, G. & Palme, H. Cosmochemical constraints on the sulfur content in the Earth's core. *Geochim. Cosmochim. Acta* **60**, 1125–1130 (1996).

[20] Campbell, A. J., Seagle, C. T., Heinz, D. L., Shen, G. & Prakapenka, V. B. Partial melting in the iron-sulfur system at high pressure: A synchrotron X-ray diffraction study. *Phys. Earth Planet. Inter.* **162**, 119–128 (2007).

[21] Morard, G. *et al.* In situ determination of Fe-Fe$_3$S phase diagram and liquid structural properties up to 65 GPa. *Earth Planet. Sci. Lett.* **272**, 620–626 (2008).

[22] Hulot, G., Eymin, C., Langlais, B., Mandea, M. & Olsen, N. Small-scale structure of the geodynamo inferred from Oersted and Magsat satellite data. *Nature* **416**, 620–623 (2002).

[23] Dobson, D. P., Vocaldo, L. & Wood, I. G. A new high-pressure phase of FeSi. *Am. Mineral.* **87**, 784–786 (2002).

[24] Terasaki, H. *et al.* Viscosity change and structural transition of molten Fe at 5 GPa. *Geophys. Res. Lett.* **29**, 68 (2002).

[25] Terasaki, H. *et al.* Effect of pressure on the viscosity of Fe-S and Fe-C liquids up to 16 GPa. *Geophys. Res. Lett.* **33**, L22307 (2006).

[26] Harper, C. L. Evidence for ^{92g}Nb in the early solar system and evaluation of a new p-process cosmochronometer from ^{92g}Nb/^{92}Mo. *ApJ* **466**, 437–456 (1996).

[27] Kleine, T., Münker, C., Mezger, K. & Palme, H. Rapid accretion and early core formation on asteroids and the terrestrial planets from Hf-W chronometry. *Nature* **418**, 952–955 (2002).

[28] Chamorro-Perez, E. M., Gillet, P. & Jambon, A. Argon solubility in silicate melts at very high pressures. experimental set-up and preliminary results for silica and anorthite melts. *Earth Planet. Sci. Lett.* **145**, 97–107 (1996).

[29] Chamorro-Perez, E. M., Gillet, P., Jambon, A., MacMillan, P. F. & Badro, J. Low argon solubility in silicate melts at high pressure. *Nature* **393**, 352–355 (1998).

[30] Bouhifd, M. A. & Jephcoat, A. Aluminium control of argon solubility in silicate melts under pressure. *Nature* **439**, 961–964 (2006).

[31] Sanloup, C. *et al.* Retention of xenon in quartz and Earth's missing xenon. *Science* **310**, 1174–1177 (2005).

[32] Sanchez-Valle, C. *et al.* Dissolution of strontianite at high P-T conditions: An in-situ synchrotron X-ray fluorescence study. *Am. Mineral.* **88**, 978–985 (2003).

[33] Bureau, H. *et al.* In situ mapping of high-pressure fluids using hydrothermal diamond anvil cells. *High Pressure Research* **27**, 235–247 (2007).

[34] Ridgen, S. M., Ahrens, T. J. & Stolper, E. M. Densities of liquid silicate at high pressures. *Science* **226**, 1071–1074 (1984).

[35] Akins, J. A., Luo, S.-N., Asimow, P. D. & Ahrens, T. J. Shock-induced melting of MgSiO$_3$ perovskite and implications for melts in Earth's lowermost mantle. *Geophys. Res. Lett.* **31**, L14612 (2004).

[36] Agee, C. B. Crystal-liquid density inversions in terrestrial and lunar magmas. *Phys. Earth Planet. Int.* **107**, 63–74 (1998).

Chapter 12
Planetary Interiors

Chrystéle Sanloup

Université Pierre et Marie Curie, Paris, France and University of Edinburgh, United Kingdom

1 Introduction

The gross composition of planets is controlled by the condensation sequence (Fig.1a) in the early solar system. Refractory materials such as silicates and iron condense at high temperatures, while the so-called planetary ices (any compound in the CH_4-NH_3-H_2O system, either in the liquid or solid state) may condense only in the colder outer parts of the solar system. Jupiter and Saturn therefore reached sufficient masses to attract H and He gases blown away from the Sun. As a result, the bulk of planets consists of three main compositions: silicates and iron for the inner terrestrial planets, H and He for Jupiter and Saturn, and planetary ices for Uranus and Neptune.

In order to get insights on the structure and dynamics of planets, a number of properties need to be measured, on the relevant materials for each planet and at the relevant P-T conditions (Fig.1b). Those properties are briefly summarized in Table 1. The type of properties to be measured can also be guided by the available observational data (e.g. accoustic velocities on terrestrial materials to be matched with seismic profiles).

Our knowledge of the internal structure of the planets, at the exception of the Earth, is based on only indirect and partial information such as astrophysical data (mass, radius, gravity field, Table 2) and chemical data (e.g. composition of atmospheres). Moreover, in the case of the giant planets, the pressure and temperature ranges largely exceed the experimental capacities, at least in the static pressure mode. One therefore needs either to extrapolate data acquired most often using shock-wave techniques, or to carry numerical simulations.

Table 1. *What we need to measure?*

Planets	Terrestrial planets	Jupiter, Saturn	Uranus, Neptune
Material	Silicates, Fe alloys	H, He	Planetary ices H_2O-CH_4-NH_3
Physical properties	Phase diagrams, equation of states $\Rightarrow$ structure and composition		
	Acoustic velocities (if seismic data available)		
	Transport properties $\Rightarrow$ planetary dynamics		
Chemical properties	Reactivity	P-T evolution of intramolecular bonds	Reactivity
	Chemical affinity	Im/miscibility	

Table 2. *Physical characteristics of the planets of the solar system.*

Planet	Equatorial radius (km/Earth=1)	Mass (kg/Earth=1)	Mean density (g·cm^{-3})	Moment of inertia I/MR^2	Gravity (m·s^{-2})
Mercury	2440/0.38	3.30 10^{23}/0.055	5.43	0.33	3.70
Venus	6052/0.95	4.87 10^{24}/0.815	5.20	0.33	8.87
Earth	6371/1	5.97 10^{24}/1	5.52	0.3308	9.78
Mars	3397/0.53	6.4 10^{23}/0.107	3.91	0.363	3.69
Jupiter	71492/11.21	1.898 10^{27}/317.71	1.33	0.254	23.12
Saturn	60268/9.45	5.685 10^{26}/95.16	0.69	0.210	8.96
Uranus	25559/4.0	8.683 10^{25}/14.535	1.318	0.225	8.69
Neptune	24766/3.88	1.024 10^{24}/17.14	1.638	?	11.0

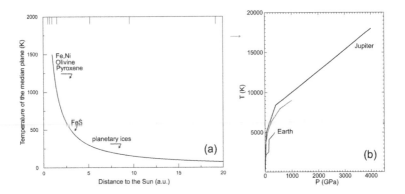

Figure 1. *(a) The sequence of condensation in the early solar system (adapted from [1]). (b) The pressure-temperature profiles of selected planets.*

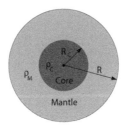

Figure 2. *The two-shell model for the terrestrial planets.*

2   Terrestrial planets

Two shells models

The terrestrial planets are by definition the planets which are similar to the Earth and include Mercury, Venus, the Earth and Mars. Their interiors are usually modeled using a two shells model, *i.e.* a silicate mantle and a metallic core (Figure 2), hereby neglecting the superficial crust. The equations for the mass of the planet, M (equation 2), and its moment of inertia, I (equation 1), can then be solved for a given mantle density, ρ_M.

$$M = M_{\text{mantle}} + M_{\text{core}} = \rho_M \times \left(\frac{4}{3}\pi R^3 - \frac{4}{3}\pi R_C^3 \right) + \rho_C \frac{4}{3}\pi R_C^3 \quad (1)$$

$$I = I_{\text{mantle}} + I_{\text{core}} = \frac{2}{5}\rho_M \left(\frac{4}{3}\pi R^5 - \frac{4}{3}\pi R_C^5 \right) + \frac{2}{5}\rho_C \frac{4}{3}\pi R_C^5 \quad (2)$$

The main results of this exercise are:

- Mercury has a large core (∼0.7 of the planet's radius), reflected in its large mean density (see Table 2) compared to the other terrestrial planets;

- Venus and the Earth have a similar internal structure;

- Mars is less density stratified, as seen in the moment of inertia which is closer to the 0.4 value of a homogeneous sphere; this is interpreted in terms of the martian mantle being denser (probably more Fe-rich, see paragraph below) and the martian core lighter (containing more volatile light elements) than their terrestrial counterparts.

The special case of Mars

A great deal of additional information can be extracted from the study of martian meteorites, the so-called SNC (for Shergotty, Nakhla and Chassigny, the three first discovered). The community has gathered over 30 of them by now, either directly from falls or by collection in deserts and in Antarctica. They share common petrological characteristics and most importantly, the same oxygen isotopic signature. Martian meteorites can be used to infer models of mantle compositions by two different ways. First the chemical and petrological analyses of martian basalts can lead to the composition of their mantle source [2], and second by matching their oxygen isotopic signature with non-differentiated meteorites of known composition [3]. Both methods predict a mantle richer in Fe. The oxygen isotopes model [3] implies mantle assemblages richer in pyroxene, as the Mg/Si ratio gets closer to 1 instead of 2 for olivine. Interestingly, an increase of the pyroxene/olivine ratio with the Sun to planet distance has also been proposed from astrophysical observations of planetary disks [4].

Models predict a S-rich martian core, mostly due to the external position of Mars in the solar system, the lower temperatures allowing more volatile elements to condense. With up to 15wt% S, the martian core could be largely if not totally molten according to the Fe-FeS binary diagram at martian core pressures [5].

3   Giant planets

Homogeneous sphere model

A quantitative way to approach the bulk composition of the planets is to calculate the mass of a homogeneous sphere model as a function of its radius for different compositions (hydrogen, hydrogen+helium, planetary ices and silicates; see Figure 3a) [6]. The obtained relationships can then be compared with the actual mass and radius of the planets. This simple modeling confirms that the composition of Jupiter and Saturn is dominated by hydrogen and

Table 3. *Models of the composition of the martian and primitive terrestrial mantle. Oxides accounting for less than 1wt% are not reported.*

Oxide	MgO	SiO$_2$	Al$_2$O$_3$	CaO	FeO	Mg/Si	Mg/(Mg+Fe)
			weight %				molar ratio
Mars [3]	27.3	47.5	2.5	2	17.7	1.17	0.72
Mars [2]	30.2	44.4	3	2.4	17.9	1.36	0.75
peridotite (Earth)	41.5	43.5	3.6	3.2	8.2	1.42	0.90

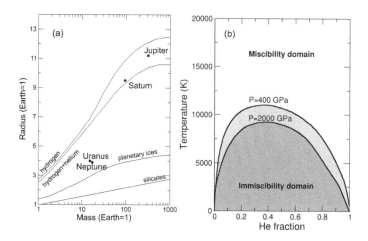

Figure 3. *(a)The homogeneous sphere model for the giant planets (adapted from [6]).(b) The H-He immiscibility domain (adapted from [6]).*

helium, being close to that of the Sun, while Uranus and Neptune are mostly made of planetary ices.

The upper atmospheres of Jupiter and Saturn are depleted in helium compared to the Sun (23 wt% and 14 wt% respectively, the Sun value being 27.5 wt%). This is attributed to a helium segregation at depth resulting from H-He immiscibility. Theoretical calculations predict a closure of this miscibility gap at extreme temperatures [6] (see Figure 3b).

Due to the large uncertainties on the equation of state of hydrogen and helium at the P-T conditions of the giant planets' interiors, it is not possible yet to investigate the interiors of Jupiter and Saturn in more detail. The existence of a rocky core for instance is still an open question.

Magnetic fields

An additional strong constraint on the interiors of giant planets comes from the study of their magnetic fields. The four of them indeed have extremely intense dynamos (Table 4), which in the case of Uranus and Neptune are dominated by quadri- and octopolar components rather than dipolar. Hydrogen is a conductive fluid on a large portion of Jupiter/Saturn interiors (Figure 4) and, as such, can generate the observed magnetic field.

Metallic hydrogen cannot be at the origin of Uranus and Neptune magnetic fields though. It should rather be searched in the high P-T properties of planetary ices. And indeed, H_2O and NH_3 become fully ionized above 50 GPa at high temperatures [11]. The predominance of small scale (quadri- and octopolar) components of the magnetic field can be explained if the movements of the ionized fluid are confined in a thin outer layer [12].

Chrystéle Sanloup

Table 4. *Characteristics of planetary magnetic fields.*

Planet	Rotational period (days)	Dipolar Moment (Earth=1)	Same polarity as Earth?	Tilt between rotational and magnetic axis
Mercury	58.65	0.0007	yes	14°
Venus	243.02	not detected	-	-
Earth	1.00	1	yes	10.8°
Mars	1.03	not detected	-	-
Jupiter	0.41	20000	no	9.6°
Saturn	0.44	600	no	< 1°
Uranus	0.72	50	no	58.6°
Neptune	0.67	25	no	47°

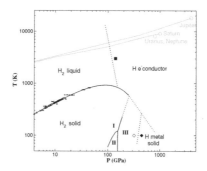

Figure 4. *Phase diagram of hydrogen; data from [7, 8, 9, 10] (adapted from [1]).*

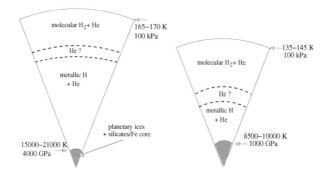

Figure 5. *Schematics of the interiors of Jupiter and Saturn (adapted from [13]).*

Models of planetary interiors

All the informations cited above have been gathered to draw the schematics of the giant planets interiors as displayed on Figures 5 and 6.

Galilean satellites and Titan

The galilean or jovian satellites are similar though smaller to terrestrial planets in terms of composition and structure, at the difference that except for Io, the closest to Jupiter, there are covered with ices. They are modeled as three-shells planets (metallic core, silicate mantle and an ice layer, see figure 7) in order to fit the astrophysical data (Table 5).

The depression in the melting curve of water at 0.2 GPa implies most likely the presence of a deep ocean under the frozen surface of Europa. The presence of liquid water at depth could explain the ice tectonics observed by the Galileo spacecraft, as well as the secondary magnetic field induced by

Table 5. *Physical characteristics of the galilean satellites.*

Planet	Equatorial radius (km/Earth=1)	Mass (kg/Earth=1)	Mean density (g·cm^{-3})	Moment of inertia I/MR^2
Io	1815/0.28	88.86 10^{21}/0.015	3.53	0.371-0.380
Europe	1569/0.25	47.85 10^{21}/0.008	3.04	0.346
Ganymede	2631/0.41	148.1 10^{21}/0.025	1.93	0.3105
Callisto	2400/0.38	107.5 10^{21}/0.018	1.83	0.358

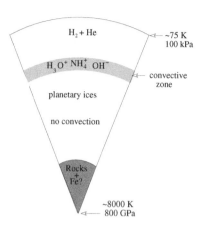

Figure 6. *Schematics of the interiors of Uranus and Neptune (adapted from [13]).*

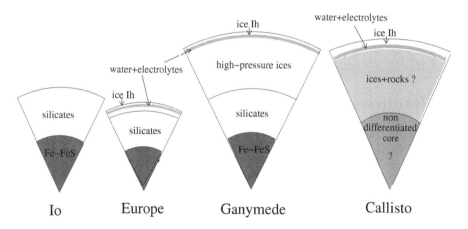

Figure 7. *Schematics of the interiors of the galilean satellites (adapted from [1]).*

Jupiter if some electrolytes are added to the water. The origin of the induced magnetic field of Callisto is more mysterious as the satellite presents no surface tectonics and as temperatures are expected to be much colder further away from Jupiter. However, antifreezes such as ammonia could help setting the conditions for an internal ocean. Pressure was shown to increase the antifreeze effect of NH_3 by a maximum of 100 K at 0.2 GPa [14]. Among the galilean satellites, only Ganymede has an internal magnetic field, probably originating from movements in the metallic core. The existence of internal oceans in Europa and Callisto, and of a molten core in Ganymede, require and internal heat source: the tide forces generated by the proximity of Jupiter.

Last but not least, Titan, the largest satellite of Saturn has received a large interest due to the presence of methane and other organic compounds in its atmosphere, raising the question of a potentially biological origin. The short lifetime of methane in Titan's atmosphere due to photodissociation implies that it must be replenished from a surface or internal reservoir. As methane clathrates are stable over Titan's internal pressure range [15], the solution of an internal methane reservoir has long been favored, until surface lakes were evidenced by the Huygens probe [16].

4   Conclusions

Planets are natural high P-T laboratories. Their interiors map an extreme range of pressure and temperature conditions to which are subjected a variety of materials. Their close observation can therefore reveal a wealth of new physical and chemical phenomena, unsuspected at ambient conditions. The search for metallic hydrogen by hard-condensed matter physicists to solve the origin of Jupiter's magnetic field is certainly the best illustration of this. But we should expect more to be discovered with the advent of new *in situ* analyses and particularly *in situ* chemical probes at extreme conditions.

References

[1] Dewaele, A. & Sanloup, C. *L'intérieur de la Terre et des planètes* (Belin, Paris, 2005).

[2] Dreibus, G. & Wänke, H. A volatile–rich planet. *Meteoritics* **20**, 367–382 (1985).

[3] Sanloup, C., Jambon, A. & Gillet, P. A simple chondritic model for Mars. *Phys. Earth Planet. Int.* **112**, 43–54 (1999).

[4] van Boekel, R. *et al.* The building blocks of planets within the "terrestrial" region of protoplanetary disks. *Nature* **432**, 479–482 (2004).

[5] Stewart, A. J., Schmidt, M. W., van Westrenen, W. & Liebske, C. Mars: A new core-crystallization regime. *Science* **316**, 1323–1325 (2007).

[6] Stevenson, D. J. Interiors of the giant planets. *Annual Review of Earth Sciences* **10**, 257–295 (1982).

[7] Weir, S. T., Mitchell, A. C. & Nellis, W. J. Metallization of fluid molecular hydrogen at 140 GPa. *Phys. Rev. Lett.* **76**, 1860–1863 (1996).

[8] Datchi, F., Loubeyre, P. & LeToullec, R. Extended and accurate determination of the melting curves of argon, helium, ice (H_2O), and hydrogen (H_2). *Phys. Rev. B* **61**, 6535–6546 (2000).

[9] Loubeyre, P., Ocelli, F. & LeToullec, R. Optical studies of solid hydrogen to 320 GPa and evidence for black hydrogen. *Nature* **416**, 613–617 (2002).

[10] Gregoryanz, E., Goncharov, A. F., Matsuishi, K., Mao, H. k. & Hemley, R. J. Raman spectroscopy of hot dense hydrogen. *Phys. Rev. Lett.* **90**, 175701 (2003).

[11] Cavazzoni, C. *et al.* Superionic states of water and ammonia at giant planet conditions. *Science* **283**, 44–46 (1999).

[12] Stanley, S. & Bloxham, J. Convective-region geometry as the cause of Uranus' and Neptune's unusual magnetic fields. *Nature* **428**, 151–155 (2004).

[13] Guillot, T. Interior of giant planets inside and outside the solar system. *Science* **286**, 72–77 (1999).

[14] Grasset, O., Sotin, C. & Deschamps, F. On the internal structure and dynamics of Titan. *Planet. Space Science* **48**, 617–636 (2000).

[15] Loveday, J. S. *et al.* Stable methane hydrate above 2 GPa and the source of Titan's atmospheric methane. *Nature* **410**, 661–663 (2001).

[16] Hayes, A. *et al.* Hydrocarbon lakes on Titan: Distribution and interaction with a porous regolith. *Geophys. Res. Lett.* **35**, L09204 (2008).

Chapter 13
Temperature Measurement and Control in High-Pressure Experiments

Paul F. McMillan[1*], **Edward Bailey**[1], **Kurt Leinenweber**[2]

[1] Department of Chemistry, Christopher Ingold Laboratory, University College London, 20 Gordon St., London WC1H 0AJ, UK
[2] Department of Chemistry and Biochemistry, Arizona State University, Tempe, Arizona 85287-1604, USA
* Email: p.f.mcmillan@ucl.ac.uk

Measurement and control of the high T variable is necessary in high P synthesis and *in situ* experiments. We introduce concepts of heat and temperature and how to apply and measure them in high P environments using piston-cylinder and multi-anvil environments as examples. These are readily adapted to other types of high-pressure apparatus. We also describe the operation of the laser-heated diamond anvil cell.

1   Introduction

Many useful high pressure experiments are carried out at ambient T; however we must also simultaneously heat samples to explore their structural and thermodynamic behaviour and to initiate chemical reactions. During his pioneering experiments P.W. Bridgman heated his apparatus with a "Bunsen" burner [1]. Modern experiments in "large-volume" high-P devices use resistive heating with a thermocouple to measure T [2, 3, 4] and related heating methods are applied in high-T diamond anvil cell (DAC) experiments [3, 4, 5]. Laser-heated DAC (LH-DAC) experiments are carried out at up to $T \approx 6000$ K using IR lasers and T is measured *via* spectroradiometry techniques [3, 5, 6, 7, 8].

The definitions of "energy", "heat" and "temperature" have evolved over several centuries and are now incorporated into modern thermodynamics

[9, 10]. The present definition of "energy" as the capacity of a body to provide heat or to do work came into use in the mid-1800's, promoted by W. Thomson (Lord Kelvin) and W. Rankine, a Scottish physicist and engineer. The concept of "heat" began from the idea that it was due to increased motion of tiny corpuscles within a body [11]. However, heat was also considered as a fluid that could flow from "hot" to "cold" regions of a system [10]. Observations of B. Thomson (Count Rumford) in Munich followed by J.R. Mayer and J.P. Joule, a Manchester brewer, demonstrated that providing heat (q) to a substance or carrying out mechanical work (w) on it were alternative means of increasing its internal energy (E): that now constitutes the "First Law" of Thermodynamics; $E = q + w$. It also began to be suggested that heat might be a form of wave vibration, like light. It is now accepted that the "internal" energy of a substance or system is altered by transferring energy into or out of it by thermal, mechanical, electrical, optical, chemical, magnetic etc. processes, and that various forms of energy storage and their transfer between systems can be interconverted. In the case of high-P,T experiments, the internal energy is conveniently increased by supplying heat derived from an electrical resistor in experiments using "large-volume" high-pressure apparatuses or resistively-heated diamond cells, or by allowing the sample to absorb photons from incident laser radiation during laser-heated DAC experiments. Other means of changing the temperature within the sample chamber can be envisaged for future experiments, such as the release of energy from exothermic chemical reactions [12]. Raising the internal energy of the sample in shock studies is achieved by mechanical impact with a high-velocity projectile or by coupling to a laser or other electromagnetic energy sources [3, 4, 5, 13, 14, 15, 16].

The concept of "temperature" as a measure of the internal energy or heat contained within a sample was devised to indicate the degree of "hotness" of a body. Observed transitions between solid, liquid and gaseous H_2O and other "standard" substances in response to specific amounts of heat gave rise to fixed reference points and development of a temperature scale. The observation that liquid volume increased regularly during heating led to the invention of alcohol and mercury thermometers, still used today, in which the T scale was divided into equal parts between fixed points such as melting and boiling of water. The Celsius (centigrade) scale of temperature took reference points at the freezing and boiling points of H_2O and marked off equal divisions between them quoted as °C. However, such instruments are not suitable for most high-pressure experiments, that use a thermocouple to determine T, described below. It was discovered that extrapolating the $V(T)$ relations for gases to low T indicated that they converged and the volume would vanish at $T = -273.15°C$. The ideal-gas-theory indicates this represents a natural limit of T and a new "absolute" scale was devised with $T = -273.15°C$ as its lowermost reference point; the new units were termed Kelvins (K). An internally consistent temperature scale is now established over a wide range by combining physical and phase equilibrium measurements [17].

2 Resistance heating and the thermocouple principle for temperature measurements

Joule heating occurs as an electrical current (I) is passed through a conductor as a function of its electrical resistance (R): the heat energy released can be used to raise the temperature of another body placed in thermal contact with the current-carrying wire ($q = I^2R$). A second relation is the Thomson effect, in which a temperature gradient ΔT across a conductor leads to an electrical potential gradient (ΔV) or electromotive force (*emf*): $\Delta V = \sigma \Delta T$, where σ is the Thomson coefficient; ΔV is on the order of several mV. In semiconductors, the sign of ΔV distinguishes between n- and p-type conduction. The Thomson effect is understood classically in terms of the average velocity and kinetic energy of charge carriers (electrons or holes) in hot *vs* cold ends of the material. At the high-T end, carriers have greater energy and drift statistically to the cold part. In band theory terms, the effect involves a greater distribution of electron energies around the Fermi level for hot *vs* cold substances: thermally excited electrons occupy higher energy levels in the hot regime and a net drift of electron density occurs to the cold part depending upon ΔT. The Peltier effect is used for thermoelectric cooling and power generation. Different electrical conductors are connected in a circuit and heat (q) is absorbed or emitted as current is made to flow in one direction or another, depending on the relative σ values of the materials. In the Seebeck effect, a similar circuit is formed by two different conductors, but the junctions between them are held at temperatures T_1 and T_2. The Peltier *emf* at each junction is different because of ΔT. One junction is set to a reference value (e.g., $T_1 = 0°C$, by placing it in an ice-water bath); T_2 at the other junction is determined by measuring the *emf*, following calibration [17]. The resulting thermocouple is readily inserted into high-P apparatus. Thermocouple referencing is now usually carried out electronically [17]. The maximum T limit mainly occurs due to melting of the thermocouple. However, high-P studies can be carried out to higher T values by extrapolating the thermocouple *emf* [18].

Choice of the thermocouple used for high-P,T experiments is driven by the target T range, ease of use and cost (Table 1). K-type thermocouples are not generally used above about 1100–1200°C, and W-Re types (C, D) should not be used below ~50°C. Another consideration is the likelihood of the thermocouple surviving the high-P experiment. As the assembly is placed under load, it deforms, the central portion undergoes compression, and outer sections are partly extruded beyond the anvils. Thermocouples often break during this process. The Pt-Rh thermocouples (R, S) are most flexible and less likely to break during initial assembly construction, but harder W-Re types are more resistant to stretching and snapping in the gasket under high-P. Thermocouple wires must be electrically insulated inside the assembly: the most common approach used is to route the wire through alumina tubes. Four-bore tubing is often used in high-P,T research as it provides a convenient

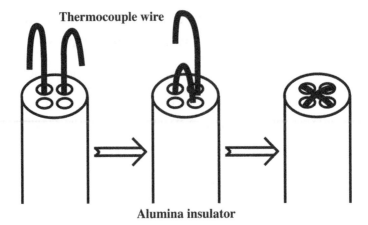

Figure 1. *Thermocouple assembly used for piston-cylinder and multianvil assemblies using four-hole ceramic tubing. Contact is made as the wires are forced together during pressurisation, although welding is also possible. Care must be taken to electrically insulate the contact from metallic sample capsules.*

way to make a thermocouple junction (Figure 1).

3   "Large volume" devices and sample assemblies

Joseph Bramah (1795) first patented a hand-pumped hydraulic press apparatus building on Pascal's principle of pressure transmission in fluids (Figure 2). One of its first scientific applications took place in 1825, when Michael Faraday used it to purify benzene. An early piston-cylinder apparatus was developed in the late 19th century to attempt to transform graphite into diamond [19]. To heat the graphite sample rods, a large current (300 A) was passed through them, previewing the modern use of graphite furnaces in modern piston-cylinder experiments. The piston-cylinder device was developed into its current form much later [20] (Figure 3). During resistivity studies of metals at high-P, Bridgman developed an opposed-anvil device in which the sample was placed between anvils made from a newly-available hard ceramic-metal (cermet) composite consisting of tungsten carbide (WC) grains held together by Co or another metallic binder. The natural material pyrophyllite was also identified at this time as a useful material for sample containment and electrical insulation [21]. In the "belt" apparatus, conical pistons compress a cylindrical pressure-transmitting medium and sample assembly confined by a "belt" of press-fit steel rings [22, 23, 24, 25]. T is not normally measured in each run, but a sample assembly calibrated for use within a given P,T range is used

Table 1. *Typical thermocouple types used for high-P,T experiments. T limits are shown for ambient P-calibrations; the T range for measurements can be extended at high-P because of the increased stability of the thermocouple.*

Thermocouple Type	+ leg	− leg	(Nominal) maximum temperature[a]	Advantages	Disadvantages
K	Ni-Cr	Ni-Al	1250°C	Inexpensive, workable	Limited T range
R	Pt-13%Rh	Pt	1450°C	Good for use under oxidising conditions	Expensive
S	Pt-10%Rh	Pt	1450°C	Good for use under oxidising conditions	Expensive
C	W-5%Re	W-26%Re	2320°C	High T use under reducing conditions	Brittle, easily oxidized
D	W-3%Re	W-25%Re	2320°C	High T use under reducing conditions	Brittle, easily oxidized

Figure 2. *Bramah hydraulic press used by Michael Faraday at the Royal Institution for expression of water from benzene crystals.*

to achieve similar conditions during subsequent runs. The split-sphere multi-anvil apparatus was introduced by von Platen [26, 27] and T. Hall developed cubic and tetrahedral devices with independently mounted anvils that pushed on the sample assembly [24]. Kawai and Endo placed a second pressurisation stage inside von Platen's split-sphere design composed of 8 carbide cubes with triangular truncations on each inside corner, to apply an equal force to each face of an octahedral sample assembly [28] (Figure 4). A split-cylinder modification made the module portable among different press frames [29]. A new type of opposed-anvil device using "toroidal" pressurisation geometry [30] was adapted to develop the "Paris-Edinburgh" cell with a lightweight frame specially designed to be transported to neutron and synchrotron beamlines [31].

Generation and control of P and T in large-volume devices are linked to sample assemblies designed for each experiment. Samples are enclosed within a metal or ceramic capsule including Pt, Au, Ag, Fe, Ta, h-BN or MgO [2] (Fig-

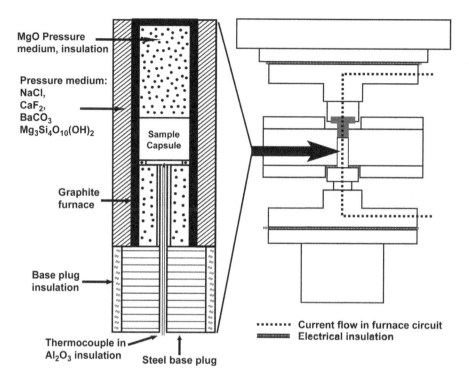

MgO Pressure medium, insulation

Pressure medium:
NaCl,
CaF$_2$,
BaCO$_3$
Mg$_3$Si$_4$O$_{10}$(OH)$_2$

Sample Capsule

Graphite furnace

Base plug insulation

Thermocouple in Al$_2$O$_3$ insulation

Steel base plug

········ **Current flow in furnace circuit**
▬▬▬▬▬▬ **Electrical insulation**

Figure 3. *Schematic of a non-end-loaded piston-cylinder assembly and press frame (for clarity, frame posts are not shown). Note that the heating current passes through the press frame, piston and furnace.*

ures 4 and 5). The capsule is surrounded by a deformable ceramic pressure-transmitting medium (PTM) that must transmit load evenly throughout the sample chamber, ideally with low compressibility and shear modulus to produce approximately hydrostatic conditions in the sample chamber. Small grain size provides low shear strength at small applied loads: initial porosity ∼30% yields uniform compression. Gasketing media placed around the PTM and between the anvils protect the anvils from entering into contact. Flow of the mechanically soft gasket materials away from high-stress regions allows the pressurising anvils to advance smoothly. Pyrophyllite tends to lose H$_2$O during high-P,T experiments that can react with the sample. Although machined pyrophyllite can be cured before use, it undergoes significant volume contraction and increase in hardness and shear strength that render it useless as a PTM or gasket material. To provide lubrication and electrical insulation between components, sheets of Mylar plastic are placed between anvil stages; epoxy composite sheets are often glued around the cubes that form the second stage in multi-anvil assemblies. In piston-cylinder experiments, the PTM is formed into cylinders that fit around the sample assembly: in multi-anvil experiments, the PTM forms an octahedron to fit inside the truncated WC cubes and that is drilled to contain the furnace and sample capsule assembly (Figures 4 and 5). Typical materials used are listed in Table 2. In such internally heated devices, pressure effects on thermocouple *emf* can affect the measured temperature [32]. Also, the thermocouple passes through regions in which considerable *P*-*T* gradients exist, but we usually make the approximation that most of the signal is generated within a few mm of the thermocouple junction. Because of the density variation of the Seebeck coefficients, low-compressibility metals based on tungsten alloys are often used for high-*P,T* research (Table 1). During studies in which *V* is held constant, especially during laser heated DAC experiments, the *P* increases during heating [33, 34]. The pressure can fall during heating experiments because of chemical reactions or relaxation within the sample chamber, PTM or surrounding components. It is usually found that *P* effects are reproducible between runs using a given apparatus and assembly and suitable calibration experiments can account for them.

Table 2. *Various materials used in large-volume (piston-cylinder, multianvil) press assemblages.*

Material	Properties and comments
NaCl	A commonly used PTM material for piston-cylinder experiments: it is cheap and easy to form into cylinders to surround the sample assemblage. It is especially good for use up to 900°C above which it melts, causing a large change in the mechanical properties, and where it becomes conducting, reducing furnace efficiency. Also, the molten salt is highly corrosive.
Pyrex glass	Used as an inner sleeve to protect NaCl in piston-cylinder experiments. The glass is an excellent thermal and electrical insulator. One disadvantage is that it undergoes cracking and shearing at low temperatures.
CaF_2 (fluorite)	Similar properties to NaCl but useful to ~1600°C.
$BaCO_3$	More difficult to shape than NaCl or CaF2 but good for use to ~1800°C. PTM materials for piston-cylinder experiments are now often doped with transition metal oxides to increase the opacity and reduce heat loss by radiative conduction.
$Mg_3Si_4O_{10}(OH)_2$ (talc)	Relatively expensive and undergoes dehydration like pyrophyllite but thought to be generally good for use to quite high-*T* although the limits are not yet reliably determined.

Pyrophyllite	Naturally occurring layer-structured hydrous alumi-nosilicate, known in early high-P,T experiments as "wonderstone" or pipestone. Low shear strength at low T. Easily machineable. Hardens significantly, dehydrates and contracts above 600°C.
MgO	Deformable ceramic material cast or moulded into octahedra as the PTM in most multianvil experiments. The first MgO-based PTM were developed in Japan and are still commonly used in conjunction with pyrophyllite gaskets. Octahedra cast from potting compounds (e.g., Ceramcast®) were developed and calibrated by Walker [26, 30]: these contain MgO mixed with other oxides (SiO_2, Al_2O_3 etc): the composition of the mixture and grain size can vary between batches causing possible variations in sample assembly performance. It is cast into octahedra using a plastic mould from an aqueous slurry that is then dried and fired at 1000°C for ∼1 hour: slight variations in the casting procedure can cause irregularities in performance. Ceramic octahedra with highly reproducible properties are available commercially from specialised suppliers. The ceramic mixture is commonly doped with transition metal oxides to increase its opacity and reduce heat loss by radiative conduction.
ZrO_2, $LaCrO_3$	Sleeves of these materials are placed around the furnace arrangement to provide thermal insulation. Mg_2SiO_4 (forsterite), and $Al_2O_3 \cdot 2SiO_2$ (mullite) can also be used to provide thermal insulation to avoid ZrO_2 for synchrotron X-ray experiments [31]. Forsterite has the advantage that its thermal conductivity becomes as low as that of ZrO_2 above 1100°C [32].

The circuits used for heating and temperature measurement need to be electrically insulated from each other and from the earthing environment in the laboratory. In high-P,T experiments, the press frame can form part of the heating circuit and thus become "live" during the experiment (Figure 2). Proper electrical insulation and controls during the experiment are required to ensure safety in the laboratory, as well as to reduce power or heat losses from the experiment. Resistance furnaces used in large-volume high-P,T devices are either (a) high-current, low-voltage heaters such as graphite or metal foils (e.g., Mo, Re) or (b) low-current, high-voltage materials such as semiconducting lanthanum chromite, $LaCrO_3$, that can also be used to provide thermal insulation around high-current, low-voltage heaters (Figure 6). Different transformers are required to switch between the two types of heater. Graphite is a typical furnace material used for piston-cylinder as well as belt and Paris-Edinburgh cell experiments. Cylinders of high purity material formed to specified size and shape are commercially available, along with end caps required to make contact with the heating circuit. Below 5 GPa, graphite can be used to $T > 2000°C$. An obvious problem in multianvil devices is the possible transformation of graphite into diamond above ~7 GPa. However, the transition occurs slowly and experiments can be performed for several hours in the lower P,T range of the diamond stability field: problems are only encountered for $T > 1200°C$ at $P > 10$ GPa or during prolonged high-P,T runs. Graphite cannot be used in experiments above 15 GPa even at low T because of its reversible transformation to an optically transparent material in this range [35]. The use of $LaCrO_3$ as a furnace material in high-P,T multianvil experiments presents special problems because of the "hard starting" nature of the semiconductor that has high resistance at room T (Figure 6). A typical voltage controller cannot be used to start the furnace because as the material heats up its resistance drops precipitously, causing spikes in the drawn current leading to T runaway (up to 1000°C or more in a few seconds). Some laboratories address this problem by using a power rather than a voltage controller. Others use a current limiter at the beginning of the run to eliminate the spikes, then increase the set limit at higher T as the furnace stabilises. Another consideration common to all large-volume high-pressure assemblies is the supply of electrical power to the furnace. The "electrodes" that carry current to the furnace are resistive conductors that dissipate power in the form of heat. In multianvil and belt-type assemblies this can be beneficial as the heat produced helps reduce thermal gradients around the sample. This is offset, however, by the relatively high thermal conductivity of the metals typically used for the purpose (e.g., Mo, W or Ti-Zr-Mo alloy). At low T, heat losses from the sample area occur *via* conduction through the sample assembly, and PTM materials containing heavy ions with low thermal conductivities (e.g. $BaCO_3$ or ceramics containing $LaCrO_3$, ZrO_2 etc.) can be used to alleviate this. Above ~800°C heat losses *via* radiative processes also become important. To counteract this, coloured components (e.g. Cr_2O_3, Mn_2O_3) are introduced as dopants in the PTM. Photons leaving the furnace assembly are absorbed within the PTM and then re-emitted randomly sending thermal energy back to the sample chamber. Temperatures up to ~2500°C at 2.2 GPa have been achieved using a graphite furnace, Mo foil leads, and a composite of $BaCO_3$ with Ca^{2+}-doped $LaCrO_3$ as the pressure-transmitting medium [36].

Large thermal gradients can exist within the sample chamber, e.g. >200°C reported over a 4 mm sample length in a multi-anvil device [37]. Thermal gradients in piston-cylinder assemblies are also a concern. Kushiro [38] introduced a tapered graphite furnace reducing the T gradient in the capsule to <15°C, compared with >75°C using a straight heater (Figure 7). Other workers have developed different multianvil furnace arrangements to reduce the influence of

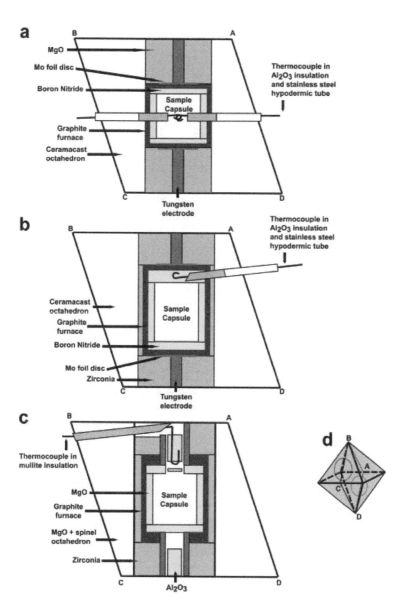

Figure 4. *Examples of multianvil assemblies, showing a variety of thermocouple arrangements a) Assembly for use up to about 1400° C. The sample must not melt, produce a gas phase or react with the thermocouple. b) and c) High-T assemblies.*

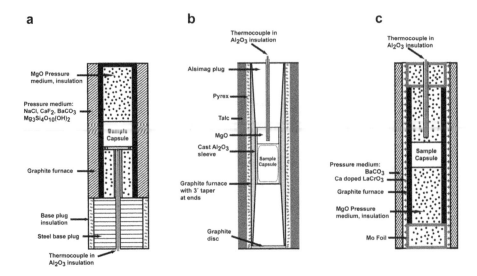

Figure 5. *Piston-cylinder assemblies. a) Low- and high-T assembly. Temperature attainable is determined by the pressure medium: NaCl up to about 1000° C, doped BaCO₃ up to 2200° C. b) Low-T gradient assembly [38]. c) Very-high-T assembly to 2500° C [36].*

T gradients in multianvil experiments. Takahashi *et al.* [39] suggested using a tapered heating furnace and Rubie *et al.* developed a three-stage furnace, with thinner sections above and below the sample and a thicker furnace section in the centre of the assembly. Another useful design feature to reduce thermal gradients is to surround the sample with materials that conduct heat efficiently: metal capsules provide good thermal conductors, and MgO or Al₂O₃ surrounding the sample are also good; diamond or c-BN "capsules" (perhaps formed *in situ* by high-*P,T* conversion from graphite or h-BN during the experiment) can have excellent properties. Thermal profiles have been modelled using numerical calculations [40] and determined experimentally [37, 41]. Sites of high electrical resistance, such as those that occur at contacts to furnace ends within multianvil assemblies, can lead to hot spots and runaway heating effects within various parts of the assembly. Such effects can also arise due to experimental design faults in the furnace circuitry leading to high current density.

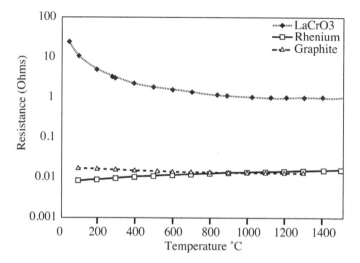

Figure 6. *Resistance versus temperature for some multianvil heaters. The sharp drop in resistance at low temperatures for the LaCrO₃ furnace can cause control problems.*

4 Blackbody radiation and laser-heated diamond anvil cell experiments

The science of thermodynamics was established with the development of the statistical (kinetic) theory of gases that leads to the Maxwell-Boltzmann distribution law [9, 10]. In a box filled with gas particles that collide and exchange energy with each other, an equilibrium (Gaussian) distribution of the kinetic energy is obtained. The most probable speed moves to higher values with increased thermal energy, and this fixes the temperature. The statistical theory was extended to polyatomic systems that have "internal" rotational and vibrational degrees of freedom, and to solids in which the thermal distribution is fixed by the degree of excitation of the lattice vibrations or phonons, as well as electronic degrees of freedom. Two descriptions of the "temperature" now existed: one as the calibrated response of a thermometer placed in the system as the internal energy was raised mechanically or by supplying heat to the system, and the "statistical" variable T defined by the equilibrium distribution of particle speeds and other excitations. With the advent of quantum mechanics the thermodynamic models used to understand the energy distribution were further refined. It was recognised that electrons as "fermions" in which the sign of the wave function is reversed upon particle interchange during scattering result in the Fermi-Dirac distribution function that applies to electronic exitations; however, the quasi-particles that include thermal vi-

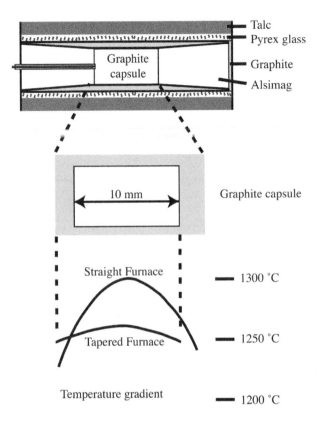

Figure 7. *Reduction of the temperature gradient in a piston-cylinder assembly is achieved with a tapered furnace (redrawn from [38]).*

brational excitations in solids obey Bose-Einstein statistics with an energy distribution related to the Maxwell-Boltzmann relation.

Understanding the wavelength distribution of light emitted from a heated body first led to the development of modern quantum mechanics [9]. It also gives rise to the optical technique used to measure T in laser-heated DAC experiments. A hypothetical blackbody is a material that absorbs and emits all wavelengths of light equally: this also applies to light emitted from a small hole in a perfectly reflecting cavity. The overall radiated intensity from heated objects increases with T^4 according to Stefan's law. As the intensity distribution of the emitted light was examined as a function of the wavelength, λ, the classical result assuming equipartition of energy between all possible degrees of freedom broke down [9]. The issue was resolved by M. Planck who suggested that energy could only be exchanged among "oscillators" present in the body *via* finite units or "quanta" whose magnitude depended on the oscillator frequency (ν) related to the light wavelength: $\Delta E = hc/\lambda = h\nu$ (c is

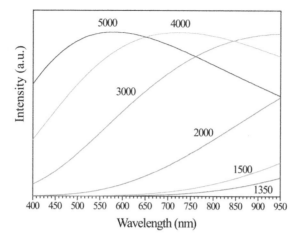

Figure 8. *Intensity* vs *wavelength functions calculated from Planck's formula for emitted radiation profiles between 400 and 950 nm at temperatures ranging from 1350 to 5000 K. Fitting data to these profiles over an experimentally defined range and assuming $\varepsilon \leq 1$ (i.e., blackbody vs greybody fit) results in T determination.*

the speed of light in vacuum; Planck's constant h has the experimentally determined value $h = 6.626 \times 10^{-34}$ Js). The resulting Planck radiation function is given by:

$$I(\lambda, T) = \varepsilon \frac{2\pi hc^2}{\lambda^5} \frac{1}{e^{hc/\lambda kT} - 1} \tag{1}$$

where $I(\lambda, T)$ is the radiant power emitted per unit surface area, and ε is the emissivity. For a perfect blackbody, $\varepsilon = 1$ at all wavelengths. In a "greybody" approximation, $\varepsilon < 1$, with a value fitted to experimental results. ε is also likely to vary with wavelength [42]. The Planck relation is used to determine T in laser-heated DAC experiments from optical emission measurements (Figure 8).

Recognising that diamonds are transparent throughout the near-IR region, Ming and Bassett directed light from a 60 W Nd^{3+}:YAG laser ($\lambda = 1064$ nm) onto a graphite sample held in the DAC and achieved temperatures of 2000–3000°C [7]. The LH-DAC technique is now the method of choice for achieving static high-P,T conditions extending T to $\sim$6000 K at pressures into the megabar range, and it has been used in studies ranging from mineralogy and planetary physics, new materials synthesis and determining phase transitions including melting at high-pressure [8, 43, 44, 45, 46, 47, 48, 49, 50]. Nd^{3+}:YAG or Nd^{3+}:YLF near-IR lasers contain Nd^{3+} ions substituted for Y^{3+} inside a solid state $Y_3Al_5O_{12}$ garnet (YAG) or yttrium lithium fluoride ($YLiF_4$) matrix: lasing involves (i) absorption of "pump" radiation provided by intense emission from lamps or solid-state diodes operating near 880 nm to cause a population inversion between two electronic energy levels and (ii) a lasing transition between these levels with population inversion with the emission of laser radiation around 1060 nm in wavelength. Using near-IR excitation, metals and dark-coloured samples can be heated efficiently at high P in the DAC. Other materials that do not absorb near-IR light can be heated by mixing an "absorber" material such as powdered Pt, Re, graphite or α-B with the sample [3, 5]. However that strategy has associated problems. First, the heat is not transferred homogeneously to the sample from the absorber material and the T determined by optical emission can vary throughout the sample chamber. Also, the absorber could react chemically with the sample under high-P,T conditions. Experimental strategies devised to minimise or control such considerations include a "hot plate" approach in which the laser beam is first absorbed by discs of material located on either side of the sample chamber, that then heat the sample by thermal diffusion or re-radiation [51]. Another approach is to use a CO_2 laser ($\lambda \approx 10.6$ μm) for sample heating [34, 52, 53, 54]. Many ceramic materials have large absorption coefficients in this range due to IR-active vibrational transitions.

The laser heating technique involves directing a focused laser beam through the diamond windows onto the sample inside the DAC held at high-pressure. Considerations concern the spatial extent of the beam and the resulting hot spot developed at the sample surface, as well as the degree of penetration of the beam and development of the T profile inside the sample [8, 42, 44, 46, 47, 48, 55, 56, 57, 58]. The near-IR beam from a Nd^{3+}:YAG or Nd^{3+}:YLF is focused using visible-light microscope objectives to provide a heated spot at the sample surface $\sim$5-10 μm in diameter. Modelling studies indicate the lateral T distribution is approximately Gaussian, decaying rapidly within a few tens of μm away from the centre [46, 58]. Usually, near-IR LH-DAC studies are carried out using double-sided heating, in which two laser beams or a single beam split into two components are directed simultaneously at opposing sides of the sample held in the DAC [51, 59, 60]. In CO_2 laser-heating experiments, the incident IR beam must be focused using ZnSe IR-transmitting lenses or Cassegrain reflecting objectives [6, 53]. The diffraction limit for the longer-wavelength beam increases the resulting laser-heated spot to $\sim$30-50 μm. The penetration depth inside the sample is on the order of several μm, and CO_2 LH-DAC runs are usually carried out *via* single-sided experiments, in which the incident laser is introduced on the back side of the sample and *in situ* T measurements and optical spectroscopy studies are carried out from the front (Figure 9).

The sample temperature in LH-DAC experiments is usually determined using spectroradiometry [44, 46, 47, 48]. Emitted radiation from the heated sample is collected and directed into a spectrometer and its intensity profile measured as a function of the wavelength over an appropriate range (generally $\sim$600–900 nm) [46, 49, 60]. The spectrometer and optical system are first calibrated against a standard source such as a W lamp to account for any

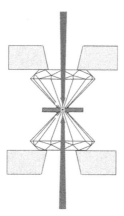

Figure 9. *Schematic diagram of a laser-heated DAC experiment. Using near-IR radiation (Nd^{3+}:YAG or Nd^{3+}:YLF lasers) the incident beam is focused on the sample held in the DAC from both sides; with IR (CO_2) heating single-sided excitation is usual.*

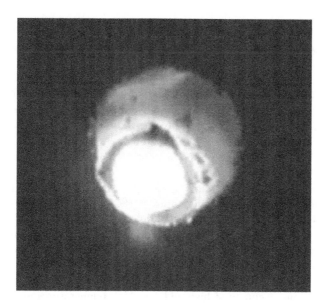

Figure 10. *View through the top diamond during a laser-heated DAC experiment.*

wavelength-dependent background signal or variations in the spectrometer and detector efficiency (Figure 10) [8]. The resulting $I(\lambda, T)$ spectrum from the sample emission is then fitted with Planck's formula (eq. 1) to obtain T.

At low T, where $hc/\lambda kT \gg 0$, the Wien approximation can be used:

$$\ln \frac{I(\lambda, T)\lambda^5}{2\pi hc^2} = \ln(\varepsilon(\lambda)) - \frac{hc}{k\lambda T} \qquad (2)$$

For a perfect blackbody, $\ln(\varepsilon(\lambda)) = 0$. Wien fitting has the advantage that plotting $J(\lambda) = \ln(I(\lambda)\lambda^5/c_1)$ with $c_1 = 3.7418 \times 10^{-16}$ Wm^2 *vs* $1/\lambda$ provides a linear relationship to estimate $1/T$. Optical emission spectroradiometry has been used throughout LH-DAC studies and it yields temperature values that are usually quoted to be reliable to within a few tens or hundreds of degrees. Various experimental considerations must be taken into account to evaluate the precision and accuracy of the measurements, including the temperature profile across the sample, the stability of the exciting laser, chromatic and other aberrations introduced by the diamond windows and other optical elements within the system, and methods used to analyse the data [6, 8, 46, 55, 61].

References

[1] Bridgman P.W., The Physics of High Pressure, G Bell and Sons, London, 1931.

[2] Holloway J.R. and Wood, B.J., Simulating the Earth: Experimental Geochemistry, Unwin Hyman, New York, 1988.

[3] Holzapfel W.D. and Isaacs, N.S., ed, High-Pressure Techniques in Chemistry and Physics: A Practical Approach, Oxford University Press, Oxford, 1997.

[4] Liu L.-G. and Bassett W., Elements, Oxides, Silicates: High-Pressure Phases with Implications for the Earth's Interior, Oxford University Press, New York, 1986.

[5] Eremets M., High Pressure Experimental Methods, Oxford University Press, Oxford, 1996.

[6] Benedetti L.R., Guignot N. and Farber D.L., Achieving accuracy in spectroradiometric measurements of temperature in the laser-heated diamond anvil cell: diamond is an optical component, J. Appl. Phys., **101**, 013109, 2007.

[7] Ming L.-C. and Bassett W.A., Laser heating in the diamond-anvil press up to 2000°C sustained and 3000°C pulsed at pressures up to 260 kilobars, Rev. Sci. Instrum., **45**, 1115, 1974.

[8] Benedetti L.R. and Loubeyre P., Temperature gradients, wavelength-dependent emissivity, and accuracy of high and very-high temperatures measured in the laser-heated diamond cell, High Pressure Res., **24**, 423, 2004.

[9] d'Abro A., The Rise of the New Physics: Its Mathematical and Physical Theories (vols. 1 and 2), Dover Press (original hardback edition published by D. Van Nostrand Co, 1939), New York, 1951.

[10] Ihde A.J., The Development of Modern Chemistry, Dover Publications, New York (original edition published by Harper and Row, 1964), 1984.

[11] Inwood S., The Man who Knew Too Much: The Strange and Inventive Life of Robert Hooke 1635-1703, Pan MacMillan, London, 2002.

[12] Gillan E.G. and Kaner R.B., Rapid solid-state synthesis of refractory nitrides, Inorg. Chem., **33**, 5693, 1994.

[13] Nellis W., Shock wave techniques, in High Pressure Phenomena, ed. Hemley R.J., et al., IOS Press, Amsterdam, 2002, p. .

[14] Ahrens T.J., Shock wave techniques for geophysics and planetary physics, in Methods of Experimental Physics, ed. Sammis, C.G. and Henyey, T.L., Academic Press, San Diego, CA, 1987, p. 185.

[15] Gupta Y.M. and Sharma S.M., Shocking matter to extreme conditions, Science, **277**, 909, 1997.

[16] Loubeyre P., Celliers P.M., Hicks D.G., Henry E., Dewaele A., Pasley J., Eggert J., Koenig M., Occelli F., Lee K.M.,Jeanloz R.,Neely D., Benuzzi-Mounaix A., Bradley D., Bastea M., Moon S., Collins G.W., Coupling static and dynamic compressions: First measurements in dense hydrogen, High Pressure Res., **24**, 25, 2004.

[17] Preston-Thomas H., The International Temperature Scale of 1990 (ITS-90), Metrologia, **27**, 3, 1990.

[18] Zhang J., Liebermann R.C., Gasparik T., Herzberg C.T., Fei Y., Melting and subsolidus relations of SiO_2 at 9-14 GPa, J. Geophys. Res., **98**, 19785, 1993.

[19] Parsons C.A., Experiments on carbon at high temperatures and under great pressures, and in contact with other substances, Proc. R. Soc., **44**, 320, 1888.

[20] Boyd F.R. and England J.L., Apparatus for phase-equilibrium measurements at pressures up to 50 kilobars and temperatures up to 1750°C, J. Geophys. Res., **65**, 741, 1960.

[21] Bridgman P.W., Bakerian Lecture: Physics above 20,000 kg/cm^2, Proc. R. Soc. London, **203**, 1, 1950.

[22] Bovenkerk H.P., Bundy F.P., Hall H.T., Strong H.M. and Wentorf R.H., Preparation of diamond, Nature, **184**, 1094, 1959.

[23] Bundy F.P., Hall H.T., Strong H.M. and Wentorf Jr R.H., Man-made diamonds, Nature, **176**, 51, 1955.

[24] Hall H.T., Some high-pressure, high-temperature apparatus design considerations: equipment for use at 100 000 atmospheres and 3000°C, Rev. Sci. Instrum., **29**, 267, 1958.

[25] Hall H.T., Ultra-high-pressure, high-temperature apparatus: the Belt. Rev. Sci. Instrum., **31**, 125, 1960.

[26] von Platen B., A multiple piston, high-pressure, high-temperature apparatus , in High Pressure Measurements, ed. Giardini, A.A. and Lloyd, E., Butterworth, London, 1962, p. .

[27] Hazen R.M., The Diamond Makers, Cambridge Univ Press, Cambridge, UK, 1999.

[28] Kawai N. and Endo S., The generation of ultrahigh-pressure by a split sphere apparatus, Rev. Sci. Instrum., **41**, 1178, 1970.

[29] Walker D., Carpenter M.A. and Hitch C.M., Some simplifications to multianvil devices for high-pressure experiments, Am. Mineral., **75**, 1020, 1990.

[30] Khvostanstev L.G., Slesarev V.N. and Brazhkin V.V., Toroid type high-pressure device: history and prospects, High Pressure Res., **24**, 371, 2004.

242   *Paul F. McMillan, Edward Bailey, Kurt Leinenweber*

[31] Besson J.-M., Hamel G., Grima T., Nelmes R.J., Loveday J.S., Hull S., Hausermann D., A large-volume pressure cell for high temperatures, High Pressure Res., **8**, 625, 1992.

[32] Getting I.C. and Kennedy G.C., Effect of pressure on the *emf* of chromel-alumel and platinum-platinum 10% rhodium thermocouples, J. Appl. Phys., **41**, 4552, 1970.

[33] Heinz D.L., Thermal pressure in the laser-heated diamond-anvil cell, Geophys. Res. Lett., **17**, 1161, 1990.

[34] Andrault D., Fiquet G., Itie J.-P., Richet P., Gillet P., Haeusermann D., and Hanfland M., Thermal pressure in the laser-heated diamond-anvil cell: An x-ray diffraction study, Eur. J. Mineral., 1998.

[35] Mao W.L., Mao H.-k., Eng P.J., Trainor T.P., Newville M., Kao C.-c., Heinz D.L., Shu J., Meng Y. and Hemley R.J., Bonding changes in compressed superhard graphite, **302**, 425, 2003.

[36] Cottrell E. and Walker D., Constraints on core formation from Pt partitioning in mafic silicate liquids at high temperatures, Geochim Cosmochim Acta, **70**, 1565, 2006.

[37] Domanik K.J. and Holloway J.R., The stability and composition of phengitic muscovite and associated phases from 5.5 to 11 GPa: Implications for deeply subducted sediments, Geochim Cosmochim Acta, **60**, 4133, 1996.

[38] Kushiro I., Changes in viscosity and structure of melt of NaA12SiO6 composition at high-pressures, J. Geophys. Res., **18**, 6347, 1976.

[39] Takahashi E., Yamada H. and Ito E., A new furnace assembly to 100 kbar and 1500°C with minimum temperature uncertainty, Geophys. Res. Lett., **9**, 805, 1982.

[40] Hernlund J., Leinenweber K., Locke D. and Tyburczy J.A., A numerical model for steady-state temperature distributions in solid-medium high-pressure cell assemblies, Am. Mineral., **91**, 295, 2005.

[41] Watson E.B., Wark D.A., Price J.D. and Van Orma J.A., Mapping the thermal structure of solid-media pressure assemblies, Can. J. Phys., **73**, 273, 2002.

[42] Heinz D.L., Sweeney J.S. and Miller P., A laser heating system that stabilizes and controls the temperature: Diamond anvil cell applications, Rev. Sci. Instrum., 1568-1575, 1991.

[43] Boehler R., Temperatures in the Earth's core from melting-point measurements of iron at high static pressures, Nature, **363**, 534, 1993.

[44] Boehler R., Melting of mantle and core materials at very high-pressures, Phil. Trans. R. Soc. A, **354**, 1265, 1996.

[45] Duffy T.S., Synchrotron facilities and the study of the Earth's deep interior, Rep. Progr. Phys., **68**, 1811, 2005.

[46] Heinz D.L. and Jeanloz R., Temperature measurements in the laser-heated diamond cell, in High-Pressure Research in Mineral Physics, ed. Manghnani M.H. and Syono Y., Amer Geophys Union, Washington D C, 1987, p. 113.

[47] Jeanloz R., Kavner A., Lazor P. and Saxena S.K., Melting criteria and imaging spectroradiometry in laser-heated diamond-cell experiments, Phil. Trans. R. Soc. A, **354**, 1279, 1996.

[48] Jephcoat A.P. and Besedin S.P., Temperature measurement and melting determination in the laser-heated diamond-anvil cell, Phil. Trans. R. Soc. A, **354**, 1333, 1996.

[49] Shen D. and Heinz D.L., High-pressure melting of deep mantle and core materials, in Ultra-High-Pressure Mineralogy: Physics and Chemistry of the Earth's Deep Interior, ed. Hemley R.J., Mineral Soc. America, Washington D C, 1998, p. 369.

[50] McMillan P.F., Chemistry of materials under extreme high pressure-high temperature conditions, Chem. Commun., **919**, 2003.

[51] Shen G., Mao H.-k. and Hemley R.J., Laser-heated diamond anvil cell technique: double sided heating with multimode Nd:YAG laser, in Advanced Materials '96 - New Trends in High Pressure Research, NIRIM, Tsukuba, Japan, 1996, 149.

[52] Yagi T., Kondo T., Watanuki T., Shimomura O. and Kikegawa T., Laser heated diamond anvil apparatus at the Photon Factory and SPring-8: Problems and improvements, Rev. Sci. Instrum., **72**, 1293, 2001.

[53] Boehler R. and Chopelas A., A new approach to laser heating in high-pressure mineral physics, Geophys. Res. Lett., **18**, 1147, 1991.

[54] Hearne G., Bibik A. and Zhao J., CO_2 laser-heated diamond-anvil cell methodology revisitied, J. Phys.: Cond. Matter, **24**, 11531, 2002.

[55] Campbell A.J., Measurement of temperature distributions across laser heated samples by multispectral imaging radiometry, Rev. Sci. Instrum., **79**, 01508, 2008.

[56] Kavner A. and Duffy T.S., Pressure-volume-temperature paths in the laser-heated diamond anvil cell, J. Appl. Phys., **89**, 1907, 2001.

[57] Kavner A. and Panero W.R., Temperature gradients and evaluation of thermoelastic properties in the synchrotron-based laser-heated diamond cell, Phys. Earth Planet Interiors, **143-144**, 527, 2004.

[58] Walter M.J. and Koga K.T., The effect of chromatic dispersion on temperature measurement in the laser-heated diamond anvil cell, Phys. Earth Planet Interiors, **143-144**, 541, 2004.

[59] Schultz E., Mezouar M., Crichton W., Bauchau S., Blattmann G., Andrault D., Fiquet G., Boehler R., Rambert N., Sitaud B. and Loubeyre P., Double-sided laser heating system for in situ high pressure-high temperature monochromatic X-ray diffraction at the ESRF, High Pressure Res., **25**, 71, 2005.

[60] Shen G., Rivers M.L., Wang Y., Sutton S.R., Laser heated diamond cell system at the Advanced Photon Source for in situ x-ray measurements at high pressure and temperature, Rev. Sci. Instrum., **72**, 1273, 2001.

[61] Panero W.R. and Jeanloz R., The effect of sample thickness and insulation layers on the temperature distribution in the laser-heated diamond anvil cell, Rev. Sci. Instrum., **72**, 1306, 2001.

Chapter 14
Solid-State and Materials-Chemistry at High Pressure

Paul F. McMillan

Department of Chemistry, Christopher Ingold Laboratory, University College London, 20 Gordon St., London WC1H 0AJ, UK
Email: p.f.mcmillan@ucl.ac.uk

1   Abstract

Solid state chemistry is concerned with understanding and developing the structures, phase relationships, properties and synthesis of crystalline and amorphous elements and compounds as well as composite materials. High pressure science began within physics and physical chemistry as new techniques were applied to increase pressures into ranges where significant changes in structures and physical properties could be observed. High-P,T studies immediately led to major advances in mineralogy, solid state chemistry and materials research that continue to be developed today.

2   Introduction

Solid state chemistry is concerned with understanding the structures and properties of substances along with their synthesis and reaction chemistry [1, 2]. Materials research combines fundamental physics and chemistry with engineering and develops the technological and commercial potential of materials [3, 4]. Beginning with the pioneering research of Bridgman it was recognised that crystalline solids undergo structural and phase transformations accompanied by changes in their properties as a function of pressure [5-8]. Modern high pressure physics and chemistry studies continue to explore the unusual structures and properties that are encountered in solids and liquids [9-27]. Some phases exhibit potentially useful properties that could be developed for commercial applications [28, 29]. The synthesis and processing of existing materials are also investigated and improved through high-P,T research.

3   Diamond and related materials

The first high-P,T experiments designed to produce materials for industrial use were carried out during the quest to achieve a laboratory synthesis of diamond [30-33]. Diamonds formed naturally within the Earth's mantle and lower crust (>100 km) are brought to the surface by explosive volcanism. As the hardest substance known since antiquity diamond is important for cutting and grinding applications ranging from machining and polishing metals and ceramics to oil and gas drilling. Diamond synthesis first reported by the GE research team has led to a worldwide industry for mass production of diamond powders for abrasive applications [31, 33, 34]. The research efforts led to advances in high-P,T technology and solid state chemistry as well as providing a model for materials technology development using high-P,T methods leading to large-scale industrial production. A second important high-P material is cubic boron nitride, c-BN, predicted and prepared shortly after the successful diamond synthesis efforts [35]. c-BN has an ordered diamond-like structure and is the second-hardest commercial material used for abrasive applications including high-speed machining of ferrous alloys (Figure 1). Research continues to improve the production and qualities of diamond and c-BN, including the role of metal catalysts and fluid environments in diamond formation, production of high-quality transparent large diamonds for specialised applications, and doping c-BN for semiconductor applications [36-43].

New studies to investigate and synthesise other known and predicted superhard compounds or semiconducting materials formed among within the B-C-N-O system also provide an active research area [11, 12, 14-16, 43-47]. These elements present a wide range of chemical and physical behaviour ranging from gases such as N_2, O_2, CO_2, CO with strong intramolecular bonding to the "soft" molecular compounds encountered in organic chemistry, polymer science and biology. They also provide some of the hardest state materials in existence (e.g., diamond, c-BN, B_6O) due to the flexibility in their chemical bonding. Recent high-P,T research has resulted in new high-P,T materials such as diamond-structured BC_2N with hardness comparable with c-BN [11, 12, 48].

An early prediction based on valency rules suggested that the compound B_2O should also form like c-BN with an ordered diamond-like structure [49]. However, despite early reports of its synthesis, that phase does not appear to exist [44, 49-51]. Instead, boron suboxide (B_6O_x) is formed with a rhombohedral structure based on that of α-B_{12}, with O atoms inserted in trigonal sites between the icosahedral B_{12} units (Figure 2) [16, 44]. Non-stoichiometric ver-

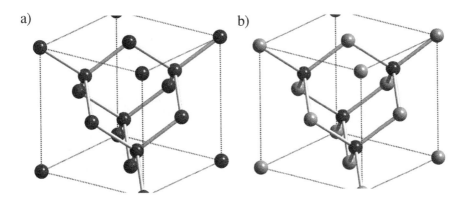

Figure 1. *The diamond and c-BN structures are materials with extremely high hardness. They are also wide bandgap semiconductors useful for high T electronic applications and potentially optoelectronic devices operating in the UV range. Diamond has high transparency throughout the electromagnetic spectrum and is used as a window material, including during high-P,T experiments in the diamond anvil cell. Diamond also has very high thermal conductivity used in semiconductor substrate applications. A new BC_2N phase has recently been discovered through high-P,T research. However, a predicted B_2O phase does not appear to exist.*

sions (B_6O_{1-x}) are developed as industrial abrasives and structural ceramics formed by hot-pressing [4]. However, high-P,T syntheses led to a nearly stoichiometric B6O phase due to the enhanced oxygen activity in the high pressure experiments [15, 16, 44, 52]. Samples recovered from 3-5 GPa showed an unusual macroscopic icosahedral morphology due either to multiple twinning or to occurrence of an unusual radial growth and packing scheme suggested by A. Mackay (Figure 2) [15, 52, 53]. The icosahedral particles could have applications as abrasives or lubricants, e.g, as nanoscale ball bearings. High-P,T synthesis experiments have also identified an analogous phase of boron subnitride, B_6N, that could exhibit metallic conductivity [16, 28, 44, 46]. Recently, new phases of elemental boron have also been demonstrated from high pressure research, and reports of superconductivity in highly B-doped diamond are under investigation [54-61].

Recent studies have been directed at producing new dense phases within the C-N system since theoretical predictions that high density sp³-bonded forms of C_3N_4 could be superhard [47, 62, 63]. Related C-N-H materials might also have applications for energy storage and release. The experimental research is being carried out using high-P,T synthesis techniques as well as thin film deposition methods, that have produced nanocrystalline materials that are not easy to characterise. Despite a large number of publications including

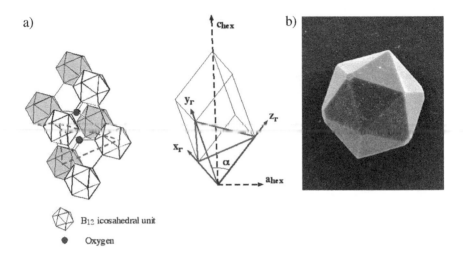

a)

b)

B_{12} icosahedral unit

Oxygen

Figure 2. *The structure of boron suboxide, B_6O_{1-x}. This provides the stable phase in the B-B_2O_3 system and it is a high hardness material used as a refractory structural ceramic applications in sintered polycrystalline forms prepared by hot-pressing. The structure is based on that of α-B containing icosahedral B_{12} units arranged into a rhombohedral unit cell (both rhombohedral x_r, y_r, z_r and hexagonal a_{hex}, c_{hex} axes are indicated)), with O atoms inserted into trigonal electron deficient sites between the B_{12} units. High-P,T synthesis results in large ($\sim$50 μm) crystals that have close to ideal composition ($x\sim0.04$: $B_6O_{0.96}$). Syntheses at P=2-3 GPa result in formation of free-standing icosahedral particles up to 40 mm in dimension described via multiple twinning of the rhombohedral cells or by radial Mackay packing of successive concentric layers of B_{12} units and O atoms.*

several reports of synthesis of superhard carbon nitride materials, there has still been no demonstration of dense crystalline C_3N_4 [45, 64]. Recent research has used molecular precursors with high N:C ratios including dicyandiamide (DCDA: $C_2N_4H_4$) in synthesis attempts. Following high-P,T treatment in a laser heated diamond anvil cell crystals with composition C_2N_3H determined to have a defective wurtzite (dwur) structure were produced and recovered to ambient conditions (Figure 3) [18]. Previously shock synthesis studies using DCDA as precursor had reported formation of the predicted β-C_3N_4 phase identified by X-ray powder diffraction [65]; recent work indicates that the reflections are mainly due to dwur-C_2N_3H (Figure 3) [66]. This structure is analogous to those of Si_2N_2O and $Si_2N_2(NH)$. The latter is known to be a precursor to Si_3N_4 ceramics under low pressure conditions [4].

There is also interest in layered graphitic carbon nitride materials produced at lower P,T conditions, e.g., in the 1-5 GPa range (Figure 4) [67-

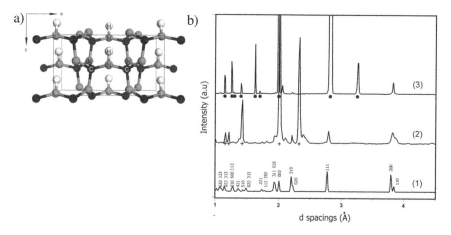

Figure 3. *The defective wurtzite (dwur-) structure of C_2N_3H contains C and N atoms in tetrahedral sites with sp^3 hybridisation. H atoms occupy a further set of tetrahedral sites but move off-centre to provide N-H bonds.*

71]. Polymeric solids derived from (C,N)-containing molecular precursors are well known and they give rise to high-strength refractory plastics such as the melamine-formaldehyde series. However, the nanocrystalline structures of compounds based on C_3N_3 triazine rings such as "melon" or "melem" are only just being elucidated [72, 73]. Extended layered C_xN_y materials have useful photoluminescence and catalysis properties including water splitting applications related to energy research [74, 75].

4 High-pressure mineralogy and solid-state materials research

Geologists took advantage of high-P,T techniques to explore the mineralogy and structure of the deep Earth [6, 33, 76-83]. There has always been a close link between mineralogy and materials research as many natural minerals and their derivatives are developed and used for technological applications [1]. High-P,T research in mineralogy continues to link with new materials development as phases such as SiO_2 stishovite are determined to have high hardness [84], mineral structures such as $CaTiO_3$ or $MgSiO_3$ perovskite provide important dielectric materials as well as geophysically important phases [85], and $MgAl_2O_4$ and γ-Mg_2SiO_4 lead to new nitride and oxynitride spinels [17, 24, 86, 87].

The spinel structure occurs among a wide range of A_2BX_4 oxides, halides and chalcogenides (Figure 5). The type phase $MgAl_2O_4$ is a well known semi-precious gem mineral with important properties as a semiconductor substrate

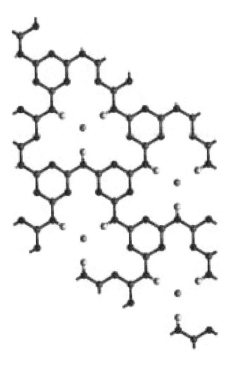

Figure 4. *Syntheses at lower P,T conditions from heterocyclic aromatic precursor molecules result in formation of graphitic layered C_xN_y structures.*

and transparent window material. The magnetic mineral phase magnetite (Fe_3O_4) leads to ferrite spinels MFe_2O_4 (M = Fe^{2+}, Ni^{2+}, Cu^{2+}, Mn^{2+} etc) with important magnetic and electronic properties used for transformer cores and in information storage applications. $Li_{1-x}(Mn,Fe)_2O_4$ spinels are used for battery and fuel cell applications. No spinel-structured compounds based on the N^{3-} anion had been reported prior to 1999 when laser heated diamond cell techniques were first used to prepare γ-Si_3N_4 and γ-Ge_3N_4 from the elements [17, 88]. Independently, high-P,T studies of Ge_3N_4 produced the spinel-structured γ-Ge_3N_4 phase using diamond cell and large press techniques [24]. The new spinel nitrides were also formed by shock synthesis [89]. Another near-simultaneous discovery of spinel-structured Sn_3N_4 did not involve high P synthesis but occurred following reaction between SnI_4 and KNH_2 in liquid NH_3 [90]; the synthesis has now been carried out under high-P,T conditions [91]. These studies led to a new area within high-P,T materials chemistry research [92]. γ-Si_3N_4 is has low compressibility and high hardness [93-95]. γ-Si_3N_4 and γ-Ge_3N_4 possess wide bandgaps in the 2-4 eV range, comparable with (Ga,In,Al)N used in optoelectronic devices [96, 97]. Spinel-structured SiAlON ($Si_{3-x}Al_xO_xN_{4-x}$) ceramics have been produced via high-P,T syn-

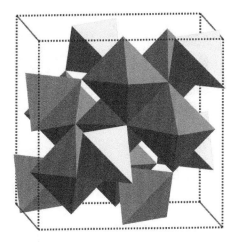

Figure 5. *The spinel structure taken by $MgAl_2O_4$, high-pressure γ-Mg_2SiO_4 important in determining the mineralogy and seismic properties of the Earth's mantle, and important magnetic and electronic materials such as Fe_3O_4 $(Fe^{2+}Fe_2\,3+O_4)$ and related ferrite spinels, and battery or fuel cell electrodes such as $Li_{1-x}(Mn,Fe)_2O_4$. The first nitride spinels based on the N^{3-} anion were created in 1999 and they include γ-Si_3N_4, γ-Ge_3N_4 and Sn_3N_4, with the group 14 cations (Si^{4+}, Ge^{4+}, Sn^{4+}) occurring on both iv- and vi-coordinated sites. γ-Si_3N_4 has high hardness and the new nitride spinels exhibit wide direct bandgaps and potentially useful optoelectronic properties. New Si-Al-O-N and Ga_3O_3N spinels are now known.*

thesis [86, 87]. The existence of Ga_3O_3N with an anion-ordered spinel structure predicted by ab initio calculations was synthesised at high-P,T [23, 25]. A large family of new spinel-structured nitrides and oxynitrides with useful technological properties now awaits discovery.

Perovskite-structured oxides based on $CaTiO_3$ continue to provide important electronic, magnetic and multiferroic materials and new phases have been explored using high-P,T techniques (Figure 6) [85, 98, 99]. Within mineralogy, a perovskite-structured polymorph of $MgSiO_3$ produced above 24 GPa was found to be stable throughout nearly the entire range of the lower mantle [100-102]. Recent high-P,T research has now indicated a further transformation into a "post-perovskite" polymorph with a layered $CaIrO_3$ structure that could explain unusual seismic properties occurring at the base of the lower mantle [103-107]. $MgSiO_3$ perovskite is recoverable to ambient P,T conditions but is unstable when heated above 500-600deg C [108]. The corresponding $CaSiO_3$ perovskite exhibits amorphisation during decompression as the SiO_6 octahedra transform back into tetrahedral silicate structures [109]. To harness this coordination instability to create new dielectric materials $SrGeO_3$

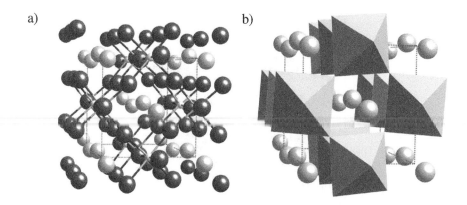

a)

b)

Figure 6. *Perovskite (CaTiO$_3$) is a common mineral that gives rise to structures that are remarkably important for geophysics and materials science. The ideal cubic structure expressed by BaTiO$_3$ contains corner-linked TiO$_6$ octahedra and Ba^{2+} cations in 12-fold coordination. Structural distortions involving off-centre displacements of the cations along with concerted rotations of the octahedra about their axes occur readily as a function of T, P and composition to yield phases with lower symmetry. The phase transitions are associated with large changes in the dielectric properties and perovskite materials are developed as radiowave transducers (e.g., in mobile telephones) and as capacitors. The high-P,T perovskite phase MgSiO$_3$ dominates the mineralogy of the Earth's lower mantle.*

was prepared at 6 GPa and decompressed to ambient P. The result was a series of perovskites on the verge of mechanical instability with a high dielectric constant [110-112]. New multiferroic materials have now been prepared using high-P,T synthesis placing lone pair cations (Se^{4+}, Te^{4+}) on the large perovskite A site and transition metal ions on the octahedral B site to achieve compositions like SeMO$_3$ (M=Co, Ni, Mn etc) [113, 114]. It is likely that new phases with useful electrconic, magnetic and optical properties will continue to be identified from high-P,T solid state chemistry and mineralogy research.

5   Superconductors, elemental alloys and high-hardness metals

Solid state physics studies have explored the electrical conductivity and magnetic properties of materials at high pressure and many high-P,T syntheses have been concerned with producing and studying materials with novel electronic and magnetic behaviour [26, 99, 115-117]. Bernal's original prediction that all materials should eventually become metallic at sufficiently high pres-

a) 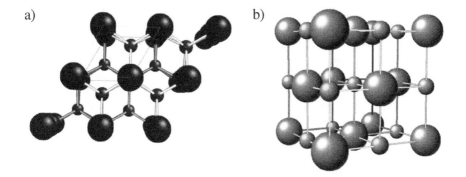 b)

Figure 7. *Hexagonal layered close-packed (a) and the cubic rocksalt structure (b) forms provide the basis for transition metal nitrides and carbides
that constitute refractory high-strength structural ceramics that are also metals
and superconductors that can achieve high T_C. values. In non-stoichiometric
phases such as TiN_x and Mo_2N the anion sites in the cubic structures are
only partly filled. Within hexagonal δ-MoN the occupancy of the nitrogen sites
can be ordered by high-P,T treatment resulting in high T_C values (12-17 K).*

sure was applied to the apparently "simple" element hydrogen [118, 119].
The quest to establish a metallic solid phase of hydrogen has now spanned
several decades, with complementary and sometimes competing results from
theoretical and experimental studies [120-129]. However, metallic conductivity has so far only been demonstrated in the dense fluid state during high-P,T
shock experiments [130]. High pressure studies have revealed surprisingly complex structures coupled with changes in the electronic properties of elements
and "simple" compounds at high pressure [21, 27, 117, 131-134]. Insulators
and semiconductors generally undergo transitions into metallic phases at high
pressure [56, 117, 133], but there are exceptions including Li [135-137]. Nearly
all the elements and materials including well known ionic compounds such
as CsI become superconducting by $\sim$1 megabar (100 GPa = 1 million atmospheres) [26, 56, 116, 137-139]. The current T_C records for elemental and
complex oxide materials are currently held at high P [13, 140]. Transition
metal nitrides, carbides and borides include refractory high hardness materials such as TiN and WC that are used as abrasives and structural materials,
including apparatus and components for high P research [6, 141-144]. They
are also metallic and compounds including cubic δ-NbN and hexagonal δ-MoN
achieve high T_C values [143, 144]. New carbides, nitrides and borides exhibit
low compressibility and high hardness values [28, 95, 145-147]. The structures are generally based on cubic and hexagonal polymorphs found within
the metallic elements, with N and C atoms occurring either interstitially or
as N^{3-} or C^{4-} ions [143, 144] (Figure 7). The compounds are often non-

stoichiometric (e.g., TiN_{1-x}) and the loss of translational symmetry causes activation of phonons away from the Brillouin zone centre so that acoustic and optic density of states are observed in optical spectra up to high pressures in the diamond anvil cell. The superconducting T_C can be related to the frequency of transverse acoustic modes and this can be used to predict pressure variations [148, 149].

In the Mo-N system, two phases are recorded at ambient P: hexagonal δ-MoN and rocksalt-structured γ-MoN$_{0.5}$ in which half the anion sites are empty [144]. It was though that if a cubic stoichiometric MoN phase could be prepared it might have a very high superconducting T_C [150]. However, despite attempts to prepare such a phase under high-P,T synthesis conditions, it does not appear to exist [151], in agreement with later theoretical predictions [152]. The superconducting T_C of Mo_2N samples prepared to date is $\sim$2-5 K. The hexagonal δ-MoN phase is usually formed at low P from mixtures of elements or reaction between $MoCl_5 + NH_3$. The structure is based on hexagonal Mo layers and the N atoms occupy one half of the trigonal prismatic sites between the layers. The ambient P synthesis procedures usually result in disordered N site occupancy and the T_C is low ($\sim$5 K). However, annealing at high pressure permits the N sites to become ordered without loss of nitrogen from the system and the T_C of δ-MoN is raised to 12-16 K [153, 154]. Recent high pressure investigations carried out for transition metal oxides also yield new high-hardness materials with the cotunnite ($PbCl_2$) structure in which the metal cation is 9-coordinate, that lead to the hardest oxide materials identified to date [84, 91].

6 Clathrates and new "light element" solids

A remarkable observation made by Bridgman was that solid H_2O underwent pressure-induced transformations into high density polymorphs [5]: over a dozen ice phases are now known to exist [133, 155-157]. Current studies are exploring the crystal chemistry and thermodynamic relations of high pressure phases of H_2O as well as NH_3 and other molecular ices important for planetary mineralogy [158-164]. That work extends to clathrate hydrates that form deep within Earth's oceans and that could play a significant role in CH_4 or H_2 storage or CO_2 sequestration as materials for energy applications [133, 158-170]. The internal structure of giant gas planets such as Jupiter is largely determined by the high pressure behaviour of hydrogen and helium, that have received considerable attention within the physics community [6, 118, 119, 121-130, 133, 134, 171-175]. These studies provide important testing grounds for extreme conditions materials exploration and also for theory.

New high pressure solid state compounds have been discovered following compression of molecular CO_2 and N_2. The linear molecule CO_2 contains very strong double O=C=O bonds. Upon initial pressurisation it freezes to form a molecular solid. Early experiments indicated that the molecular structure

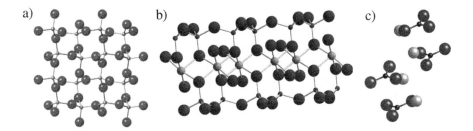

Figure 8. *(a) Molecular CO_2 is predicted to transform at high-P,T into a three-dimensional polymeric network structure analogous to the SiO_2 polymorphs quartz or cristobalite. (b) Carbonates such as $CaCO_3$ appear to form pyroxene-like chain structures based on CO_4 tetrahedral units at high-P,T conditions. (c) N_2O and NO_2 form molecular-ionic solids NO^+ NO^{3-} analogous to carbonate structures following high-P,T treatment (the NO^+ cations are represented as single large balls).*

remained intact up to at least 80 GPa. However, it was found that heating to T~1000-1500 K at P > 25 GPa results in polymerisation into a tetrahedral network structure containing corner-linked CO_4 groups analogous to the SiO_2 polymorphs cristobalite or tridymite [10, 14, 176, 177] (Figure 8). The new structure ($CO_2 - V$) could undergo further transitions at high pressure, and formation of an amorphous high-coordinate polymerised CO_2 material has been described from high-P,T experiments [176, 177]. Now carbonate phases including $CaCO_3$, $SrCO_3$ and $MnCO_3$ have been shown to transform to new pyroxene-like structures containing chains of CO_4^{4-} units containing bridging and non-bridging O^{2-} ions under high-P,T conditions [178-180] (Figure 8). This observation is likely to give rise to a new solid state chemistry of tetrahedrally bonded carbonate-based phases at high pressure. However, none of the phases investigated to date have been recoverable to ambient conditions that constitutes a barrier to their development as useful technological materials [181]. Among phosphosilicate materials the P^{5+} ion replaces Si^{4+} in tetrahedral coordination. The molecule N_2O (N=N=O) is isoelectronic with CO_2 and it was thought that it might form a tetrahedrally-bonded network analogous to the cristobalite-structured SiO_2 analogue PON at high pressure [182, 183]. However, N_2O breaks down into N_2 and NO_2 components to form calcite- or aragonite-structured compounds NO^+ NO^{3-}, like NO_2 (N_2O_4) [184, 185] (Figure 8).

Triply bonded N_2 contains one of the strongest covalent bonds known among molecular compounds (954 kJ/mol). However, theoretical calculations indicated that a transition into a singly-bonded dense polymeric polymorph could occur at high pressure and the new material with interesting semiconducting properties might be recoverable [186-188]. In view of the structural chemistry of elemental phosphorus this would lead to a new solid state chemistry based on nitrogen. Early experimental studies observed opacity of N_2 samples above 150 GPa with the appearance of new Raman and infrared active bands [189, 190]. A transition into a polymeric phase occurs at $\sim$100 GPa, that remains semiconducting up to at least 240 GPa [9]. The new high pressure structure was initially described as the "cubic gauche" (cg-) form predicted from theoretical studies. A further orthorhombic "ζ-nitrogen" structure was later identified providing a link between molecular and cg polymorphs [9]. Such studies indicate that the high-P,T solid state chemistry of nitrogen-based dense polymorphs is just beginning. The complex structural chemistry of elemental phosphorus itself is not yet well understood [2, 191]. The stable but highly reactive "white" phosphorus structure is based on isolated tetrahedral P4 groups, and the high pressure metallic black polymorph contains interwoven zigzag chains and sheets of singly-bonded $(P_3)_n$ pyramidal units [2, 6, 191]. Amorphous "red" varieties are formed by partly polymerised P_3 and P_4 units with different degrees of P-P bonding between them, but little is known about their structural or thermodynamic relationships. Following observation of an unusual density-driven first order phase transition in liquid phosphorus at high pressure [192, 193], new studies are investigating the high pressure solid state as well as liquid and fluid polymorphism and the thermodynamic phase relations [194]. Zaug et al described the structural behaviour of amorphous red phosphorus at high pressure [195]. Density-driven liquid-liquid and "polyamorphic" transitions as well as pressure-induced amorphisation are resulting in new forms of amorphous materials [19, 196-200].

Zintl phases such as NaSi and LiSi contain oligomeric polyanions based on the semimetallic or semiconducting group of elements along with highly electropositive cations [201]. The Si^- anion is formally isoelectronic with elemental P and the NaSi structure contains isolated Si_4^{4-} tetrahedral units like those in white phosphorus; the LiSi structure instead contains zigzag chain and sheet units of $(Si^-)_n$ analogous to the high pressure black P structure [202-204] (Figure 9). Although it is thermodynamically stable, formation of the LiSi structure is kinetically impeded relative to phases such as $Li_{12}Si_7$: only high-P,T synthesis could demonstrate the existence of this phase. Recent studies have investigated the high pressure behaviour of Zintl phases such as $BaSi_2$ that contain corrugated sheets of the silicide polyanion with three-fold coordinated silicon species [205]: the crystalline structure is lost at high pressure indicating further crystalline phases existing at high density. Pressure-induced amorphisation also occurs for NaSi that is likely accompanied by loss of Na from the structure. The resulting redox reaction $(Na^+ + Si^- = Si^0 + Na^0)$ indicates formation of new Si-Si bonds between the Si_4^{4-} tetrahedral groups to produce partly polymerised oligomeric units within the amorphous structures [206]. The Zintl phases also provide precursors and models for the high-pressure behaviour of semiconductor clathrate phases that provide new thermoelectric, superconducting and electronic materials [207-211].

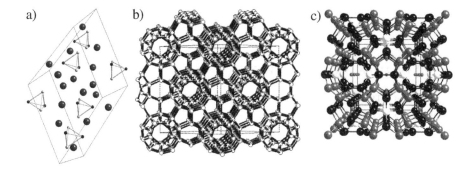

Figure 9. *(a) The NaSi structure contains tetrahedral $Si_4{}^{4-}$ units analogous to those found in white phosphorus. (b) The Type II semiconductor clathrate Si_{136} contains empty cages that are usually filled with metal atoms. (c) The Type I Na_8Si_{46} clathrate. These compounds are isostructural with H_2O-based clathrate structures in which the tetrahedral framework is constituted by H-bonded OH_2 units and the cages are filled with molecules such as CH_4 or CO_2, providing materials for energy applications and planetary ices.*

7   Summary

Solid state chemistry involves the study of the structures, phase transformations, properties and synthesis of crystalline and amorphous solids. High pressure science permits access to a wide range of thermodynamic conditions in which new crystalline and amorphous structures and physical or chemical behaviour of elements and compounds are identified and studied. Some of the materials and synthesis routes to them will lead to development of industrially relevant materials and processes. An exciting new area concerns nanoscale materials in which the properties and reactivity are determined by the nanoparticle dimensions and its surface properties. High pressure studies of free-standing and agglomerated nanoparticles and as well as dense nanocomposites are providing a new understanding of their physical and chemical behaviour and it is likely that new materials will result. New structural changes and phase transformations are also being discovered among amorphous solids and liquids as a function of the density, and some of these could result in new materials developments [19, 198]. Studies are also being carried out to establish the limits of biological function under high-P,T conditions along with changes in biomolecular systems that are important for discussions of the origins and existence of life under extreme conditions. These studies are also concerned with design of new biologically inspired materials and nanocomposites for device applications. The field of high pressure solid state chemistry and new materials development is only just beginning and it will lead to a new range of elemental polymorphs, compounds, nanocomposites and biologically-inspired or polymeric macromolecules for existing and new technologies as well as synthesis and processing routes to them.

8 Acknowledgements

Work within PFM's group is supported by EPSRC through Portfolio grant EP/D504782 (in collaboration with P. Barnes (Birkbeck) and C.R.A. Callow (UCL)) and Senior Research Fellowship EP/D07357X.

References

[1] West A.R., Solid State Chemistry and its Applications, John Wiley & Sons, Chichester, 1984.

[2] Wells A.F., Structural Inorganic Chemistry, Clarendon Press, Oxford, 1984.

[3] Anderson J.C., et al., Materials Science, Chapman & Hall, London, 1990.

[4] Riedel R., ed, Handbook of Hard Ceramic Materials, Wiley VCH, New York, 2000.

[5] Bridgman P.W., The Physics of High Pressure, G Bell and Sons, London, 1931.

[6] Liu L.-G. and Bassett W., Elements, Oxides, Silicates: High-Pressure Phases with Implications for the Earth's Interior, Oxford University Press, New York, 1986.

[7] McMillan P.F., Nature Materials, **4**, 2005.

[8] Bridgman P.W., Proc. R. Soc. London, **203**, 1, 1950.

[9] Eremets M., et al., Nature Materials, **3**, 558, 2004.

[10] Iota V., Yoo, C.-S. and Cynn, H., Science, **283**, 1999.

[11] Solozhenko V.L., et al., Appl. Phys. Lett., **78**, 1385, 2001.

[12] Solozhenko V.L., Dub, S.N. and Novikov, N., Diamond Related Mater, **10**, 2228, 2001.

[13] Struzhkin V.V., et al., Nature, **390**, 1997.

[14] Yoo C.-S., et al., Phys. Rev. Lett., **83**, 5527, 1999.

[15] Hubert H., et al., Nature, **391**, 376, 1998.

[16] Hubert H., et al., Chem. Mater., **10**, 1530, 1998.

[17] Zerr A., et al., Nature, **400**, 1999.

[18] Horvath-Bordon E., et al., Angew. Chem. Int. Ed., **46**, 1476, 2007.

[19] McMillan P.F., et al., Nature Materials, **4**, 680, 2005.

[20] Bhat M.H., et al., Nature, **448**, 787, 2007.

[21] McMahon M.I., Degtyareva O. and Nelmes R.J., Phys. Rev. Lett., **85**, 2002.

[22] Mao W.L., et al., Science, **302**, 425, 2003.

[23] Kinski I., et al., Z. Naturforsch., **60b**, 832, 2005.

[24] Leinenweber K., et al., Chem. Eur. J., **5**, 3076, 1999.

[25] Soignard E., et al., Chem. Mater., **17**, 5465, 2005.

[26] Amaya K., Shimizu K. and Eremets M., Int. J. Mod. Phys., **13**, 3623, 1999.

[27] Nelmes R.J., Phys. Rev. Lett., **88**, 155503, 2002.

[28] McMillan P.F., Nature Materials, **1**, 19, 2002.

[29] McMillan P.F., High Press Res., **23**, 7, 2003.

[30] Bundy F.P., et al., Carbon, **34**, 141, 1996.

[31] Bundy F.P., et al., Nature, **176**, 51, 1955.

[32] Hazen R.M., The New Alchemists, Times Books, Random House, New York, 1993.

[33] Hazen R.M., The Diamond Makers, Cambridge Univ Press, Cambridge, UK, 1999.

[34] Bovenkerk H.P., et al., Nature, **184**, 1094, 1959.

[35] Wentorf R.H., J. Chem. Phys., **26**, 956, 1957.

[36] Demazeau G., Biardeau, G. and Vel, L., Mater. Lett., **10**, 1990.

[37] Pal'yanov,Y.N., et al., Am. Mineral., **87**, 1009, 2002.

[38] Sokol A.G., et al., Diamond Related Mater, **10**, 2131, 2001.

[39] Solozhenko V.L., et al., Phys. Chem. Chem. Phys., **4**, 5386, 2002.

[40] Solozhenko V.L., et al., J. Phys. Chem. B, **106**, 6634, 2002.

[41] Sumiya H., Toda N. and Satoh S., New Diamond Frontier Carbon Technol., **10**, 233, 2000.

[42] Sung C.-M. and Tai M.-F., Int. J. Refractory Metals Hard Mater., **15**, 237, 1997.

[43] Taniguchi T., et al., Jpn. J. Appl. Phys., **41**, L109, 2002.

[44] Hubert H., et al., MRS Symp. Proc., **410**, 191, 1996.

[45] Malkow T., Mater. Sci. Eng. A, **302**, 309, 1996.

[46] Solozhenko V.L., Le Godec Y. and Kurakevych O.O., C. R. Chimie, **9**, 1472, 2006.

[47] Teter D. and Hemley R.J., Science, **271**, 1996.

[48] Solozhenko V.L., High Press Res., **22**, 519, 2002.

[49] Hall H.T. and Compton, L.A., Inorganic Chemistry, **4**, 1213, 1965.

[50] Endo T., Sato T. and Shimida M., J. Mater. Sci. Lett., **6**, 683, 1987.

[51] Grumbach M.P., Sankey O.F. and McMillan P.F., Phys. Rev. B, **52**, 15807, 1996.

[52] McMillan P.F., et al., J. Solid State Chem., **147**, 281, 1999.

[53] Mackay A., Acta Crystallogr., **15**, 916, 1962.

[54] Dubrovinskaia N., et al., J. Appl. Phys., **99**, 033903, 2006.

[55] Ekimov E.A., et al., Nature, **428**, 2004.

[56] Eremets M.I., et al., Science, **293**, 272, 2001.

[57] Hassermann U., et al., Phys. Rev. Lett., **90**, 2003.

[58] Oganov A.R., et al., Nature, **457**, 2009.

[59] Papaconstantopoulos D.A. and Mehl M.K., Phys. Rev. B, **65**, 172510, 2002.

[60] Yokoya T., et al., Nature, **438**, 647, 2005.

[61] Zarechnaya E.Y., et al., Phys. Rev. Lett., **102**, 185501, 2009.

[62] Cohen M.L., Phys. Rev. B, **32**, 7988, 1985.

[63] Liu A.L. and Cohen M.L., Science, **245**, 841, 1989.

[64] Horvath-Bordon E., et al., Chem. Soc. Rev., **35**, 987, 2006.

[65] Liu J., Sekine, T. and Kobayashi, T., Solid State Commun., **137**, 21, 2006.

[66] Salamat A., et al., in prep, 2009.

[67] Demazeau G., et al., Rev. High Pressure Sci. Technol., **7**, 1345, 1998.

[68] Montigaud H., et al., Diamond and Related Materials, **8**, 1707, 1999.

[69] Alves I., et al., Solid State Commun., **109**, 697, 1999.

[70] Wolf C.H., et al., in Frontiers of High Pressure Research II: Application of High Pressure to Low-Dimensional Novel Electronic Materials, ed. Hochheimer H.D., Kluwer Academic, The Netherlands, 2001, p. 29.

[71] Zhang Z., et al., J. Am. Chem. Soc., **123**, 7788, 2001.

[72] Lotsch B.V., et al., Chemistry: A European Journal, , 2007.

[73] Jrgens B., et al., J. Am. Chem. Soc., **125**, 10288, 2003.

[74] Wang X., et al., Nature Materials, **8**, 76, 2009.

[75] Guo Q., et al., Solid State Commun., **132**, 369, 2004.

[76] Coes L., J. Am. Ceram. Assoc., **38**, 333, 1955.

[77] Ringwood A.E., Am. Mineral., **44**, 1959.

[78] Ringwood A.E., Earth Planet. Sci. Lett., **5**, 401, 1969.

[79] Stishov S.M., High Press Res., **13**, 245, 1995.

[80] Stishov S.M. and Popova, S.V., Geochemistry, **10**, 923, 1961.

[81] Anderson D.L., Theory of the Earth, Blackwell, Boston, 1989.

[82] Jeanloz R. and Thompson, A.B., Rev Geophys Space Physics, **21**, 51, 1983.

[83] Hemley R.J., ed, UltraHigh-Pressure Mineralogy, Mineral Soc America, Washington DC, 1998.

[84] Dubrovinsky L.S., Nature, **410**, 653, 2001.

[85] Navrotsky A. and Weidner, D.J., ed, Perovskite: A Structure of Great Interest to Geophysics and Materials Science, American Geophysical Union, Washington DC, 1989.

[86] Schwarz M., et al., Angew. Chem. Int. Ed., **41**, 788, 2002.

[87] Sekine T., et al., Chem. Phys. Lett., **344**, 395, 2001.

[88] Serghiou G., et al., J. Chem. Phys., **111**, 1999.

[89] Sekine T., et al., Appl. Phys. Lett., **76**, 2000.

[90] Scotti N., et al., Z. Anorg. Allg. Chem., **625**, 1435, 1999.

[91] Lundin U., Phys. Rev. B, **57**, 4979, 1998.

[92] Schnick W., Angew. Chem. Int. Ed., **38**, 3309, 1999.

[93] Sekine T., et al., J. Phys. Cond. Matter, **13**, L515, 2001.

[94] Jiang J.Z., et al., J. Phys. Cond. Matter, **13**, L515, 2001.

[95] Soignard E., et al., J. Phys. Cond. Matter, **13**, 557, 2001.

[96] Dong J., et al., Phys. Rev. B, **61**, 11979, 2000.

[97] Leitch S., et al., J. Phys. Cond. Matter, **16**, 6469, 2004.

[98] Demazeau G., Eur J. Solid State Chem., **34**, 759, 1997.

[99] Goodenough J.B., Kafalas J.A. and Longho J.M., in Preparative Methods in Solid State Chemistry, ed. Hagenmuller, P., Academic Press, New York, 1972.

[100] Hemley R.J. and Cohen R.E., Ann. Rev. Earth Planet. Sci., **20**, 553, 1992.

[101] Knittle E. and Jeanloz R., Science, **235**, 668, 1987.

[102] Liu L.-g., Geophys. Res. Lett., **1**, 277, 1974.

[103] Akber-Knutson S., Steinle-Neumann G. and Asimow P.D., Geophys. Res. Lett., **32**, L14303, 2005.

[104] Hernlund J.W., Thornas C. and Tackley P.J., Nature, **434**, 882, 2005.

[105] Iitaka T., et al., Nature, **430**, 442, 2004.

[106] Murakami M., et al., Science, **304**, 855, 2004.

[107] Oganov A.R. and Ono S., Nature, **430**, 445, 2004.

[108] Durben D.J. and Wolf G.H., Am. Mineral., **77**, 890, 1992.

[109] Hemmati M., et al., Phys. Rev. B, **51**, 14841, 1995.

[110] Grzechnik A., et al., Integrated Ferroelectrics, **15**, 191, 1997.

[111] Grzechnik A., et al., Eur. J. Solid State Inorg. Chem., **34**, 269, 1997.

[112] Grzechnik A., et al., J. Phys. Cond. Matter, **10**, 221, 1998.

[113] Munoz A., et al., Dalton Trans., **4936**, 2006.

[114] Munoz A., et al., Phys. Rev. B, **73**, 104442, 2006.

[115] McMahon M.I. and Nelmes, R.J., Chem. Soc. Rev., **35**, 943, 2006.

[116] Amaya K., et al., J. Phys. Cond. Matter, **10**, 11179, 1998.

[117] Schwarz U., Z Kristallogr, **219**, 376, 2004.

[118] Ashcroft N.W., Phys. Rev. Lett., **21**, 1748, 1968.

[119] Wigner E. and Huntington H.B., J. Chem. Phys., **3**, 764, 1935.

[120] Chen N., Sterer E. and Silvera I.F., Phys. Rev. Lett., **76**, 1663, 1996.

[121] Eggert J.H., et al., Phys. Rev. Lett., **66**, 193, 1991.

[122] Goncharov A.F., et al., Proc Nat Acad Sci USA, **98**, 14234, 2001.

[123] Hemley R.J. and Mao H.-k., Phys. Rev. Lett., **61**, 857, 1988.

[124] Lorenzana H.E., Silvera I.F. and Goettel K.A., Phys. Rev. Lett., **63**, 2080, 1989.

[125] Loubeyre P., Nature, **383**, 702, 1996.

[126] Loubeyre P., Occelli F. and LeToullec R., Nature, **416**, 613, 2002.

[127] Mao H.K. and Hemley R.J., Am. J. Sci., **80**, 234, 1992.

[128] Mao H.K. and Hemley R.J., Rev. Mod. Phys., **66**, 671, 1994.

[129] Mazin I.I., et al., Phys. Rev. Lett., **78**, 1066, 1997.

[130] Weir S.J.M., A C and Nellis W.J., Phys. Rev. Lett., **76**, 1860, 1996.

[131] McMahon M.I., Rekhi S. and Nelmes R.J., Phys. Rev. Lett., **87**, 055501, 2001.

[132] Schwarz U., et al., Phys. Rev. Lett., **83**, 4085, 1999.

[133] Hemley R.J., Ann. Rev. Phys. Chem., **51**, 763, 2000.

[134] Hemley R.J., et al., Nature, **369**, 384, 1994.

[135] Hanfland M., et al., Nature, **408**, 174, 2000.

[136] Neaton J.B. and Ashcroft N.W., Nature, **400**, 141, 1999.

[137] Shimizu K., Ishikawa H. and Amaya K., J. Phys. Cond. Matter, **14**, 10433, 2002.

[138] Shimizu K., et al., Nature, **393**, 767, 1998.

[139] Shimizu K., et al., Physica B, **194-196**, 1959, 1994.

[140] Gao L., et al., Phys. Rev. B, **50**, 4260, 1994.

[141] Eremets M., High Pressure Experimental Methods, Oxford University Press, Oxford, 1996.

[142] Holzapfel W.D. and Isaacs N.S., ed, High-Pressure Techniques in Chemistry and Physics: A Practical Approach, Oxford University Press, Oxford, 1997.

[143] Oyama S.T., The Chemistry of Transition Metal Carbides and Nitrides, Blackie Academic, Glasgow, 1996.

[144] Toth L.E., Transition Metal Carbides and Nitrides, Academic Press, New York, 1971.

[145] Kaner R.B., Gilman J.J. and Tolbert S.H., Science, **308**, 1268, 2005.

[146] Brazhkin V.V., Lyapin A.G. and Hemley R.J., Phil. Mag. A, **82**, 231, 2002.

[147] Haines J., Leger J.M. and Bocqillon J., Ann. Rev. Mater. Sci., **31**, 1, 2001.

[148] Chen, X.-J., et al., Phys. Rev. B, **70**, 2004.

[149] Shebanova O., McMillan P.F. and Soignard E., High Press Res., **26**, 87, 2006.

[150] Papaconstantopoulos D.A., et al., Nature, **308**, 494, 1984.

[151] Machon D., et al., Phys. Stat. Solidi (a), **203**, 831, 2006.

[152] Hart G.L.W. and Klein B.M., Phys. Rev. B, **61**, 3151, 2000.

[153] Bezinge A., et al., Solid State Commun., **63**, 141, 1987.

[154] Bull C.L., et al., J. Solid State Chem., **177**, 2004.

[155] Bowron D.T., et al., J. Chem. Phys., **125**, 194502, 2006.

[156] Salzmann C.G., et al., Science, **311**, 1758, 2006.

[157] Loerting T. and Giovambattista N., J. Phys. Cond. Matter, **18**, R919, 2006.

[158] Kuhs W.F., in High-Pressure Crystallography, ed. Katrusiak, A. and McMillan, P.F., Kluwer Academic, Dordrecht, The Netherlands, 2004, p. 475.

[159] Loveday J.S., et al., Phys. Rev. Lett., **87**, 215501, 2001.

[160] Loveday J.S., et al., Nature, **410**, 661, 2001.

[161] Loveday J.S., et al., Can. J. Phys., **81**, 539, 2003.

[162] Loveday J.S., et al., Phys. Rev. Lett., **76**, 74, 1996.

[163] Loveday J.S., et al., Physica B, **241-243**, 240, 1998.

[164] Datchi F., et al., Phys. Rev. B, **73**, 174111, 2006.

[165] Chou I.M., J. Phys. Chem. A, **105**, 4664, 2001.

[166] Hirai H., J. Phys. Chem. B, **104**, 1429, 2000.

[167] Max M.D., ed, Natural Gas Hydrate in Oceanic and Permafrost Environments, Kluwer Academic, Dordrecht, The Netherlands, 2000.

[168] Schmitt B., de Bergh C. and Festou M., ed, Solar System Ices, Kluwer Academic, Dordrecht, The Netherlands, 1998.

[169] Vos W.L., et al., Phys. Rev. Lett., **71**, 3150, 1993.

[170] Scandolo Science, .

[171] Duffy T.S., et al., Science, **263**, 1590, 1994.

[172] Hanfland M., Hemley R.J. and Mao H.K., Phys. Rev. Lett., **70**, 3760, 1993.

[173] van Straaten J. and Silvera I.F., Phys. Rev. Lett., **37**, 1989, 1988.

[174] Hemley R.J. and Mao H.K., Phys. Rev. Lett., **63**, 1393, 1989.

[175] Mao H.K., et al., Phys. Rev. Lett., **60**, 2649, 1988.

[176] Santoro M., et al., Nature, **441**, 857, 2006.

[177] Tschauner O., Mao H.-k. and Hemley R.J., Phys. Rev. Lett., **87**, 075701, 2001.

[178] Ono S., et al., Am. Mineral., **90**, 667, 2005.

[179] Ono S., Kikegawa T. and Ohishi Y., Am. Mineral., **92**, 1246, 2007.

[180] Oganov A.R., Glass C.W. and Ono S., Earth Planet. Sci. Lett., **241**, 95, 2006.

[181] Brazhkin V.V. and Lyapin A.G., Nature Materials, **3**, 1, 2004.

[182] Kingma K.J., Pacalo R.E.G. and McMillan P.F., Eur. J. Solid State Inorg. Chem., **34**, 679, 1997.

[183] Leger J.M., et al., Phys. Chem. Minerals, **28**, 388, 2001.

[184] Somayazulu M., et al., Phys. Rev. Lett., **87**, 135504, 2001.

[185] Jones L.H., Swanson B.I. and Agnew S.F., J. Chem. Phys., **82**, 4389, 1985.

[186] Mailhiot C., Yang L.H. and McMahan A.K., Phys. Rev. B, **46**, 14419, 1992.

[187] McMahan A.K. and LeSar R., Phys. Rev. Lett., **54**, 1929, 1985.

[188] Zhang T., et al., Phys. Rev. B, **73**, 094105, 2006.

[189] Eremets M., et al., Nature, **411**, 170, 2001.

[190] Goncharov A.F., et al., Phys. Rev. Lett., **85**, 1262, 2000.

[191] Greenwood N.N. and Earnshaw A., Chemistry of the Elements, Pergamon Press, Oxford, 1984.

[192] Katayama Y., et al., Science, **306**, 848, 2004.

[193] Katayama Y., et al., Nature, **403**, 170, 2000.

[194] Monaco G., et al., Phys. Rev. Lett., **90**, 255701, 2003.

[195] Zaug J.M., Soper A.K. and Clark S.M., Nature Materials, **7**, 890, 2008.

[196] Brazhkin V.V., et al., ed, New Kinds of Phase Transitions: Transformations in Disordered Substances, Kluwer Press, Dordrecht, the Netherlands, 2002.

[197] Brazhkin V.V., Popova S.V. and Voloshin R.N., High Press Res., **15**, 267, 1997.

[198] McMillan P.F., J. Mater. Chem., **14**, 1506, 2004.

[199] Ponyatovsky E.G. and Barkalov O.I., Mater. Sci. Rep., **8**, 147, 1992.

[200] Poole P.H., et al., Science, **275**, 322, 1997.

[201] Kauzlarich S.M., ed, Chemistry, Structure and Bonding of Zintl Phases and Ions, VCH, New York, 1996.

[202] Evers J., Oehlinger G. and Sextl G., Angew. Chem. Int. Ed., **32**, 1442, 1993.

[203] Evers J., Oehlinger G. and Sextl G., Eur. J. Solid State Inorg. Chem., **34**, 773, 1997.

[204] Stearns L.A., et al., J. Solid State Chem., **173**, 251, 2003.

[205] Nishii T., et al., Phys. Stat. Solidi (b), **244**, 270, 2007.

[206] Cabrera R.Q., et al., J. Solid State Chem., submitted, 2009.

[207] Bobev S. and Sevov S.C., J. Solid State Chem., **153**, 92, 2000.

[208] Yamanaka S., et al., Inorg. Chem., **39**, 56, 2000.

[209] San Miguel A., Chem. Soc. Rev., 2006.

[210] Machon D., et al., Phys. Rev. B, 79 184101

[211] San Miguel A. and Toulemonde P., High Press Res., **25**, 159, 2005.

Chapter 15
Liquids and Amorphous Materials

Martin C. Wilding

Institute of Mathematics and Physical Sciences, Aberystwyth University, Aberystwyth, United Kingdom, SY23 3BZ

1 Introduction

The liquid state dominates many physical, chemical and biological processes many of which occur at pressures greater than one atmosphere [1, 2, 3]. Yet understanding liquid structure has proved to be challenging even at ambient pressure. The structure and structure-dependent properties of liquids can be investigated by studying amorphous or vitreous forms of the same material, although this connection is not straightforward. Understanding of the formation of glass and other amorphous solids has been improved by new synthesis techniques which have expanded the range of glass formation from the traditional, commercially important realm to include a variety of exotic glass forming materials, including metallic glassy systems.

The structure of liquids, glasses and other amorphous materials is determined by diffraction using neutrons and more recently high energy X-rays. Recent advances include the ability to focus neutrons and X-rays and these provide new opportunities including the ability to study liquid and amorphous structures *in situ*, at high pressure and temperature [4, 5].

In this contribution the distinction between liquids and amorphous materials will be made and an introduction to techniques used to evaluate there structure given. The observed and predicted changes in liquid and glass structure with pressure will be highlighted and the concept of polyamorphism discussed.

2   Definitions

At low pressures, matter exists as a dense solid or as a diffuse vapour [2, 1, 3]. For each of these states there is an idealised model, which although may be a simplified approximation of real crystals or gases, allows a theoretical discussion. For solids and gases this is respectively the ideal crystal lattice and the ideal gas. In an ideal crystal for example there is focus on the structural order which is modified by thermal motion. An ideal gas in contrast describes the thermal motion of atoms based on their random positions.

At high pressure a third state, the liquid state appears and this occurs over a temperature range that is intermediate between the crystalline and vapour states (Figure 1).There are simple and well-established definitions of liquid using P-T and P-V phase diagrams. However there are no general theories of the liquid state in the sense that there are ideal gas laws or models of the crystalline state and most properties of liquids are calculated or determined from the sums of pair interactions and inter-atomic forces, by this token most liquids can be viewed as classical systems [2, 1].

If we consider a liquid, we can describe it in terms of a number of molecules, which in a vapour will have large entropy (S) per molecule that reflects a high disorder. If this vapour is cooled then the entropy per molecule is reduced, there is an implied increase in molecular order and an increase in density (the packing fraction increases to 0.3 to 0.5). On further cooling there is an increase of long-range order, shown by the development of a periodic crystal lattice, regularly spaced crystal planes that allow unambiguous characterisation of the crystal structure by X-ray or neutron diffraction. In this simple scenario, with the solid, vapour and liquid states being simple molecular or atomic systems, phase diagrams determine the stability of each of the three states, liquids are only stable over a limited part of the P-T phase diagram, between the triple point temperature (where all three phases coexist) and the critical temperature above which the liquid and gas merge into a single fluid phase. There are however metastable regions in these phase diagrams, these are regions where there is separation into coexisting stable phases such as the nucleation and growth of liquid droplets in a supersaturated vapour. Glasses for another important category of metastable materials, these will be discussed later.

The distinction between the spatial arrangements of gases, liquids and crystals is provided by the radial or pair distribution function. This is a quantitative measurement of the modulation in local density around a molecule as a function of distance. In the gas (and in the limit of an ideal gas) the local density seen from from one fixed molecule is equal to the average density (ρ). In a liquid however the positions of neighbouring molecules are strongly correlated and the radial distribution function (which we will call $\rho(r)$) shows a series of maxima associated with neighbouring shells of molecules, these oscillations are however damped and tend to a macroscopic value at large radial distances. At the transition to the crystalline state, the radial distri-

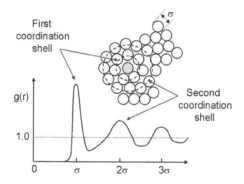

Figure 1. *Atomic configurations for a liquid of hard spheres of diameter σ showing the first and second coordination shells [1, 2, 3]. The pair distribution function (PDF) shows peaks that correspond to the first and second coordination shells.[100]*

bution function is characterised by well-defined intermolecular distances and coordination numbers that extend to macroscopic scales (at zero temperature these would be a sequence of δ-functions with the peaks defined by the lattice geometry)(Figure 2).

The macroscopic properties of a liquid distinguish it from both a vapour and a solid. Cohesion is a property shared by both liquids and solids and is a consequence of intermolecular interactions which leads to a lowering of internal energy via clustering. In gases the (thermal) kinetic energy of molecules overcomes this short-range attraction. The fluidity of liquids and gases distinguishes them from solids. In the application of an external stress there is a fluid flow rather than elastic displacement and the elastic relation that characterises crystals is replaced by the shear viscosity (*eta*) which characterises the internal friction of the fluid. On a microscopic scale this reflects the molecular diffusion.

The simple definitions outlined above represent simple substances comprising quasi spherical molecules and may be used to describe for example Ar, NH_3 CH_4 or N_2. Argon atoms are spherical and interact by short-range repulsion and can be modelled by convenient semi-empirical potential models such as the Lennard Jones potential. However for many liquids, molecules can be strongly polar (H^+) and bonds formed can be highly directional as is the case of water. This can lead to more complex behaviour, for example multi-component systems which are mixtures of several molecular species give rise to compositional ordering and complicated phase equilibria. In addition ionic fluids that consist of at least two chemical species of opposite charge can result in complicated attractive and repulsive interactions.

3   Exploring the liquid state

The properties of liquids and amorphous materials are explored on both the macroscopic and molecular level. Thermodynamics are used to characterise the equilibrium properties and to determine the equation of state and the compressibility of the liquid. Calorimetric techniques are used to evaluate enthalpy and free energy changes during phase transitions. The transport coefficients, which include thermal and electrical conductivity are also determined and the bulk and shear viscosities used to characterise internal friction. On a molecular scale, the local structure of fluids is determined by diffraction which provides the radial distribution function of the liquid via Fourier transform of the (static) structure factor. Mesoscopic structures can be explored using small angle techniques and inter facial properties can be explored in confined geometries. Individual or collective motions can be examined using dynamic or frequency dependent probes for example using inelastic neutrons [4, 6, 7, 8, 9].

4   Amorphous materials

Glasses and amorphous materials are non-crystalline and posses some degree of randomness, in the form of topological or spin disorder but it is not unique. Such a definition for example necessarily describe a non-crystalline state as many crystals are imperfect containing vacancies, substitution disorder etc. Topological disorder is the form of randomness in which there is no translational periodicity and this provides us with a definition of an amorphous material [10, 11]. An amorphous material is synonymous with a non-crystalline material by this definition, in other words an amorphous material does not posses the long-range order that is characteristic of a crystal. The term glass is more restrictive. A glass is a solid amorphous phase that exhibits an abrupt change in thermodynamic properties (heat capacity or thermal conductivity) from crystal-like to liquid-like values with temperature. This definition distinguishes glasses (or the vitreous state) from amorphous materials which can include liquids, glasses are restricted in definition to those materials that can be obtained in a reproducible state even after thermal recycling.

The oldest and most well-established method of forming amorphous solids is to cool the liquid sufficiently quickly. Although materials produced in this way invariably show glassy behaviour the feature of this melt-quenching process is continuous hardening, in other words and increase in viscosity. A prerequisite to formation of an amorphous material by this technique is that the cooling must be sufficiently fast to avoid nucleation. This depends on the speed at which the crystal-liquid interface moves and scales with viscosity.

It is easy to identify liquids that might be expected to be good glass formers; liquids with high viscosity and which reflect competing timescales that result from the free energy difference between crystals and liquids. This is a classical view of crystallisation that has been developed hand-in-hand with the glass industry. For silicates and other good glass-forming liquids such as B_2O_3 the nucleation of crystals is so slow that glass formation is very easy, indeed it is hard to crystallise B_2O_3 even if seed crystals are added to the supercooled liquid. Glass can be therefore viewed as a kinetic phenomenon and this can be used in useful definitions of the glass transition itself. A kinetic model of the glass transition does not however address the thermodynamic differences between glasses and liquids which implies an underlying thermodynamic glass transition hidden beneath the kinetic trasnsition [10].

5 The glass transition

The concept of a thermodynamic glass transition can be supported if the thermodynamic properties of liquids and glass are considered [12, 13] and this dates back to the observation of Kauzmann in 1948 [14]. By integrating the heat capacity data for glass forming liquid and determining the change in entropy of a liquid with temperature Kauzmann noted that the decrease in the entropy of a liquid with temperature was greater that the decrease of the equivalent crystal[14]. If extrapolated, the curve of liquid entropy would drop below that of the crystal, in other words the excess entropy would drop below zero. This is referred to as the *Kauzmann paradox* [10, 15], conveniently this is never reached because the kinetic glass transition occurs at higher temperature but it does point a thermodynamic basis for the transition. The existence of glasses is not therefore dependent on kinetic phenomena and the situation implied by the Kauzmann paradox, a transition where $S_{ex} = 0$ at T=0 must imply an underlying first order transition.

Kauzmann himself [14]put forward a solution to the Kauzmann paradox and suggested that molecular motion becomes progressively constrained and that free energy barriers to molceular rearrangements increases. At the same time free energy barriers to crystallisation are reduced because supercooling causes a decrease in the size of the critical nucleus. Spontaneous crystallisation therefore prevents the entropy crisis that would exist were liquid entropy to be extrapolated to low temperature.

An alternative solution proposed by Angell [13, 16] is that the Kauzmann temperature can be viewed as an absolute limit below which a liquid cannot exist. The supercooled liquid therefore escapes the entropy crisis by undergoing a sharp glass transition at Tk to an ideal glass. The ideal glass has the same entropy as a crystal and corresponds to a state in which the configuration of the system has settled to the deepest minimum of all amorphous potential energy minima. Although an ordered crystal will have an equally deep minimum the two are mutually inaccessible.

An important theory that has attempted to link the relaxation aspects with entropy was proposed in 1965 by Adam and Gibbs [17]and this is a model for relaxation that is similar to the Volger-Fulcher form but which formally links thermodynamics with relaxation, with the viscosity expressed as a function of configurational entropy, S_c,

$$\eta = \eta_0 \exp(B/TS_c) \tag{1}$$

with B and η_0 constants. The configurational entropy is established form the magnitude of the jump in the heat capacity at the glass transition.

$$S_c = \int_{T_0}^{T} \Delta d \ln T \tag{2}$$

The mode coupling theory views vitrification process of amorphous materials as a transition from the *ergodic* to *non-ergodic* behavior in the relaxation dynamics of desnity fluctuations.

Ergodic behaviour refers to the availability of microscopic configurations in phase space, when a system is ergodic all configurations are accessible but when the system is non-ergodic configurations become inaccessible due to structural arrest. This transition is not accompanied by singularities in any thermodynamic quantity and accordingly this is a dynamic viewpoint.

Mode-coupling theory is based on a structural relaxation (viscosity) mechanism and has three basic components [18, 19, 20]. Shear stress relaxation occurs through diffusive motion which leads to a feedback mechanism with viscosity. Shear stress relaxation (η)is written as the sum of vibrational and structural contributions,

$$\eta \approx G_\infty \tau_{vib} + G_\infty c(T)D^{-1}$$
$$= \eta_0(T) + b(T)D^{-1} \tag{3}$$

In which the structural relaxation time is inversely related to the diffusion coefficient, D, b is a constant. Using the Stokes-Einstein equation this becomes

$$\eta = \eta_0(T) + [6\pi ab(T)/kT]$$
$$= \eta_0(T) + B(T)\eta \tag{4}$$

with b(T) a non-increasing function of temperature and B(T) increases monotonically as temperature is decreased. This predicts an increase in viscosity on cooling and a divergence, identified with vitrification.

Central to the mode coupling theory is F, the Fourier transform of the van Hove density-density correlation function, G.

$$F_k(t) = N^{-1}\langle \rho_k(t)\rho_{-k}(0)\rangle$$
$$= \int G(r,t)\exp(-ik.r)dr \tag{5}$$

$$G(r.t) = \rho^{-1}\langle \rho(r,t)\rho(0,0)\rangle \tag{6}$$

The bulk number density is ρ and $\rho(r,t)$ is the density at distance r and time t, The zero time value of F is the static structure factor, S

$$S_k = N^{-1}\langle \rho_k(0)\rho_{-k}(0)\rangle \tag{7}$$

Central to the theory is the differential equation for F which has a solution that yields the time evolution of the decay of density fluctuations. For certain values of density and temperature, F decays to zero, a condition identified as ergodic. For other values, the solution decays to a finite non-zero value in which density functions cannot relax and there is structural arrest; non-ergodic behavior. The viscosity feedback mechanism enters through a memory function term (Γ) that causes the instantaneous rate of density fluctuations to depend on their own history[21, 22, 23].

A mode coupling model of the dynamics of a supercooled liquid requires only the static structure factor at a given temperature and the density which can be calculated from the interaction potential between molecules. This theory then yields a locus of $\rho(r,t)$ which changes sharply from erodic to non-ergodic behaviour. Numerical solutions to the mode-coupling equations for binary Lennard-Jones fluids shows a two-step relaxation and this is often equated with the two-step relaxation process that is observed spectroscopically in supercooled liquids; α and β relaxation.

In deeply supercooled liquids the dynamics are described in terms of two well-separated process [24], slow collective motions which are associated which are the exploration of deep configurational energy minima and exploration of local minima by faster non-collective motions. The slow collective motions are termed a-relaxation which the faster motion is referred to as β relaxation. The time scale for β relaxation is Arrehnius but for a relaxation progressively non-Arrhenius behavior is observed as the glass transition is approached and at the glass transition itself is the point at which a relaxation is arrested, β motion however persists. At high temperatures ($T > Tg$) the two types of relaxation are indistingushible. Goldstein has argued that when a liquid is supercooled the normal diffusive motion gradually becomes activated and relaxation increasingly occurs as space- and time localised events.

Mode-coupling predicts (essentially) a temperature of dynamic arrest and the two relaxation timescales. The sharp transition predicted by mode-coupling (Tc) occurs 10-40K higher than the calorimetric glass transitions which itself occurs at a greater temperature than the Kauzmann temperature. Mode-coupling theory does make insightful predictions into liquid behavior.

Clearly there are many issues to be resolved in the variety or ambient pressure liquids . As pressure is increased there may be further compliction, how can this be reconciled with the current state of liquid theory and, more importantly, how can high pressure behavior be explored?

6   The influence of pressure

Whilst the majority of glass made commercially is through the quenching of viscous silicates or borosilicate the range of glass-forming behaviour has been extended to include systems that have not been considered as glass-formers. Rapid quenching techniques including splat-quenching and melt spinning that have allowed for example amorphous metals ($Au_{75}Si_{25}$ in 1960) to be made, with quench rates of up to 10^6 K s ^{1}. In the case of oxide glass-formers, containerless levitation techniques have been used with success; these allow the liquid to be supercooled well below the melting temperature and cooled to form a glass, the absence of heterogeneous nucleation sites in these examples preventing crystallisation. As a consequence, it is becoming increasingly apparent that the liquid compositions that have been important for commercial glass-making only form a small subset of the range of glass-forming liquids and a range of amorphous structures and behaviour can be sampled as different temperature and composition domains are explored.

It was the work of Bridgman [25] in establishing that many systems, including H_2O demonstrate a negative dT_m/dP. It is generally expected that the slopes of melting curves, dTm/dP should be positive as indicated by the Clausius-Clapeyron relation:

$$\frac{dT_m}{dP} = \frac{\Delta V_m}{\Delta S_m} = \frac{V_{liquid} - V_{crystal}}{S_{liquid} - S_{crystal}} \tag{8}$$

Liquids are less ordered than the corresponding crystal, so that ΔS_m is always assumed to positive. Melting is usually associated with an increase in volume (positiveΔV_m). However, many simple systems show a negative melting slope and there can be one or more maxima in the melting curves. Perhaps the best known compound with a negative initial melting slope is H_2O (from the ice Ih phase) [26], as discussed below. Cs, Ba, Eu, Pu, Si and Ge also have negative dTm/dP slopes to the melting curve. Si and Ge are of additional interest in that a maximum to the melting curve is expected at negative pressure[27, 28].

7   Metastable melting

For systems with a a negative Clausius-Clapeyron relation (8), the negative melting slopes of materials under pressure have important implications for the behaviour of low pressure crystalline polymorphs. Metastable extensions of the melting curves can be intercepted and an amorphous material produced irreversibly. This is pressure-induced amorphisation. This was reported by Mishima for H_2O, when ice Ih was compressed and the *melting line* crossed [29, 26]. The amorphous H_2O produced by pressure-induced amorphisation is structurally distinct form of amorphous ice (termed high density amorphous ice; *HDA*), with a higher density that the amorphous forms of ice produced, for example, by condensation from vapour (low density amorphous ice; *LDA*).

In more complicated phase diagrams, such as SiO_2, the melting curves do not necessarily become negative but show incipient maxima in the melting curve that are intercepted by polymorphic crystal-crystal transitions. The high pressure crystalline phase may have a different dT_m/dP curve and intercept at a triple point. If the melting curve for the lower pressure crystalline polymorphs is extrapolated then these too can form metastable melting curves which are intercepted and pressure-induced amorphisation can occur.

Pressure-induced amorphisation can be considered in terms of metastable melting[27, 28, 29, 26]. In stable melting, the transformation between crystalline and liquid phases occurs when the Gibbs free energies of the two phases are equal. In the metastable case, melting (amorphisation) will likewise occur when the Gibbs free energies of the crystal and extrapolated liquid phase are equal. A solid amorphous material results with thermodynamic properties such as volume, enthalpy and entropy that can be mapped onto a non-crystalline state that is in a state of metastable thermodynamic equilibrium. Metastable melting is used to suggest a mechanism for pressure induced amorphisation. As low pressure, low density crystalline phases are compressed equilibrium structural changes include changes in short range order such as changes in coordination number. Potential energy barriers must be overcome for the low density crystalline phase to transform to the stable high density crystalline states. If there is sufficient thermal energy to overcome barriers to intermediate metastable states then amorphous forms can be produced. These intermediate states will not be crystalline and there may be several intermediate states separated by low potential energy barriers, each accessed by thermal motion. This series of related amorphous states or energy landscape is similar to that produced by quenching a supercooled liquid to a glass; the exact structural configuration is a reflection of the relaxation history, i.e. thermally activated jumps between closely related metastable, non-crystalline states.

One of the most important results from the study of pressure induced amorphisation of simple crystalline substances is that the amorphous forms produced have macroscopic thermodynamic properties that are different from amorphous materials produced at lower pressure (ΔV, ΔH and ΔS). This is the origin of the term *polyamorphism*; different amorphous forms of the same substance can be produced by different pressure-temperature routes.

From thermodynamic arguments, the Gibbs free energy of these amorphous forms will have a different pressure and temperature dependence and there may be a transition between the amorphous forms of the same material. This may be continuous or discontinuous and may be indicative of a first-order transition between liquids in the supercooled regime. The close relation between pressure induced amorphisation and changes in the structure of amorphous states implies, in a one-component system, that there are differences in density in the liquid. The presence of two species and differences in density and entropy between them can be used to construct two-state models for phase transitions that are analogous to liquid-gas transitions.

8   Two-state models

Two state models were developed from the late 1950s onwards to explain the
unexpectedly complex melting curves observed at high pressure for substances
such as Rb, Cs, Te, Ba and Eu [30, 31]. These systems display maxima in their
melting curves which may be attributed to the presence of different local environments in the liquid state. Changes in dT_m/dP slope reflect a change in
density through the Clausius-Clapeyron relation (8). In a one component system, this increased density of the liquid is assumed to reflect the presence of a
high density liquid species. High and low density species exist in the stable liquid, according to the two-state model and the relative proportion of each varies
as a function of pressure and temperature. In the original versions of the two
state model, developed by Rapaport [30, 31], the high and low-density melt
species were assumed to be domains with local packing (short-range order)
similar to those in high- and low-pressure crystal polymorphs. The increase
in liquid density, evidenced by the overturn in melting curve, is a reflection
of the increased abundance of the high-density species. The arbitrary high-
and low-density species in the two-state model are treated as thermodynamic
components. The equilibrium fraction of each species is a function of pressure
and temperature and reflects the minimisation of free energy. Rapaport applied the regular solution mixing model of Guggenheim[32] to the liquid, for a
low density species (A) and a high density species (B) the equilibrium molar
free energy for the liquid is defined by,

$$G = X_A G_A + X_B G_B \tag{9}$$

with X_A and X_B the mole fractions of the low- and high-density species. The
partial molar free energy of each species is defined in terms of the specific
volume contribution,

$$
\begin{aligned}
G_A =& G_A^0 + V_A^0(P - P_0) \\
&+ RT\ln(X_A) + W(1 - X_A)^2 \\
G_B =& G_B^0 + V_B^0(P - P_0) \\
&+ RT\ln(X_B) + W(1 - X_B)^2
\end{aligned}
\tag{10}
$$

here G_A^0 and G_B^0 are the standard state molar free energies associated with the
low- and high-density liquid species. The standard state molar volumes are
V_A^0 and V_B^0 respectively. The standard state pressure is P_0 and the absolute
temperature is T. R the universal gas constant. W is the regular solution
interaction parameter. The total molar free energy of the liquid is:

$$
\begin{aligned}
G =& X_A(H_A - TS_A) + (1 - X_A)(H_B - TS_B) \\
&+ P[X_A V_A + (1 - X_A)V_B] \\
&RT(X_A\ln X_A + (1 - X_A)\ln(1 - X_A)) \\
&+ X_A(1 - X_A)W
\end{aligned}
\tag{11}
$$

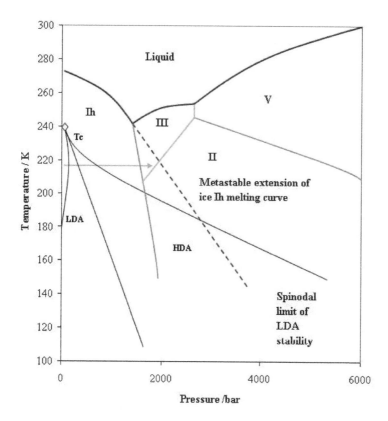

Figure 2. *The metastable phase diagram of H_2O. Two state models are used to calculate the stability and spinodal limits to the low- (LDA) and high-density amorphous (HDA) forms of water[36]. The phase boundaries between the different crystalline polymorphs of ice are shown schematically.*

The regular solution interaction parameter W in will be non-zero if there is a mixing contribution to the excess enthalpy of the liquid. This parameter is the key to interpreting liquid-liquid transitions in terms of the two-state model. In Rapaport model [30, 31]a non-zero value of W can be thought of as reflecting the direct interface energy between two structural species, or more generally as a contribution from the cooperativity of bonding arrangements if anomalous changes in bonding or coordination occur as a function of density.

One consequence of the non-ideal interaction parameters is that a second critical point (Figure 8), in addition to that terminating the liquid/gas boiling curve and can be defined according to $T_c = W/2R$. The consequence of this formalism is seen when the temperature is decreased. The equilibrium concentration of each species will vary as a function of pressure and temperature[33, 34, 35, 36]. At high temperatures, in the stable liquid, the change in species abundance is a smoothly varying function of pressure and at higher pressures a single phase liquid with an increased abundance of the high-density species is stable. This single phase liquid is stable at temperatures above the second critical point, but in the supercooled regime it is possible for sub-critical behaviour to be encountered. This can be illustrated by considering the minimisation of free energy. These functions show a series of minima with the minima associated with an excess of the HLA state becoming more favourable as the pressure is increased. As a result, as the pressure is increased, there will be a gradual increase in the abundance of the high density species as and the higher pressure liquid will have an increased density and, because of the differences in entropy and enthalpy between different species, different thermodynamic properties. At lower temperatures the variation in abundance of the high density species is less smooth this would be the regime of *critical like* fluctuations observed by Brazhkin and co-workers[37, 38]. At lower temperatures still a transition between two *supercooled* liquids occurs [39, 28, 40, 41, 42]. There are two spinodal lines defined in the subcritical region, these mark the extreme limits of stability of the two species. Transitions between liquids dominated by high and low-density species can occur in the supercooled region, above the calorimetric glass transition. If low pressure glasses or amorphous materials are compressed then an amorphous form with lower free energy could be accessed provided there was a relaxation process allowing these more stable structural configurations to be achieved. This would be equivalent to a glass quenched from the supercooled high pressure liquid. The two-state models described above have been used with success in describing the stability fields of different amorphous forms of water. In addition these types of model can be used to describe the anomalous thermodynamic properties of water, including anomalous contributions to volume and heat capacity.

There are apparently anomalous thermodynamic properties in polyamorphic systems; these include excess contributions to thermal expansion, isothermal compressibility and the specific heat capacity. The anomalous contributions to volume in H_2O, based on the differences in macroscopic thermodynamic properties and the non ideal mixing model[43, 39, 44, 33, 34, 35, 45, 26, 46] , result in the characteristic density maximum in H_2O. These excess contributions also change with pressure and reflect the increasing stability of the high-density species as the system is compressed. There are also anomalous contributions to the temperature-dependence of heat capacity. Changes in heat capacity as a function of pressure, implied by the increase in the abundance of the high density liquid species indicate that the rheological properties of the liquid will change as a function of pressure. This is a change in the liquid fragility.

9 Liquid fragility

The concept of *liquid fragility* was introduced by Angell [44, 16], building on earlier work by Uhlmann[47, 48]. Liquid fragility is a measure of departure from Arrhennius Law viscosity-temperature behaviour. A fragility plot (Figure 9)shows the viscosity-temperature relations for different liquids are scaled against the calorimetric glass transitions (Tg). SiO_2 is typically used to define the *strong* Arrhennian limit. More *fragile* liquids show increasing degrees of curvature in their viscosity when scaled to Tg. Fragile liquids therefore show non-linear increases in viscosity in the supercooled liquid regime. The relationship between the thermodynamic properties of a liquid and the viscosity is considered to be a reflection of the contribution of configurational entropy. This is the basis of the *Adam-Gibbs model* of viscosity[17] (1). The entropy differences between the liquid species in the two-state models should, therefore, correspond to differences in the rheological properties of the liquids. Liquids dominated by the high density species will be more fragile. Since the higher density species will be stable at greater pressures then higher pressure liquids will be more fragile and will have increased configurational entropy. However, the exact structural changes though are unclear and has led to Angell and others to develop versions of the two state model that are not based on specific liquid species but on the degree of excitation of the liquid structure (bond-breaking)[33, 34, 49]. In the two state model formulated by Rapaport[30, 31] and applied to systems with negative dT_m/dP slopes such as Cs, the two different liquid species have structures that are similar to the high- and low-pressure crystalline polymorphs. Such implied structural changes may be applicable to simple elemental substances but one of the surprising things about systems with reported polyamorphic behaviour is that they are not restricted to simple systems but include systems that are structurally complex such as H_2O[45, 46, 50] , BeF_2[51], triphenylphosphite (TPP)[52], Y_2O_3-Al_2O_3 [53, 54, 100, 56, 57]and traditionally *stong*, network-forming systems including GeO_2, SiO_2. Structural studies indicate that, for example in the clearly demonstrable case of a liquid-liquid transition in super-cooled Y_2O_3-Al_2O_3, the changes in structure are mid- and not short-range (coordination number) order, even though there are difference in short-range order in crystalline polymorphs in these systems. Angells version of the two-state model emphasises the configurational change and departure from *ideal configuration* rather than the presence of specific structural species. Tanaka [58, 59]has also used the two-state model as the basis for explaining polyamorphic trends again based on departure from ideal configurations, although in this case the two-state model is based on the competition between density-ordering and bond-ordering (directional, strong covalent bonds). These modified two state models have identical formulism to the version of Rapaport11[30, 31]. Critical like behaviour and transition between low- and high-density liquids is a reflection of the non-ideal mixing or clustering of the high- and low-density components, referred to as cooperativity.

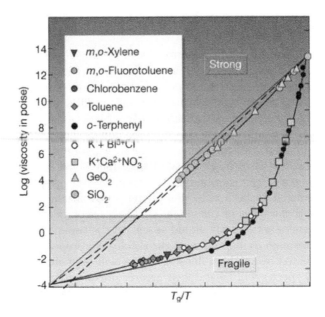

Figure 3. *The liquid fragility or so-called Angell plot[?]. This show the decrease of departure from Arrhennius Law viscosity behavior of liquids as the configruational entropy increases. A variety of liquids can be compared by scaling the viscosity temperature relations to the calorimetric glass transition. As pressure is applied, strongliquids such as SiO_2 may be expected to become more fragile.*

Although the these two versions of the two-state liquid models are very simplified and are based on the differences in macroscopic the thermodynamic properties of amorphous forms of the same substance and phase equilibria, they serve to indicate some of the expected behaviour that may occur if polyamorphism is encountered. Specific, crystal like clusters are avoided and the models require cooperative rearrangement of amorphous networks. The stability of amorphous networks is strongly dependent on temperature and pressure. Increasing pressure will favour increased density and density-ordering and so liquid fragility and cooperative clustering; possibly leading to liquid-liquid transition may be expected at moderate pressure.

10 Polyamorphic systems

The two-state models, while avoiding the exact mechanism, do predict certain type of behaviour. These behaviours should be observed in candidate polyamorphic systems. To summarise, candidate polyamorphic systems will have some or all of the following properties; overturn of the melting curve or a negative dT_m/dP slope (8), non-ideal mixing such that these slopes are no described simple by the ideal mixing of two *species* of different volume (co-operatively), pressure-induced amorphisation, a variety of structural motifs in the amorphous or liquid state, different amorphous forms produced by different synthesis routes with measurable thermodynamic differences between them, changes in macroscopic properties such as viscosity and electrical conductivity with pressure and rich phase diagrams with numerous crystalline polymorphs.

High pressure studies of liquid and amorphous structure are therefore crucial in identifying systems that may be candidates for liquid-liquid and amorphous-amorphous transitions. Such transitions may involve changes in volume, enthalpy and entropy (ΔV, ΔH and ΔS) and, if volume changes are small, the transitions can be intercepted at relatively low pressures or even under ambient pressure conditions. *in situ* observation of polyamorphic changes is difficult, involving high temperature, if the stable liquids are to be observed. Furthermore, supercooling and quenching high pressure liquids to a glass is also experimentally difficult and the high pressure amorphous phases may not be recoverable.

Extensive studies using toroid-type pressure cells [60, 61, 62, 63, 64], which can generate pressures to 0.3-13 GPa and temperatures of up to 2000K, have suggested the occurrence of phase transitions in elemental liquids such as Se, S, Te, I_2 and P , as well as in As_2Se_3, A_2S_3 and Mg_3Bi_2 [37, 65, 66]. These liquids show abrupt changes in the electrical conductivity of the stable liquids analogous to those associated with insulator/semiconductor-metal transitions in crystalline solids. These changes are associated with volumetric changes and viscosity changes, inferred from the quenching behaviour of melts under pressure. In liquid selenium, for example the electrical conductivity of the liquid increases by two orders of magnitude at pressures of approximately 4GPa within a transition width of 0.3 GPa . Changes in the properties of liquid sulphur are reported at 8 and 12 GPa. At 8 GPa there is a change in volume while at 12 GPa there is an increase in electrical conductivity of 1-2 orders of magnitude, consistent with a change from a semi-conductor to metallic liquid. In both selenium and sulphur the location of the changes in electrical properties depend on the rate of change of pressure and temperature, this hysteresis resulting in apparent regions of coexistence of different liquid states. Like selenium [38], liquid phosphorus shows an increase in electrical conductivity consistent with a semiconductor to metal transition, accompanied by a decrease in liquid viscosity. These transformations are coincident with a volume change of $\Delta V/V of 40\%$. The abrupt transformation in phosphorous may result from the same mechanism that causes bonding changes in equivalent crystalline polymorphs. As such density ordering is suggested as a mechanism for a transition between different phosphorous liquids. Direct observation of such a transition in phosphorous has been reported by Katayama [67, 68].

The work by Brazhkin [65, 38, 66] and others has shown that the abrupt transition in electrical conductivity and boundaries between semiconductor and metallic liquids are associated with changes in the slope of the melting curves, dP/dT. This means that in the stable liquids the changes in electronic properties correspond to changes in density. The transitions between liquids can occur over 0.3 to 0.5 GPa, but can be more abrupt, for example over a range of 0.01 GPa for phosphorous. The boundaries in these transition regions show negative Clapeyron (dP/dT) slopes which means that the entropy of the denser, metallic liquids is higher. A possibility is therefore raised that stable liquids can undergo transitions from one stable liquid phase to another with density and entropy as the order parameters. The mechanisms for such transitions are elusive (electrical conductivity measurements are not a direct probe of the liquid structure) and there may be fluctuating micro- and nano-scale domains as well as regions of coexistence of the high- and low-density liquids. As such the apparent transition between one liquid and another occurs over an interval of pressure and can be interpreted as critical-like fluctuations, a critical-like point occurring at lower temperature in the supercooled liquid regime. In the case of liquid phosphorus however, X-ray scattering and radiography describe a liquid-liquid transition between low-density (LDL) and high-density (HDL) stable liquid phases at 1 GPa and 1000 K.

11 Experimental techniques

The changes in physical properties of liquid and amorphous materials with pressure are of interest to inorganic and organic chemists, mineralogists and solid state physicists. Interest in this highly disciplinary field of study follows the pioneering work of Bridgman who observed changes in physical properties and behaviour of crystalline materials under pressure [25]. Studies of crystal structure have recently been performed *in situ* up to pressures of 1-2 Mbar, and the results used to interpret changes in properties such as electrical conductivity and magnetism, and also to establish the phase relationships between different crystalline phases. There have howvere been fewer studies of liquids and amorphous solids, and the interpretation of the results is less direct.

The study of materials at high pressure involve two approaches. Optical and spectroscopic experiments can be carried out with high pressure cells made using materials such as silica glass, sapphire and diamond that are ideally transparent to optical and infrared radiation and is resistant to the high pressures and temperatures required for *in situ* study. Of most importance is the diamond anvil cell (DAC). In the diamond anvil cell the sample is placed between the flattened tips of two gem quality single crystal diamonds and contained within a hole drilled in a gasket (usually made of metal). The sample chamber is brought to high pressure by applying mechanical force to the diamonds. High temperatures can be generated by resistance heating or laser heating. Diamond is transparent to radiation over a wide range of the electromagnetic spectrum, and various optical spectroscopy experiments to probe crystal and glass structures at high pressures in the DAC have been carried out. Substantial X-ray transmission occurs above 112 keV, so that X-ray diffraction and amorphous scattering experiments can be most readily carried out at X-ray synchrotron radiation sources. However the sample chamber and the sample size are generally very small, usually between 50-200 μm in diameter, depending upon the pressures require. Several studies of amorphous solids, including glasses, have however been carried out using synchrotron X-ray scattering. An alternative sample environment is to use of presses of larger volume equipped with multi- or toroidal type anvils [69, 70, 71]. While these are not transparent to visible, infrared or ultraviolet radiation, these pressure cells have been used with success at synchrotron or neutron sources. The multi- anvil or toroidal anvil pressure devices have larger sample volumes and the thermal regime is much more easily controlled and allow simultaneous *in situ* measurements of physical properties such as electrical conductivity and liquid viscosity can be made. One type of toroidal cell has been used extensively in high pressure research by Russian groups [63, 64]and also by groups from Paris and Edinburgh [60, 61, 72, 73], in a cell that was specially designed for neutron crystallography. This design utilises low- or null scattering gasket material and allows diffraction data of low scattering materials such as liquids and glasses to be obtained to high values of the scattering vector (Q).

12   The role of diffraction

The local chemical bonding and intermediate range order interactions that are seen in the pair distribution function play an important role in determining structure-property relationships [7, 2, 4, 6]. For example, the local coordination number of liquid silicon is associated with a metal/insulator transition. The pair distribution function is obtained by Fourier transformation of the total structure factor, S(Q), obtained from diffraction data from liquids, glasses and amorphous materials probed by radiation with wavelengths comparable to the inter-atomic separation. Total neutron scattering and more recently high energy X-rays are experimental probes that provide a direct measure of these nuclear arrangements over the wide range of length scales. These are highly penetrating and a powerful bulk probe, which can provide high resolution information at the atomic level which is needed for the study of liquid, structures. In addition, neutrons are sensitive to light elements, for example hydrogen (deuterium) and oxygen, which varies from isotope to isotope depending on their nuclear spin. Isotopic substitution is therefore a powerful probe in determining detailed atomic structure of specific atom pairs.

Structural studies of liquid or amorphous states generally involve *elastic* or *inelastic scattering* of electromagnetic radiation. As with crystalline (elastic) diffraction, a beam of radiation is directed at the sample and the intensity of the signal measured as a function of scattering angle 2θ and the wavelength of the incident radiation. The intensity is expressed as a function of the momentum transfer of the scattered particle, Q, where Q is defined as $Q = k_{incident} - k_{scattered}$, with $k_{incident}$ and $k_{scattered}$ the incident and scattered wave vectors respectively [74, 8, 9]. In a neutron diffraction experiment, for example, the diffraction pattern is obtained from the counts per second measured by a detector placed at a solid angle $d\Omega$ and expressed as the differential cross section. If the energy exchange between incident and scattered neutrons is small, a *static approximation* is made which assumes that scatted neutrons are counted in a detector regardless of their energies. The double differential cross section is defined by $\frac{d^2\sigma}{d\Omega dE}$ and for an incident flux of intensity I_0 this is related to the observed intensity through

$$I = I_0 \frac{d^2\sigma}{d\Omega dE} d\Omega dE. \tag{12}$$

If the energy of the scattered neutrons (in this case) is not analysed and hte neutron counts are intergrated ar a given solid angle the differential cross section and the total cross section are obtained

$$\frac{d\sigma}{d\Omega} = \int_0^\infty \frac{d^2\sigma}{d\Omega dE} dE \tag{13}$$

$$\sigma = \int_0^\infty \frac{d\sigma}{d\Omega} d\Omega \tag{14}$$

The scattered neutron intensity is expressed as a function of the momentum trasnfer, Q. The coherent and incoherent scattering cross sections are given by $\sigma \equiv 4\pi b^2$ where the scattering length b, chracterises the strenght of the neutron-nucleus interaction The quantity b can be complex and has a value that depends on particular isotope and the spin state of the neutron-nucleus system. Scattering lengths can be positive or negative and can fluctuate from one isotope to another, this is the basis of the technique of neutron diffracion with isotopic substitution. The coherent scattering law is defined as,

$$S(Q,\omega) = \frac{1}{N} \frac{k_{incident}}{k_{scattered}} \frac{4\pi}{\sigma_{coherent}} \frac{d^2\sigma}{d\Omega dE}\Big|_{coh} \tag{15}$$

where $S(Q,\omega)$ is the dynamic structure factor. The *static structure factor*, S(Q) is obtained by integrating $S(Q,\omega)$ with respect to ω at constant Q.

In a polyatomic system, the total scattering comprises the sum of several partial atom-pair contributions. The measured differential cross section has two parts, the self scattering part which is the incoherent scattering from individual scattering centres and the distinct part, the coherent interference term from different atom pairs. This is written as

$$\frac{1}{N}\left[\frac{d\sigma}{d\Omega}(Q)\right] = F(Q) + \sum_\alpha^n c_\alpha \overline{b_{\alpha,inc}^2} \tag{16}$$

where F(Q) is the total interference function (the distinct scattering) and c_α the concentration of the chemical species α. It is convenient to use the

Faber-Ziman convention to express F(Q) in terms of the dimensionless static structure factor, S(Q),

$$F(Q) = \sum_{\alpha,\beta}^{n} c_\alpha c_\beta \overline{b_\alpha b_\beta} \left[S_{\alpha,\beta}(Q) - 1 \right] \tag{17}$$

and for n components there are $n(n+1)/2$ partial structure factors, $S_{\alpha,\beta}(Q)$ labelled for species α and β.

The Sine Fourier trasnform of $S_{\alpha,\beta}(Q)$ leads to the partial pair distribution function through

$$g_{\alpha,\beta}(r) - 1 = \frac{1}{2\pi^2 r \rho_0} \int_0^\infty Q \left[S_{\alpha,\beta}(Q) - 1 \right] \sin(Qr) dQ \tag{18}$$

where the total desnity of atoms is ρ and $g_{\alpha,\beta}$ is the probability of finding an atom β at a distance r from an atom α. A Fourier transform of the total muli-component F(Q) defines the toal pair distribution function G(r), which is the weighted sum of all partial values for neutron diffraction data

$$G(r) = \frac{1}{2\pi^2 r \rho_0} \int_0^\infty Q F(Q) \sin(Qr) dQ \tag{19}$$

$$\sum_{\alpha,\beta}^{n} C_\alpha C_\beta \overline{b_\alpha b_\beta} \left[g_{\alpha,\beta}(r) - 1 \right] \tag{20}$$

With the development of third generation synchrotron sources, there has recently been huge progress in producing instrumentation for using highly penetrating neutron-like x-rays of $\approx 100 keV$ for the study of liquid and glass structure. These high energy X-rays act as a bulk probe and cover a wide Q-range and the same principles of scattering and the equivalent Faber-Ziman formalism can be used. Neutron and high energy X-ray diffraction can be viewed as complementary techniques and are particularly useful for studying oxide or hydrogenous systems, as while neutron scattering lengths (b) vary erratically across the periodic table, the equivalent x-ray form factors vary as a function of atomic number.

13   Glass and liquid structure

The simplest structural model of a glass, such as the continuous random network (CRN) indicates that a simple glass such as an alkali silicate would consist of a framework of SiO2 tetrahedra and randomly arrange network modifying cations. However, the local atomic arrangements (and intermediate range) are far from random and governed by strong inter-atomic forces. Several studies using diffraction techniques have shown that glass structures are more ordered than the CRN model and in the case of alkali silicates, domains rich in the modifying cations exist. Such modified random network models form a good basis for interpreting diffraction data but the interpretation of the full glass structure is difficult, based solely on the pair distribution function measured in a neutron diffraction experiment. It is common therefore to combine neutron data with the results spectroscopy studies and diffraction modelling techniques such as Reverse Monte Carlo and Empirical Potential Structure Refinement. Raman spectroscopy is often used to study glasses and is very sensitive to changes in local structure; the results are only qualitative however. Nuclear magnetic resonance spectroscopy has long been used in combination with diffraction to interpret silicate glass structure, usually in terms of the modification of the large polymerized silicate, aluminosilicate and aluminate frameworks (Q speciation). NMR is isotope specific and can be used for example to probe ^{11}B, ^{29}Si, ^{27}Al and ^{18}O environments providing information on bond angle distribution and ring statistics as well as the modification of network. Other spectroscopic techniques such as Extended X-ray Absorption Fine Structure are also element specific and act as a local structural probe.

The structure factor (17) and associated real space distribution function for a multi-component system (20)can be complex and peaks past the nearest neighbour can be difficult to interpret. The data do however provide hard constraints for any structural model. The use of techniques such as isotopic substitution, isomorphic substitution or combined neutron and X-ray data sets usually enable more detailed information to be obtained, than any single data set can yield.

A useful technique, used to interpret diffraction data of amorphous and liquid structures is to delimit characteristic distance ranges, a methodology developed by Wright and Price. Range I relates to the nearest neighbour distances corresponding to the basic structural unit i.e., the short-range order. This peaks are the sharpest in G(r) and contain information on the first neighbour coordination number and bond length. For SiO_2 for example, short range order refers to the Si-O and O-O correlations that reflect average size and shape of the basic intra-molecular tetrahedron. Chemical knowledge is usually assumed so that likely atom pair bond lengths are obtained by either comparison with known crystal structures or by comparison with bond valence theory. There are various methods to calculate the coordination number once the atom pair is identified. Range II is associated with the connectivity of these structural units in other words the connectivity of the main structural units and will overlap with Range I. In the case of SiO_2, Range II distances are characteristic of next nearest neighbour Si-Si distance and reflect inter-tetrahedral distances and torsional bond angles. Normally these angles cannot be determined directly from a single neutron measurement and are combined with (for example) an x-ray dataset on the same sample. Range III relates to the presence of larger correlations of several structural units and usually referred to as *intermediate range order* and is associated with the presence of a first sharp diffraction peak in the neutron spectra. Range III is characteristic of the formation of cages, rings, layers, chains or other structures through the connection of the basic structural units. In real space this topology is usually characterised by a distance range that extends up to 10-20 Å, although order may also extend beyond that (Salmon, 2004). The most prominent signature of intermediate range order in the measured structure factor S(Q)is the presence of a *first sharp diffraction peak* (FSDP), that is a peak or feature below Q.r1¡3 Å^{-1} for tetrahedral materials. The FSDP in network glasses can be attributed to structural correlations on a length scale in real space of periodicity $2\pi/Q_P$, which decreases in magnitude with increasing r. Where Q_P is the position of the FSDP. The FSDP peak height reflects both the degree of periodicity and is also a function of the packing of the structural elements

14 Case studies

14.1 Amorphous forms of H$_2$O

One of the many intriguing feature of amorphous ices is that they show an apparent amorphous-amorphous transition. A feature observed originally by Mishima [45] and the subject of much controversy. It is only relatively recently, however, that *in situ* diffraction studies have been carried out.

When water vapour is deposited on a cooled plate an amorphous form can be produced which has a glass transition temperature at 130 K. Amorphous ice produced in this way is referred to as low density amorphous ice (*LDA*). This differs in density from the high density form (*HDA*) produced by pressure-induced amorphisation by 20% . When heated, samples of recovered *HDA* will transform to lower density *LDA*. Similarly, when *LDA* is compressed at 177K it will transform to *HDA* over a narrow interval in pressure [45, 46, 50]. Transformation to *HDA* occurs at 3.2 kbar on compression and *HDA* transforms back to LDA at 0.5 kbar Differential scanning calorimetry experiments on *HDA* at atmospheric pressure show a glass transition and in the relaxed, supercooled regime an exothermic signature of a transition to the more stable *LDA* phase. These data are used in two-state models in combination with volumetric data from the phase diagram to indicate the presence of a second critical point and stable liquid structures that resemble the low and high pressure amorphous forms, i.e. *HDL* and *LDL*.

The HDA form of ice can be produced in sufficiently large quantities to allow its structure and vibrational properties to be studied. The mechanism of formation, collapse of the ice Ih lattice, would suggest that it may be an amorphous metastable state related to the underlying stable crystal structure, in this case ice XII. As noted by Klug [75] , there are similarities in the $g_{OO}(r)$ of *HDA* and ice XII. Vibrational properties determined by Raman Spectroscopy and inelastic neutron spectroscopy are strong functions of O-H bond length and provide further insight into the nature of the amorphous *HDA* form. *HDA* ice has an excess in the vibrational density states. Infrared and incoherent inelastic neutron scattering techniques and lattice dynamics suggests and origin of this excess in low frequency vibrational modes from several sources including damped acoustic modes, interacting soft harmonic oscillators and quasi localised vibrations. This excess in the vibrational density of states is absent in LDA. These low frequency modes are the origin of the excess in entropy responsible for the increased fragility, i.e. the HDA amorphous form is consistent with a more fragile glass-forming liquid[75, 76].

A comparative study of LDA and HDA, using neutron diffraction with isotopic substitution and combined with empirical potential structural refinement (EPSR) has been used to ascertain the differences in the pair-distribution function of the two forms. Both forms of amorphous ice are fully hydrogen-bonded tetrahedral networks. The structure of HDA resembles that of liquid water at high pressure [77] while LDA is similar to ice Ih [77, 78]. The pair distribution functions for the two forms differ most notably because of the presence of an interstitial water molecule in the HDA form, which lies just beyond the first O-O coordination shell. The presence of this molecule results in HDA being less ordered than LDA [79]. The diffraction data and resulting pair-correlation functions show limited change in the O-H and H-H partial contributions, with a sharpening of the main peaks as LDA is transformed to the HDA form. In contrast there are distinct changes in the $g_{OO}(r)$ [79]. The O-O coordination number for the LDA form is 3.7 comparable to the value for low pressure water (4.3). In the HDA form the O-O coordination number is increased to 5.0 and suggests an additional water molecule present in the first neighbour shell. Spatial density functions, obtained from EPSR models of the diffraction data suggest that, on compression, the second neighbour shell of water molecules collapses and water molecules can become interstitial. Interstitial molecules increase in abundance as water is compressed and the HDA form of ice may be regarded as being related to the high pressure form of liquid water [77, 80, 78]. The potential relationship between liquid and amorphous forms is however further complicated by the report of an additional amorphous form of ice.

When the LDA form of ice is compressed to form HDA at 77K, an additional form can be produced and recovered by isobarically heating the HDA to 140 K [80]. This form has a higher density and is termed very high density amorphous ice (VHDA) . Diffraction data for VHDA show significant differences in the $g_{OO}(r)$ when compared to that of HDA and LDA. The most obvious changes are increasing intensity in the second neighbour O-O region between 3.1 and 3.4 Å, this is a minimum in $g_{OO}(r)$ for HDA. In the VHDA form, there is a peak that appears as a shoulder to the first O-O peak. This is distinct from the second neighbour peak in HDA which occurs at a greater radial distance and is separated by a minimum between 3.1 and 3.4 Å, indicative of more directional bonding. The VHDA form may, therefore, be viewed as having more disorder in the second neighbour shell. Both HDA and VHDA forms have interstitial molecules which secure the amorphous structure and inhibiting relaxation back to an LDA form. It is postulated that VHDA is more representative of the high pressure liquid and has more interstitial molecules present. What it is not clear is how the HDA and VHDA forms are related and whether there is a sharp transition between them. Some authors suggest that the VHDA form is more stable form and that HDA is an intermediate phase. If this was the case then any two-state or similar model would have a second critical point that should be based on the thermodynamic differences between the LDA and VHDA forms.

The change in structure during the transformation between HDA and LDA forms of ice has been studied *in situ* by neutron diffraction with isotopic substitution. Far from clarifying the nature of this polyamorphic change, however, different studies suggest two alternative transition mechanisms; continuous and discontinuous. In a study by Klotz and others [81, ?] diffraction data show a shift in the position of the principal peak in the structure factor as samples are compressed from 0 to 0.7 GPa and on to 2.2 GPa. The Fourier transform of these data show changes in $g_{OO}(r)$ with the second neighbour peak moving to shorter radial distance. The highest pressure data resembles that of the VHDA form confirming the close relationship between the two forms and indicating that the mechanism of formation of the high density amorphous forms is the collapse of the second neighbour shell and formation of an interpenetrating network of water molecules[82]. The three diffraction data sets indicate three different structures and a potential transition from LDA to VHDA by an intermediate HDA form [83].

The presence of intermediate forms of amorphous ice has been suggested by Tulk and others[84, 85] . In the region of transition, diffraction studies using both neutrons and high energy X-rays show changes in the position and shape of the first peak in the diffraction pattern. In addition the relaxation to these intermediate amorphous forms has been monitored by annealing *HDA* at different temperatures. The formation of intermediate structures over the completed transition from *HDA* to *LDA* has been shown by Guthrie and others[86] . The change from *HDA* to *LDA* represents a shift in the first peak in the structure factor from 2.1 to 1.7 Å^{-1} and there are similar dramatic changes in the real space transform of these data, i.e. the $g_{OO}(r)$ (Figure 14.1). The changes in O-O correlation in the 2.75 to 4.5 Årange are seen as the depletion of the interstitial oxygen in the 3.6 Åregion. This is seen as the shoulder to the first O-O peak becoming more distinct and moving to a greater radial distance through the transition from HDA to LDA.

A study of the transition from *LDA* to *HDA* at 130 K and 0.3 GPa has, by contrast, been interpreted as a first order transition . The neutron data in this study has been interpreted as a linear combination of the *HDA* and *HDA* components[83, 81]. This study suggests the nucleation and growth of the HDA phase in a matrix of the LDA assuming crystal-like behavior and using an arbitrary shift parameter to model the shift in the first diffraction peak. This does not account for the dramatic changes in intermediate-range order demonstrated by Guthrie and others [86].

The current debate on *LDA-HDA* transition focuses mainly on the presence of the second critical point that is suggested by two-state and similar

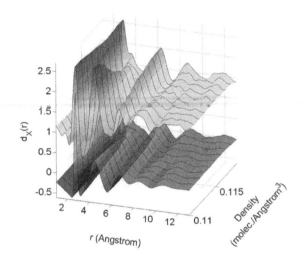

Figure 4. *Oxygen-oxygen partial differential distribution function for amorphous ice. The X-ray diffraction data is shown at the top, while the results from molecular dynamics simulation are shown at the bottom. The collapse of the second shell can be clearly observed as the density increases and interstitial molecules are pushed into the first O-O shell.*

models. The data of Tulk, Guthrie and others[84, 85, 86] argues against it's presence since the transition is continuous (fig 14.1). From versions of the two-state models currently favoured by Angell, Tanaka and others [33, 59, 87] , however, a second critical point does not have to be present, the liquid or supercooled liquid needs only to show strong cooperativity. These intermediate states would have different relaxational properties and fragilities but would be highly cooperative systems. Without recourse to complicated interpretations it can be seen that the behavior of amorphous forms of ice can be interpreted in these terms.

14.2   Amorphous silicon

Crystalline silicon has a semi-conducting tetrahedrally-coordinated diamond-structured polymorph that is stable at low pressure. At high pressure the tetrahedral structure collapses and a metallic phase with octahedral coordination of silicon is stable, the β-Sn phase . The melting curve of the low pressure polymorph has a negative dT_m/dP slope which indicates an increase in liquid density on melting and suggests that silicon might be a candidate *polyamorphic system*. Amorphous forms of silicon can be made at atmospheric pressure by chemical vapour deposition and similar synthesis techniques but these forms are semi-conducting and have a tetrahedral structure, while the liquid at atmospheric pressure is metallic. The low pressure amorphous forms are not, therefore, quenched representatives of the low pressure liquid and may suggest that there is more than one form of amorphous silicon and a possible transition between different phases.

One consequence of the negative dT_m/dP slope is the potential for pressure-induced amorphisation and when porous nanophase diamond-structure silicon (π)is compressed an amorphous phase can be produced. In the demonstration of *pressure-induced amorphisation*, Raman spectra and X-ray diffraction data were collected as π-silicon was compressed in a diamond anvil cell. The diamond structure is shown by a strong (111) reflection which is present up to pressures of 7-8 Gpa. The structure is completely amorphous at 12 GPa. Raman spectra collected simultaneously show a red luminescence that shifts to increasingly longer wavelengths with pressure. At high pressures, coincident with the amorphisation of the sample, this band disappears and is replaced by a weak broad feature between 200 and 400 cm^{-1}. The high pressure Raman spectrum is different from that of amorphous phases produced at ambient pressure but does resemble that of β-Sn. As a result the high density amorphous form of Si is tentatively assigned an *HDA* form[88].

On decompression the Raman spectrum of the amorphous form changes. The *HDA* Raman signal persists until 10 GPa at which point a broad amorphous band at 470 cm^{-1} appears, a signal characteristic of the low pressure amorphous form i.e. an LDA form [88].

Based on the available thermodynamic and computer simulation data, a two-state model can be constructed for Si and predict an amorphous-amorphous transition in the pressure range where the Raman signal changes[88]. An interesting feature of this simple model is that the position of the second critical point occurs at a negative pressure (under tension) and means that if the liquid stable at atmospheric pressure is supercooled then it will intercept a liquid-liquid transition in the supercooled regime and an *LDA* form would result different in structure and electronic properties from the low pressure liquid [88, 89].

As noted from the Raman study, the optical properties of the amorphous forms of silicon change on compression. At high pressure the reflectivity of the *HDA* form is greater than the metallic gaskets used in the diamond anvil cell and suggests that the *HDA* form is metallic. Electrical resistance measurements also change dramatically in the vicinity of the proposed *HDA-LDA* transition. The two-state model predicts a transformation between LDL and HDL supercooled liquids at approximately 1060 K. This temperature is coincident with the *unusual* melting transition reported when amorphous (*LDA*) silicon is heated to the crystalline melting temperature. Molecular dynamic simulations using the Stillinger-Weber potential [90] have been used to explore this region in temperature. Above the proposed *LDL-HDL* transition region, the equilibrated volumes in the simulation show fluctuations consistent with thermal fluctuations. Closer to the transition however, the fluctuations in volume are much greater and the magnitude consistent with the density differences between the *LDL* and *HDL* liquids. There are changes in mean coordination number associated with these fluctuations, the higher density fluctuations showing a greater proportion of 5- and 6 coordinated silicon. These simulations suggest critical-like fluctuations in the supercooled regime . The vibrational properties calculated from the simulations show distinct low- and high-frequency peaks associated with stretching and bending of tetrahedral silicon in the *LDA* network. The *HDA* spectrum has a broad feature associated with an increase in 5- and 6-fold domains and is consistent with the increased fragility of the *HDL* supercooled liquid. [89]This indicates that the behavior of supercooled liquid silicon is consistent with a strong to fragile liquid transition accompanying the *LDL-HDL* transition, increased low frequency modes contribute to the increased configurational entropy of the *HDL* liquid[90].

14.3   Liquid phosphorous

Liquid forms of phosphorous have complicated structures. Metallization in the liquid state has been reported at pressures of between 0.7 and 1.2 GPa, at which point the electrical conductivity is observed to increase. Grain size differences in recovered samples are taken to indicate that there are rheology changes in this region too. The crystalline phase diagram for phosphorous is rich. White phosphorous, which has a low melting point (44oC) is tetrahedral consisting of P4 molecules. Red phosphorous has a polymeric structure with a correspondingly higher melting point (¿600oC). Black phosphorous has a layered structure and consists of three-coordinated atoms. The melting curve of black phosphorous shows a maximum at 1 GPa. This is the region where electrical conductivity is seen to change and is the point at which the liquid density is greater than that of the crystalline phase.

  X-ray diffraction studies of the liquid performed at high pressures between 0.77 and 1.38 GPa show a dramatic and sudden change in structure(Fig 14.3) . At pressures of 0.77 and 0.98 GPa the structure factor shows a prominent first peak at 1.4 . At pressures of 1 GPa this first peak is reduced in intensity and a new maximum is developed at 2.45 Å^{-1}. The Fourier transform of these data shows, at low pressures, peak centred on 2.2Å, corresponding to the P-P distance in P4 molecules . The intensities of the next-nearest neighbour P-P peaks are low and the low pressure liquid structure is interpreted as comprising an open tetrahedral framework. At pressures greater than 1GPa, the P-P peak shifts to a greater radial distance and there is an appearance of pronounced next-nearest peak at 3.5. This peak is interpreted as being characteristic of an increasingly polymeric liquid. The two different liquids have different densities, estimated from the pdfs as 2.0 gcm-3 and 2.8 gcm-3 for the low and high density liquids respectively. Following the initial observation, subsequent X-ray diffraction studies have concentrated on characterizing the changes in liquid structure at higher temperatures, effectively mapping the

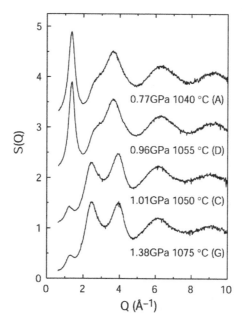

Figure 5. *X-ray diffraction data for liquid P_2O_5 showing the dramatic change in liquid structure between 0.96 and 1.10 GPa. The change in structure is demonstrated by the absence of the peak at 1.4 $Å^{-1}$ at high pressure. This study provides unequivocal evidence for first order transitions in the stable liquid regime at high pressure*

suggested LDL-HDL transition curve as a function of pressure and temperature (Figure 14.3). The *in situ* data show that the lowest pressure and highest temperature at which there is a transition between the two liquids occurs at 0.3 GPa and 2200°C. The transition between the low density molecular form and the high density polymeric form is also seen as changes in the first peak in the diffraction pattern. A similar trend is observed in computer simulations which also predict a change in electrical conductivity. The transition between the two liquids would be expected to terminate in a critical point. What is surprising about liquid phosphorous is that the transition emerges into the stable liquid fields and no critical point or critical-like fluctuations have been reported. Radiography data from Katayama clearly demonstrate the nucleation and growth of one liquid in the matrix of another as predicted by two-state and similar models. The occurrence in the stable liquid field is unusual but can be thought of as consistent with Tanaka's two-state model, that is, if a system shows strong directional bonding that acts in competition with density-driven ordering then the melting temperature based on density-ordering (close packing) may be much higher than the experimental melting curve. If this situation were applicable to liquid phosphorous then the LDL-

HDL transition is in effect in the regime below the density ordered melting curve because of the strong directional bonds. A second critical point in this interpretation could again occur at slightly negative pressure. More recently it has been noted that the transition between the LDL and HDL forms of liquid phosphorous occurs above the critical point for the white form of P, which melts at 44oC ; i.e., the *liquid* produced in the decompression experiments is a molecular tetrahedral fluid and the transition is actually between LDL and polymeric HDL fluid phases.

15 Transitions in the strong amorphous network

GeO_2 and SiO_2 are classic network-forming glasses with corner-shared tetrahedral networks and *strong* behavior. GeO_2 glasses, when compressed, are believed to show an amorphous-amorphous transition from a glass with an open network structure based on corner-linked tetrahedra, at low pressure, to a glass structure dominated by GeO_6 octahedral units at higher pressure[91]. This conclusion is based on XAS measurements that show a change in Ge-O distance consistent with the analogous tetrahedral-octahedral change in crystal phases and Raman spectroscopy data using a diamond anvil cell. In situ neutron diffraction studies of GeO_2 (combined with high energy X-ray diffraction studies and molecular dynamics simulations) have been used to investigate the nature of the change in short- and intermediate-range order[92]. It has also been suggested that vitreous GeO_2 may undergo a first order amorphous-amorphous transition. As GeO_2 glass is compressed the height and position of the first peak in the structure factor changes and indicating a decrease in intermediate range order through the shrinkage and collapse of the open network structures[93]. Prior to a coordination change there are changes in the O-O correlations as oxygen atoms move closer to central germanium atoms. Between 6 and 10 GPa the nearest neighbor coordination number increases and a mixture of 4, 5 and 6 coordinate germanium-centered polyhedra co-exist. This is again an intermediate state and not simply a mixture of 4 and 6 coordinate Ge. As the pressure is increased to above 15 GPa a high pressure octahedral glass forms, which is not recoverable. This network comprises of a mixture of edge- and face-shared GeO_6 octahedral units.

More recently there has been *in situ* diffraction measurements made on vitreous B_2O_3, [94]which along with GeO_2 and SiO_2 is one of the archetypal network-forming glasses. The structure of B_2O_3 comprises boron atoms surrounded by three oxygen atoms and these can be linked to for boroxyl (B_2O_6) rings, although the fracion of these rings remains contetnious. It has been known from recovered B_2O_3 samples that a residual density is retained but it is only recently that *insitu* diffraction measurements have been made. X-ray diffracion studies using a cubic, multi anvil press [94] and energy dispersive X-rays at the Sping-8 synchrotron X-ray source shows that pressure results in two forms of structural change. B_2O_3 shows a smooth change in the total structure factor with a change in intermediate range ordered evidenced by the changes in teh first shapr diffracion peak. As pressure is increased to above 5 GPa,changes in structure are more dramatic, these are short range ordering changes, changes in the B-O distance which suggests an increase in the fraction of four-coordinate B-O. The reposnse to pressure has two compoenents, first there is the distortion of the BO_3 triangles which partly reversible at higher pressures there is a completely reversible transition in which a fraction of the boron becomes four-coordinate.

16   Non-oxide glasses: $GeSe_2$

Network glasses of AX_2 stoichiometry exhibit a variety of structures, depending on constituent atoms and the character of bonding. Short-range order is reflected in well-defined structural units such as AX_4 tetrahedra, which are linked to form networks and rings with varying members. $GeSe_2$ is considered an archetypal network-forming glass. Unlike AX_2 oxides glasses such as GeO_2 and SiO_2, however, $GeSe_2$ has a considerable number of homo-nuclear bonds and consequently, there are a greater variety of different packing arrangements that can be made in response to changes in pressure. This is reflected in the amplitude of the first sharp diffraction peak (FSDP) in the diffraction pattern [95]. The structure of $GeSe_2$ has been extensively studies by neutron diffraction with isotopic substitution and by *ab initio* computer simulation [95]. The basic structural unit is the $Ge(Se_{1/2})_4$ tetrahedron and the diffraction data imply a large number of different arrangements of these polyhedra. The ambient pressure glass structure comprises both edge- and corner-shared tetrahedral arranged in a open framework with a non-uniform arrangement of Ge and Se atoms in which chemical order is broken by homo-nuclear (homopolar) bonds. estimated from the pdfs as 2.0 gcm-3 and 2.8 gcm-3 for the low and high density liquids respectively.

The structure of $GeSe_2$ liquids has been shown to change as a function of both temperature and pressure. In situ studies of liquid $GeSe_2$ under pressure show changes in the intermediate range order as evidenced by changes in the FSDP and these are interpreted as a change from a two-dimensional network to three dimensional fluid. This has led to the suggestion that $GeSe_2$ may show a first-order liquid-liquid transition under the application of pressure. There are additional characteristics of the $GeSe_2$ system that suggest polyamorphism[96, 97]. There is an increase in density on melting indicating a negative dT/dP slope to the melting curve and the different amorphous structures that can be produced mimic the structures of crystal polymorphs. In addition there are changes in electrical properties as the pressure is increased. The low pressure semi-conducting form transforms to a metallic amorphous form at 9 GPa. Recent *in situ* studies of amorphous $GeSe_2$ using high energy X-rays and a diamond anvil cell show changes in structure as samples are compressed. These changes are seen as a decrease in the intensity of the first sharp diffraction peak, which also shifts in position from 1.01 to 1.23 Å^{-1} and an increase (by a factor of 1.46) in the intensity of the principle peak in the X-ray S(Q). The response to pressure, an increase in density, is accomplished by changes in both intermediate- and short-range order. The changes in $GeSe_2$ are qualitatively similar to those in GeO_2. For $GeSe_2$ the changes in intermediate range order are a conversion from edge- to corner shared $Ge(Se_{1/2})_4$ tetrahedra up to pressures of 3GPa. Above 3GPa the response to pressure is an increase in coordination number from a mean Ge coordination number of 3.98 at ambient pressure [97], increasing from 4.15 to 4.52 between 3.9 and 9.3 GPa. The mechanisms differ in detail between GeO_2 and $GeSe_2$, with the intermediate range order changes in GeO_2 reflecting the greater ionicity in the oxide glass[98, 95]. Tetrahedral GeO_4 units can only be corner shared and intermediate order changes reflect a decrease in void space which becomes accompanied by short-range changes increasing the coordination number from 4 to 5 through intermediate 5-coordinate polyhedra. In $GeSe_2$, because of the homo-nuclear bonding the connections between structural units is very different and density can increase by a change from edge- to corner shared tetrahedral units. The in situ study is consistent with Raman spectroscopy data where the ratio of edge- to corner-shared tetrahedral units reduces from 34% at ambient pressure to 20% at 3 GPa. This change is apparently continuous and the reported disappearance of the FSDP does not correlate with an amorphous-amorphous transition. Densification apparently occurs by stabilizing a series of intermediate structures and does not occur over a narrow pressure range, although in the relaxed liquid the change in structure occurs between 4.1 and 5.1 GPa. It has been further suggested that the changes in intermediate range order are similar for other tetrahedral systems . A comparison of the peaks in the structure factor and mean inter-atomic spacing as a function of pressure show similar trends towards a limit, which is the dense packing of random spheres. This would favour an increase in disorder as pressure increases.

17   Future directions

In this contribution the complexity of glass structure and the potential for glass structures to become even more complicated at high pressures has been discussed. The structures of chemically complex liquids and glasses can be studied in detail if the partial structure factors can be determined either by neutron and X-ray diffraction techniques. The pair distribution function is the starting point for interpreting amorphous structure and diffraction data used in combination with computer simulation and spectroscopy provide a means for interpreting the short- and intermediate-range structure, reproduced in the S(Q) or G(r). These techniques are important if the characteristics of *polyamorphism* and the changes in the rheology (*fragility*) of liquids are to be understood. Since these changes in the structure and structure-related properties are invariably not recovered in quench samples *in situ* diffraction and associated spectroscopic studies are desired. Developments in the specialised sample environments for use in combination with neutron diffraction mean that the change in liquid or glass structure with pressure and temperature can now be ascertained [99]. There are few studies on the changes in amorphous structure with pressure however those that have been perfomred on a range of candidate *polyamorphic* system show a wealth of structural changes, some of which occur over narrow intervals in pressure. The nature of these changes remains controversial and the high pressure liquid regime is as yet largely unexplored.

As the sample environments and experimetnal facilities become developed there are opportunities to probe extremes of temperature and pressure are offered. New neutron and X-ray sources and instruments such as the Spallation Neutron Source (SNS) at Oak Ridge national Laboratory (US) and the new second target station at ISIS, Rutherford Laboratory (UK) for example, will offer high neutron fluxes and there is the opportunity to examine small samples such as those contained in high pressure cells. Disordered and isotopically-substituted materials such as $^{11}B_2O_3$, can be examined and structural changes determined. Chemically complex liquids and glasses can also be studied if the partial structure factors can be determined either by using isotopic substitution techniques, or more cheaply, by combining data from neutron and X-ray diffraction experiments. Such an approach is ideal for oxide materials since the scattering of neutrons from oxygen in the sample may account for up to 60% of the scattered signal. Combined neutron and high energy X-ray measurements can be used to identify different partial contributions to the $S(Q)$ and the real space transforms.

References

[1] Barrat J.-L. and Hansen J. P., Basic concepts for simple and complex liquids, Cambridge University Press, 2003.

[2] Egelstaff P. A., An introduction to the liquid state, Academic Press, 1967.

[3] Zallen R., The physics of amorphous solids, Wiley, 1983.

[4] Bacon G. E., Neutron diffraction, Clarendon Press, 3d edn, 1975.

[5] Parise J. B., Introduction to neutron properties and applications, Neutron Scattering In Earth Sciences, **63**, 1, 2006.

[6] Bacon G. E. and I. U. of Crystallography, Fifty years of neutron diffraction: the advent of neutron scattering, A. Hilger, published with the assistance of the International Union of Crystallography , 1986.

[7] Egelstaff P. A., Thermal neutron scattering, Academic Press, 1965.

[8] Fischer H. E., Barnes A. C., and Salmon P. S., Neutron and x-ray diffraction studies of liquids and glasses., Rep. Prog. Phys., **69**, 233, 2006.

[9] Fischer H. E., Salmon P. S., and Barnes A. C., Neutron and x-ray diffraction for structural analysis of liquid and glass materials, J. Phys. Iv, **103**, 359, 2003.

[10] Debenedetti P. G., Metastable liquids: concepts and principles, Princeton University Press, 1996.

[11] Elliott S. R., Medium-range structural order in covalent amorphous solids., Nature, **354**, 445, 1991.

[12] Angell C. A., Liquid fragility and the glass transition in water and aqueous solutions., Chem. Rev., **102**, 2627, 2002.

[13] Angell C. A. and Goldstein M., Dynamic aspects of structural change in liquids and glasses, Annals of the New York Academy of Sciences, v. 484, New York Academy of Sciences, 1986.

[14] Kauzmann W., The nature of the glassy state and the behavior of liquids at low temperatures., Chem. Rev., **43**, 219, 1948.

[15] Debenedetti P. G. and Stillinger F. H., Supercooled liquids and the glass transition., Nature, **410**, 259, 2001.

[16] Angell C. A., Formation of glasses from liquids and biopolymers, Science, **267**, 1924, 1995.

[17] Adam G. and Gibbs J. H., On temperature dependence of cooperative relaxation properties in glass-forming liquids., J. Chem. Phys., **43**, 139, 1965.

[18] Bengtzelius U., Theoretical calculations on liquid-glass transitions in lennard-jones systems., Phys. Rev. A, **33**, 3433, 1986.

[19] Bengtzelius U., Gotze W. and Sjolander A., Dynamics of supercooled liquids and the glass-transition., J. Phys. C, **17**, 5915, 1984.

[20] Geszti T., Pre-vitrification by viscosity feedback., J. Phys. C, **16**, 5805, 1983.

[21] Gotze W. and Sjogren L., Alpha-relaxation near the liquid glass-transition., J. Phys. C, **20**, 879, 1987.

[22] Gotze W. and Sjogren L., The glass-transition singularity, Z. Phys. B, **65**, 415, 1987.

[23] Gotze W. and Sjogren L., Alpha-relaxation in supercooled liquids, Slow Dynamics In Condensed Matter, AIP Conf. Proc., **256**, 95, 1992.

[24] Johari G. P., Glass-transition and secondary relaxations in molecular liquids and crystals., Ann. N.Y. Acad. Sci., **279**, 117, 1976.

[25] Bridgman P. W., Recent work in the field of high pressures., Rev. Mod. Phys., **18**, 1, 1946.

[26] Mishima O. and Stanley H. E., Decompression-induced melting of ice iv and the liquid-liquid transition in water., Nature, **392**, 164, 1998.

[27] McMillan P. F., High pressure synthesis of solids., Curr. Opin. Solid State Mater. Sci., **4**, 171, 1999.

[28] McMillan P. F., Liquid state polymorphism. Hemley, R. J., Chiarotti, G. L., Bernasconi, M., and fisica., S. i. d., eds.),, High pressure phenomena: proceedings of the International School of Physics "Enrico Fermi": course CXLVII:Varenna on Como Lake, Villa Monastero, 3-13 July 2001, pp. 511–543., IOS Press;Ohmsha, 2002.

[29] Mishima O., Calvert L. D., and Whalley E., Melting ice-i at 77-k and 10-kbar - a new method of making amorphous solids., Nature, **310**, 393, 1984.

[30] Rapaport E., Model for melting-curve maximum at high pressure, J. Chem. Phys., **46**, 2891, 1967.

[31] Rapaport E., Melting-curve maximum at high pressure. ii. liquid cesium. resisitivity, hall effect and compositon of molten tellurium., J. Chem. Phys., **48**, 1433, 1968.

[32] Guggenheim E. A., Modern thermodynamics by the methods of Willard Gibbs, Methuen & co. ltd, 1933.

[33] Angell C. A. and Moynihan C. T. Ideal and cooperative bond-lattice representations of excitations in glass-forming liquids: Excitation profiles, fragilities, and phase transitions., Metall. Mater. Trans. B, **31**, 587, 2000.

[34] Angell C. A., Moynihan C. T. and Hemmati M., 'strong' and 'superstrong' liquids, and an approach to the perfect glass state via phase transition., J. Non-Cryst. Solids, **274**, 319, 2000.

[35] Moynihan C. T., Two species/non ideal solution model for amorphous/amorphous phase transitions., Mat. Res. Soc. Proc., **455**, 411, 1997.

[36] Ponyatovsky E. G., Belash I. T. and Barkalov O. I., Pressure-induced amorphous phases., J. Non-Cryst. Solids, **117**, 679, 1990.

[37] Brazhkin V. V., Buldyrev S. V., Rhzhov V. N. and Stanley H. E., New kinds of phase transitions: transformations in disordered substances, NATO science series. Series II, Mathematics, physics, and chemistry; v. 81, Kluwer Academic Publishers, 2002.

[38] Brazhkin V. V., Popova S. V. and Voloshin R. N., Pressure-temperature phase diagram of molten elements: selenium, sulfur and iodine., Physica B, **265**, 64, 1999.

[39] Angell C. A., Energy landscapes for cooperative processes: nearly ideal glass transitions, liquid-liquid transitions and folding transitions, Philos. Trans. R. Soc. London, Ser. A, **363**, 415, 2005.

[40] McMillan P. F., Polyamorphic transformations in liquids and glasses. , J. Mat. Chem., **14**, 1506, 2004.

[41] Poole P. H., Grande T., Angell C. A., and McMillan P. F., Polymorphic phase transitions in liquids and glasses., Science, **275**, 322, 1997.

[42] Poole P. H., Grande T., Sciortino F., Stanley H. E. and Angell C. A., Amorphous polymorphism., Comput. Mater. Sci., **4**, 373, 1995.

[43] Angell C. A., Amorphous water., Annu. Rev. Phys. Chem., **55**, 559, 2004.

[44] Angell C. A., Glass transition dynamics in water and other tetrahedral liquids: 'order-disorder' transitions versus 'normal' glass transitions, J. Phys. Cond. Matter, **19**, 0953, 2007.

[45] Mishima O., Calvert L. D. and Whalley E. An apparently 1st-order transition between 2 amorphous phases of ice induced by pressure, Nature, **314**, 76, 1985.

[46] Mishima O. and Stanley H. E. The relationship between liquid, supercooled and glassy water., Nature, **396**, 329, 1998.

[47] Uhlmann D. R., Relaxation in glass-forming liquids., Am. Ceram. Soc. Bull., **58**, 384, 1979.

[48] Uhlmann D. R., Nucleation, crystallization and glass-formation, J. Non-Cryst. Solids, **38-9**, 693, 1980.

[49] Moynihan C. T. and Angell C. A., Bond lattice or excitation model analysis of the configurational entropy of molecular liquids., J. Non-Cryst. Solids, **274**, 131, 2000.

[50] Mishima O. and Suzuki Y., Propagation of the polyamorphic transition of ice and the liquid-liquid critical point., Nature, **419**, 599, 2002.

[51] Hemmati M., Moynihan C. T. and Angell C. A., Interpretation of the molten bef2 viscosity anomaly in terms of a high temperature density maximum, and other waterlike features., J. Chem. Phys., **115**, 6663, 2001.

[52] Ha A., Cohen I., Zhao X. L., Lee M. and Kivelson D. Supercooled liquids and polyamorphism., J. Phys. Chem., **100**, 1, 1996.

[53] Aasland S. and McMillan P. F. Density-driven liquid-liquid phase separation in the system al2o3-y2o3., Nature, **369**, 633, 1994.

[54] Greaves G. N. et al. Detection of first-order liquid/liquid phase transitions in yttrium oxide-aluminum oxide melts., Science, **322**, 566, 2008.

[55] Wilding M. C. and McMillan P. F. Polyamorphic transitions in yttria-alumina liquids., J. Non-Cryst. Solids, **293**, 357, 2001.

[56] Wilding M. C. and McMillan P. F., liquid polymorphism in yttrium aluminate liquids, Brazhkin, V. V., Buldyrev, S. V., Rhzhov, V. N., and Stanley, H. E., eds.),, New kinds of phase transitions: transformations in disordered substances, pp. 57–73, NATO science series. Series II, Mathematics, physics, and chemistry; v. 81, Kluwer Academic Publishers, 2002.

[57] McMillan P. F. and Wilding M. C., Direct density determination of low- and high-density glassy polyamorphs following a liquid-liquid phase transition in y2o3-al2o3 supercooled liquids., J. Non-Cryst. Solids, **354**, 1015, 2008.

[58] Tanaka H., Two-order-parameter description of liquids. i. a general model of glass transition covering its strong to fragile limit., J. Chem. Phys., **111**, 3163, 1999.

[59] Tanaka H., General view of a liquid-liquid phase transition, Phys. Rev. E, **62**, 6968, 2000.

[60] Besson J. M., Klotz S., Hamel G., Makarenko I., Nelmes R. J., Loveday J. S., Wilson R. M. and Marshall W. G., High pressure neutron diffraction. present and future possibilities using the paris-edinburgh cell. , High Pressure Res., **14**, 1, 1995.

[61] Besson J. M. and Nelmes R. J., New developments in neutron-scattering methods under high-pressure with the paris-edinburgh cells., Physica B, **213**, 31, 1995.

[62] Loveday J. S., Nelmes R. J., Marshall W. G., Besson J. M., Klotz S., Hamel G. and Hull S., High pressure neutron diffraction studies using the paris-edinburgh cell., High Pressure Res., **14**, 303, 1996.

[63] Khvostantsev L. G., Sidorov V. A. and Tsiok O. B. High pressure toroid cell: Applications in planetary and material sciences., Properties Of Earth And Planetary Materials At High Pressure And Temperature, **101**, 89, 1998.

[64] Khvostantsev L. G., Slesarev V. N. and Brazhkin V. V., Toroid type high-pressure device: History and prospects., High Pressure Res., **24**, 371, 2004.

[65] Brazhkin V. V. and Lyapin A. G., High-pressure phase transformations in liquids and amorphous solids., J. Phys. Cond. Matter, **15**, 6059, 2003.

[66] Brazhkin V. V., Voloshin R. N., Popova S. and Lyapin A. Transitions in liquids: Examples and open questions. Brazhkin, V. V., Buldyrev, S. V., Rhzhov, V. N., and Stanley, H. E., eds.),, New kinds of phase transitions: transformations in

disordered substances. NATO science series. Series II, Mathematics, physics, and chemistry; v. 81, vol. 81 of NATO science series. Series II, Mathematics, physics, and chemistry, pp. 239–251, Kluwer Academic Publishers, 2002.

[67] Katayama Y., In situ observation of a first-order liquid-liquid transition in phosphorus., J. Non-Cryst. Solids, **312**, 8, 2002.

[68] Katayama Y., Inamura Y., Mizutani T., Yamakata M., Utsumi W. and Shimomura O., Macroscopic separation of dense fluid phase and liquid phase of phosphorus., Science, **306**, 848, 2004.

[69] Bailey I. F., A review of sample environments in neutron scattering. , Z. Kristallogr., **218**, 84, 2003.

[70] Le Godec Y., Dove M. T., Redfern S. A. T., Tucker M. G., Marshall W. G., Syfosse G. and Klotz S., Recent developments using the paris-edinburgh cell for neutron diffraction at high pressure and high temperature and some applications., High Pressure Res., **23**, 281, 2003.

[71] Le Godec Y., Strassle T., Hamel G., Nelmes R. J., Loveday J. S., Marshall W. G. and Klotz S. Techniques for structural studies of liquids and amorphous materials by neutron diffraction at high pressures and temperatures., High Pressure Res., **24**, 205, 2004.

[72] Besson J. M., Nelmes R. J., Hamel G., Loveday J. S., Weill G. and Hull S., Neutron powder diffraction above 10-gpa., Physica B, **180**, 907, 1992.

[73] Loveday J. S., Nelmes R. J., Marshall W. G., Besson J. M., Klotz S. and Hamel G. Structural studies of ices at high pressure., Physica B, **241**, 240, 1997.

[74] Barnes A. C., Fischer H. E. and Salmon P. S., Structure of disordered systems and liquid metal alloys., J. Phys. Iv, **111**, 59, 2003.

[75] Klug D. D., How ice 'melts' below its melting point., Physics World, **18**, 25, 2005.

[76] Klug D. D., Tulk C. A., Svensson E. C. and Loong C. K., Dynamics and structural details of amorphous ice by incoherent inelastic neutron scattering., Abstracts Of Papers Of The American Chemical Society, **218**, U298, 1999.

[77] Ricci M. A. and Soper A. K., Jumping between water polymorphs, Physica A, **304**, 43, 2002.

[78] Soper A. K. and Ricci M. A., Structures of high-density and low-density water., Phys. Rev. Lett., **84**, 2881, 2000.

[79] Finney J. L., Hallbrucker A., Kohl I., Soper A. K. and Bowron D. T., Structures of high and low density amorphous ice by neutron diffraction, Phys. Rev. Lett., **8822**, 5503, 2002.

[80] Finney J. L., Bowron D. T., Soper A. K., Loerting T., Mayer E. and Hallbrucker A., Structure of a new dense amorphous ice., Phys. Rev. Lett., **89**, 0031, 2002.

[81] Klotz S., Hamel G., Loveday J. S., Nelmes R. J., Guthrie M. and Soper A. K., Structure of high-density amorphous ice under pressure, Phys. Rev. Lett., **89**, 285502, 2002.

[82] Strassle T., Saitta A. M., Le Godec Y., Hamel G., Klotz S., Loveday J. S. and Nelmes R. J., Structure of dense liquid water by neutron scattering to 6.5 gpa and 670 k., Phys. Rev. Lett., **96**, 067801, 2006.

[83] Nelmes R. J., Loveday J. S., Strassle T., Bull C. L., Guthrie M., Hamel G. and Klotz S., Annealed high-density amorphous ice under pressure. , Nature Physics, **2**, 414, 2006.

[84] Tulk C. A., Benmore C. J., Urquidi J., Klug D. D., Neuefeind J., Tomberli B. and Egelstaff P. A., Structural studies of several distinct metastable forms of amorphous ice., Science, **297**, 1320, 2002.

[85] Tulk C. A., Hart R., Klug D. D., Benmore C. J. and Neuefeind J., Adding a length scale to the polyamorphic ice debate., Phys. Rev. Lett., **97**, 0031, 2006.

[86] Guthrie M., Tulk C. A., Benmore C. J. and Klug D. D., A structural study of very high-density amorphous ice., Chem. Phys. Lett., **397**, 335, 2004.

[87] Tanaka H., Thermodynamic anomaly and polyamorphism of water, Europhys. Lett., **50**, 340, 2000.

[88] Deb S. K., Wilding M., Somayazulu M. and McMillan P. F. Pressure-induced amorphization and an amorphous-amorphous transition in densified porous silicon., Nature, **414**, 528, 2001.

[89] McMillan P. F., Wilson M., Daisenberger D. and Machon D., A density-driven phase transition between semiconducting and metallic polyamorphs of silicon., Nature Materials, **4**, 680, 2005.

[90] Sastry S. and Angell C. A., Liquid-liquid phase transition in supercooled silicon., Nature Materials, **2**, 739, 2003.

[91] Itie J. P., Polian A., Calas G., Petiau J., Fontaine A. and Tolentino H., Pressure-induced coordination changes in crystalline and vitreous geo2., Phys. Rev. Lett., **63**, 398, 1989.

[92] Sampath S., Benmore C. J., Lantzky K. M., Neuefeind J., Leinenweber K., Price D. L. and Yarger J. L., Intermediate-range order in permanently densified geo2 glass, Phys. Rev. Lett., **9011**, 5502, 2003.

[93] Guthrie M., Tulk C. A., Benmore C. J., Xu J., Yarger J. L., Klug D. D., Tse J. S., Mao H. K. and Hemley R. J., Formation and structure of a dense octahedral glass., Phys. Rev. Lett., **93**, 115502, 2004.

[94] Brazhkin V. V., Katayama Y., Trachenko K., Tsiok O. B., Lyapin A. G., Artacho E., Dove M., Ferlat G., Inamura Y. and Saitoh H., Nature of the structural transformations in b2o3 glass under high pressure, Phys. Rev. Lett., **101**, 0031, 2008.

[95] Salmon P. S. and Petri I., Structure of glassy and liquid gese2, J. Phys. Cond. Matter, **15**, S1509, 2003.

[96] Crichton W. A., Mezouar M., Grande T., Stolen S. and Grzechnik A., Breakdown of intermediate-range order in liquid gese2 at high pressure, Nature, **414**, 622, 2001.

[97] Mei Q., et al., Topological changes in glassy gese2 at pressures up to 9.3 gpa determiend by high-energy x-ray and neutron diffraction measurements. , Phys. Rev. B, **74**, 2006.

[98] Salmon P. S., Martin R. A., Mason P. E. and Cuello G. J., Topological versus chemical ordering in network glasses at intermediate and extended length scales., Nature, **435**, 75, 2005.

[99] Wilding M. C., Guthrie M., Bull C. L., Tucker M. G. and Mcmillan P. F., Feasibility of in situ neutron diffraction studies of non-crystalline silicates up to pressure of 25 gpa., J. Phys. Cond. Matter, **20**, 244122, 2008.

[100] Wilding M. C and Benmore C J Neutron Scattering in Earth Science, Reviews in Mineralogy and Geochemistry **63** 275-311 (2006).

Chapter 16
Dense Hydrogen

Russell J. Hemley

Geophysical Laboratory, Carnegie Institution of Washington, Washington, DC 20015 USA

> "Let's start at the very beginning, a very good place to start."
> R. Rogers & O. Hammerstein, *The Sound of Music*

1 Introduction

Many developments in high-pressure research have been driven by questions surrounding the nature of the first element in the Periodic Table under extreme conditions. Discovered by Cavendish and reported in 1766 [1], hydrogen is the starting point for our understanding of much of physical science. As such, the element has been a crucial testing ground for theory in atomic, molecular, solid state, and plasma physics. The unique and simple electronic structure of the hydrogen atom gives rise to an elemental dichotomy as a halogen or an alkali metal (Figure 1). Pressure thus serves to illuminate the nature of that chemical duality, as pointed out by Wigner and Huntington in 1935 [2]. Moreover, deeper theoretical inquiry some 40 years ago into the behavior of hydrogen at very high pressures indicated that the element under extreme conditions could exhibit high-temperature superconductivity [3] and a liquid ground state [4]. These propositions in turn led to the prediction of additional novel behavior, including dissipationless properties as a combined superfluid and superconductor: indeed, an altogether new state of matter [5]. Highly compressed hydrogen is also of interest because it is a high energy density material and central to a potential hydrogen-based fuel economy [6]. Finally, hydrogen has long been recognized as the major component of large planets [7]. Approximately 90% of the atoms in the solar system are hydrogen and most of those atoms experience conditions of ultrahigh pressures and temperatures. Indeed, hydrogen is the most abundant element in the visible cosmos, accounting for about 75% of its observable mass.

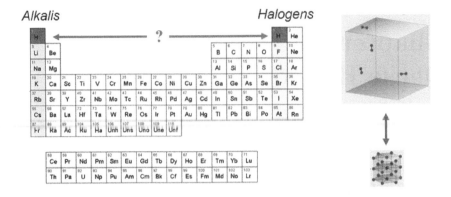

Figure 1. *The dual nature of hydrogen as alkali-metal and halogen.*

Intriguing questions about the behavior of dense hydrogen are juxtaposed with the considerable technical challenges of studying the material under extreme conditions. As a result, the study of hydrogen at high densities has driven technical developments in the field of high pressure research [8]. These challenges include the high compressibility of the material (which results in large deformation of gaskets and the high-pressure chambers in which samples are contained), high reactivity with metals (diffusion through containers), and weak x-ray scattering power. On the other hand, the molecular form has a strong Raman cross-section and variable infrared absorption and the nuclei have a large neutron cross-sections (both coherent and incoherent). The material has now been investigated by a broad range of high pressure techniques in different regimes of pressure and temperature. These include Raman, infrared (IR), nuclear magnetic resonance (NMR), x-ray diffraction, x-ray scattering, neutron diffraction, neutron scattering, optical, electron, sound velocity, Brillioun scattering, shock compression, and isentropic compression. Challenges associated with theoretical studies of hydrogen arise from the large quantum motion of the nuclei and the need to treat the electronic and nuclear degrees of freedom on the same footing. Addressing the latter challenge has spurred the development of both fundamental theory and computational techniques [3, 4, 5, 9].

This chapter comprises highlights in the study of hydrogen at high density. Rather than an exhaustive review, which is beyond the scope of this forum, the paper attempts to provide useful background on the fundamentals while at the same time featuring selected recent developments, primarily from the standpoint of compression experiments. This introduction will be followed by a review of the properties of the isolated molecule and hydrogen in condensed phases at zero (or ambient) pressure. The high pressure prop-

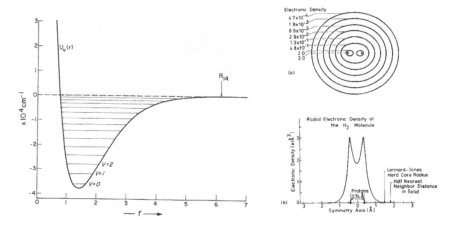

Figure 2. *Properties of the isolated molecule. Left: Intramolecular potential [16]. Right: Charge distribution of p-H$_2$ [13]*

erties are then discussed, with a focus on the behavior of the material up to approximately two megabars (200 GPa) or roughly ten-fold compression at (and below) room temperature. Selected results that illustrate key phenomena are discussed. This is followed by discussion of hydrogen at more extreme conditions, including combined high pressure and both low and high temperatures. The chapter ends with a brief summary of major conclusions and the outlook for future work.

2 The isolated molecule and low-density solid

Different scales of interactions characterize the hydrogen molecule, both in isolation and in condensed phases. Our understanding of the nature of the chemical bond began with H$_2$ through the development of valence bond [10] and molecular orbital [11] theory. We now know, based on over 80 years of calculations and experiments, that the isolated molecule is essentially spherical in terms of its charge density. The intramolecular interaction potential has been the subject of a vast number of quantum mechanical calculations and is now well understood. There are 14 bound vibrational states and a strong covalent bond of 4.53 eV (Figure 2). The zero-point energy is approximately 0.25 eV. The magnitude of these energies has important implications for the high-pressure behavior of hydrogen.

The molecular character of the system persists into condensed phases, first produced at the turn of the last century by Dewar [12]. Indeed, even the spherical state of the molecule is retained over a broad range of conditions. Bringing together a pair of molecules gives rise to a set of weak intermolecular

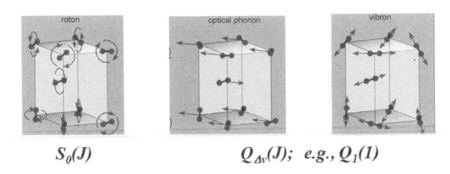

$$S_0(J) \qquad\qquad Q_{\Delta v}(J); \ e.g., \ Q_1(1)$$

Figure 3. *Principal Brillouin zone-center (k=0) vibrational excitations of solid* H_2 *[100].*

interactions that increase in magnitude with decreasing distance. Well understood dispersion interactions produce a binding energy between two molecules of 3.0 meV (35 K). The pure pairwise interactions can be broken down into isotopic and anisotropic parts. The leading anisotropic term at low densities is the electric quadrupole-quadrupole (EQQ) interaction (for odd J) that scales with the coupling parameter $\Gamma'(r)=(6/25)e^2Q^2/r^5$, where Q is the EQ moment, e is the electronic charge, and r is the intermolecular distance [13]. In condensed phases, the intermolecular interactions are typically described using an isotropic effective potential with higher order many-body terms (usually unconstrained by experiment) that are spherically averaged. The shape of the effective intramolecular potential has been the focus of numerous theoretical calculations, including those constrained by experiment (see Refs. [14, 15, 16] and references therein).

Another important aspect of the physics of hydrogen is the ortho-para state of the molecules. This distinction arises from the coupling of nuclear spin and rotational states and gives rise to molecules with distinct properties for a given isotope. The spin I_{NJ} and rotational states denoted by quantum number J contribute to the total molecular spin I_{mol}; for H_2 and D_2, $I_N = 1/2$ and 1, respectively. This distinction arises from the Pauli Exclusion Principle, which requires that the molecular wavefunction be antisymmetric with respect to inversion. This requirement gives rise to the separation of molecular hydrogen into two types, depending on the isotope and the allowed combinations of nuclear spin and rotational states. The rate of conversion between the two forms is very low at ambient pressure in the absence of catalytic effects [13].

Characteristic features of the principal intermolecular excitations need to be identified. By convention, intermolecular and intramolecular excitations involving ΔJ = -2,-1,0,1,2 are labeled O,P,Q,R,S with energies $F_v = J(J+1)B$, where B is the rotational constant and neglecting centrifugal distortion terms (Figure 3). Combined vibrational-rotational excitations involve changes in quantum numbers ΔJ and Δv; excitations for the Q branch ($\Delta J=0$) are labeled $Q_{\Delta v(J)}$. Notably, because H_2 and D_2 are homonuclear diatomic molecules, there is no dipole-allowed infrared absorption corresponding to vibrational excitations for the isolated molecule. Collective excitations are known as vibrons, rotons, and librons; the excitations at the Brillouin zone center ($\mathbf{k}=0$) are shown in Figure 3. Pressure affects rotational ordering (breakdown of J as a good quantum number), interactions between molecules (via molecular coupling), and molecular stability (where the lattice mode frequencies become comparable to that of the internal modes or vibrons). The model for the coupling of vibron excitations developed by Van Kranendonk [16] for the zero-pressure solid has been used successfully over a range of densities. The vibron coupling parameter ϵ' scales with van der Waals interaction; *i.e.*, $\sim 1/r^6$, where r is the intermolecular distance (Figure 4).

Detailing the electronic structure of the molecule is crucial for understanding bonding in highly compressed states. The electronic excitations have been extensively studied, particularly for the isolated molecule. To first order, it is the evolution of the strong covalent bond of H_2, including its eventual cleavage, that we wish to characterize and understand under pressure. Ionization to form H_2^+ occurs above about 14 eV, with a complex manifold of excited states with precisely determined Born-Oppenheimer potential surfaces [17]. The vaccuum ultraviolet absorption spectra of the zero-pressure solid has been measured, but these excitations are not well resolved [18]. The refractive index in the visible spectrum is determined by these higher energy electronic excitations, a concept that has been used to predict the pressure dependence of the electronic properties including the band gap.

Finally, the crystal structure is a crucial property. X-ray diffraction measurements first reported for crystalline hydrogen by Keesom et al. [19], found the structure of p-H_2 to be hexagonal closed packed (hcp). The lattice is expanded due to the large zero-point motion, $V_0=23.0$ cm^3/mol for H_2 versus $V_0=19.9$ cm^3/mol for D_2 for $J=0$ solids at $T=$k (see Ref. [13]. Note that for the intermolecular H_2-H_2 distance, r is 3.8 Å as compared with the H-H bond distance R of 0.74 Å under ambient conditions. The molecules are orientationally ordered in o-H_2, which has the $Pa3$ structure [13].

3   Hydrogen under pressure

We now turn to the question of what happens to hydrogen under pressure. The first breakthrough in studies at very high pressures came with the application of the diamond anvil cell to study dense hydrogen [20]. The ability to view samples of hydrogen at tens of kilobars pressure for the first time was a key step experimentally. A surprising pressure induced freezing transition at room temperature (5.4 GPa at 298 K) was observed [20]. To probe the state of the material under pressure, vibrational spectroscopy was later measured. Specifically, *in situ* Raman scattering measurements showed the persistence of

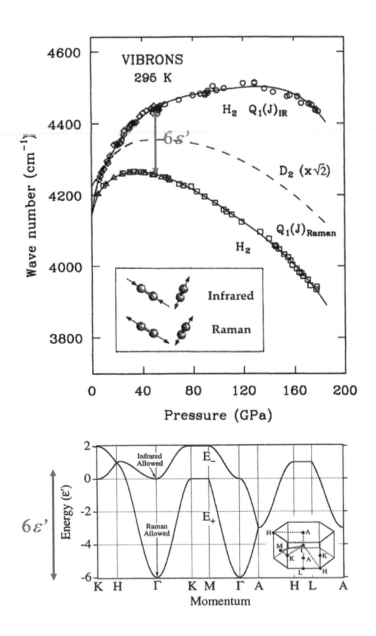

Figure 4. *Vibron excitations. Left: Splitting of the Raman and IR vibron frequencies. Right: Dispersion of the exciton in the hexagonal close packed structure. The energy scale can be represented in terms of the coupling parameter ϵ' (see Ref. [8])*

molecular bonding to at least 63 GPa and room temperature [21]. Diffraction measurements carried out later indicated that hydrogen undergoes a transition from a fluid to a molecular solid in which the molecular centers form an hcp lattice while the molecular orientations are dynamically disordered (phase I) [22, 23].

A pressure-induced softening of the vibron in the Raman measurements was observed above 30 GPa [21]. This finding was of immediate interest because it suggested weakening of the H_2 covalent bond. In fact, comparison of the IR and Raman spectra indicated that the softening at 30 GPa is a consequence of vibrational coupling (Figure 4). The pure vibron fundamental excitation is induced in one phonon IR absorption. It is useful to look at the band structure calculated from the van Kranendonk model, according to which the vibrons in solid hydrogen are collective excitations whose wave functions are analogous to the linear combination of atomic states of electron energy band theory. The corresponding (tight binding) vibron Hamiltonian has both on-site and hopping terms: basis states are vibron excitations localized on single molecules. Accordingly, the vibron Hamiltonian can be written as a tight-binding representation for single vibron excitations in the solid,

$$H = \sum_i W_i - \sum_{ij} \epsilon'(i,j)|\phi_i > < \phi_j| \tag{1}$$

where ϕ_i are the basis states employed for one-vibron excitation: the ϕ_i correspond to N-molecule Hartree functions of molecular wave functions, in which molecule i is in a vibrationally excited state, and W_i is the on-site term [16, 24, 25, 26, 27]. Measurements reveal a dramatic enhancement of vibrational coupling ϵ' with pressure [8, 97].

Low temperature vibrational spectroscopy has been used to probe a variety of phenomena. The pressure effects on the ortho-para conversion rates were found to be striking and non-monotonic (Figure 5) [31, 32, 33]. The rates were obtained by measuring the time dependence of the intensity of NMR spectra [33] and of the $S_0(0)$ and $S_0(1)$ rotational Raman bands, which were used for the measurements to the highest pressure (70 GPa) [31, 32]. The increase at the highest pressures is attributed to the increase in EQQ interactions [34]. Low temperature roton spectra give information about pressure effects on orientational ordering. Measurements of the rotational Raman spectra revealed a pressure-induced broken symmetry transition in the $J=0$ solids, first identified as a broadening of o-D_2 rotons around 28 GPa [28]. This broken symmetry phase is also called phase II. There is a large isotope effect [28 GPa (o-D_2); 110 GPa (p-H_2)]. At these pressures, libron excitations with higher J are mixed into the ground state. The vibrational spectroscopy revealed that the material undergoes orientational ordering in phase III and the libron spectrum becomes "classical" [29, 30].

A series of breakthroughs in the late 1980s in high-pressure x-ray [22] and neutron [23] scattering techniques led to the first diffraction measurements at tens of gigapascal pressures. Phase I showed a monotonic decline in the axial

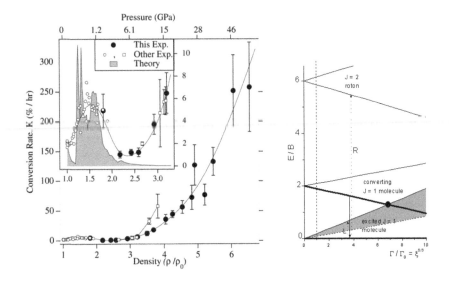

Figure 5. *Ortho-para conversion rate of H_2. Left: Conversion rate as a function of pressure and density [32]. Right: Energy scheme of conversion transition involving J=1 and J=2 excitations as a function of EQQ interaction energy [34].*

ratio c/a; the result seemed to be sensitive to the orientation of the single crystal in the diamond anvil cell, and therefore dependent on residual strength of the sample; no measurable isotope effect could be discerned. Most notably, these results clarified the equation of state (EOS) for the solid, which had been a longstanding problem for predicting high-pressure behavior from the earliest calculations [2]. The compressibility of the molecular solid was found to be remarkably high, with $\rho/\rho_0 = 5$ at 30 GPa and reaching $\rho/\rho_0 \approx 14$ on extrapolation to 300 GPa, though this is dependent on pressure calibration [35]. The P-V relations can be analyzed in terms of phenomenological EOS functions or in terms of effective intermolecular potentials (Figure 6). The analysis revealed a need for softening of the EOS relative to the predictions of pure pair potentials; *i.e.*, effective two body potentials are required to describe the P-V relations at high densities [36], an effect also observed for other simple molecular systems [35]. In general, these effective potentials do not reproduce other properties such as phonon frequencies, while at the same time they do satisfy the EOS [8]. These models can, however, be helpful in uncovering phenomena and providing physical insight [37]. The crystal structure at low temperature and high pressure has been challenging. Nevertheless, in recent years elegant attempts have been made using both x-rays and neutrons. For example, changes in intensities were followed on cooling of D_2 at 60 GPa and

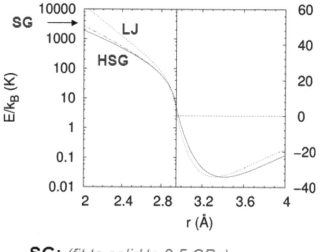

SG: *(fit to solid to 2.5 GPa)*

$$V_{SG}(r) = \exp(\alpha - \beta r - \gamma r^2) - \left(\frac{C_6}{r^6} + \frac{C_8}{r^8} + \frac{C_{10}}{r^{10}}\right)f(r) + \frac{C_9}{r^9}f(r)$$

LJ

$$V_{RRY}(r) = V_{SG}(r), \qquad r \geq r_C,$$
$$= A\exp\left[-B(r-r_C) - C(r-r_C)^2 - D(r-r_C)^3 - E(r-r_C)^3(r-r_1)\right], \qquad r < r_C$$

HSG: *(fit to x-ray/solid to 26 GPa)*

$$V_{HSG}(r) = V_{SG}(r) + V_{SR}(r)$$
$$V_{SR}(r) = a_1(r-r_c)^3 + a_2(r-r_c)^6, \qquad r \leq r_c$$
$$= 0, \qquad\qquad r > r_c.$$

Figure 6. *Effective H_2-H_2 intermolecular potentials. LJ (Leonard-Jones) [37]; SG (Silvera-Goldman) [101]; HSG (Hemley-Silvera-Goldman) [37]*

passage from phase I to phase III [38].

3.1   Phase diagram

Vibrational spectroscopy has given the best constraints to date on the P-T phase diagram. The phase lines have been identified for H_2, D_2, and HD to approximately 200 GPa and 10-300 K. The phase lines meet at triple points with the possibility of other invariant points (an additional triple point or a critical or tri-critical point). As mentioned above, the character of the Raman and infrared spectra indicate that the orientational order in phases II and III are qualitatively different, with increasing order in phase III [29, 30, 39]. There is evidence for a 'charge transfer' at the transition to phase III near 150 GPa. Understanding the transitions between phases I, II, and III has been an important test of theory (i.e., intensities and frequencies) [40, 41, 42, 43, 44, 45]. The series of rich measured vibrational spectra turn out to be the primary constraints on the crystal structures, as shown by comparisons between measured and calculated modes.

The existence of phase I′ was suggested based on measured changes in the slope of the I-III phase boundary and the observation of an apparent $\sim$10 cm^{-1} discontinuity in vibron frequency as a function of pressure [46]. The change suggested either a second triple point or a tricritical point in the system at these pressures [46]. Supporting evidence for the I′ phase was obtained from path-integral Monte Carlo simulations [47], which predict a transition to an orientationally ordered phase at 145 GPa and 300 K. Baer et al. [48] reported frequency shifts based on coherent anti-Stokes Raman spectroscopy (CARS) of D_2 to 187 GPa at room temperature, which they interpreted as supporting evidence for I′. Measurements of the Raman vibron frequency shift through the I-II transitions combined with data published in Refs. [46, 49] explain the bulk of the vibron frequency difference between 77 and 300 K reported in Ref. [48]. Several phase lines have been refined in recent measurements and through reanalysis of earlier data [50].

3.2   Constraints on crystal structures

Changes in the intensities of neutron scattering peaks measured for D_2 as a function of temperature down to 1.5 K provide additional constraints on the structure of phase II [38]. The results indicate a superstructure in the *a-b* plane. It is important to consider that the structure may depend on *o-p* content (e.g., residual J=1 molecules) [51]. The number of possible structural symmetries of phases II and III is reduced by symmetry considerations, specifically from the character of the vibrational spectra [52, 53, 46]. In addition, the lack of major discontinuities in d spacings reported to date across the I → II and II → III transitions provide additional constraints [38, 63]. The results suggest that group-subgroup relationships are preserved between the three phases. Accordingly, the transitions to higher pressure phases would involve distinct symmetry-breaking order parameters. The vibrational properties of phase II imply the presence of an inversion center at the molecules, and a multiplication by at least a factor of 2 of the number of molecules in the primitive cell relative to phase I. The spectral changes observed in phase III indicate that its primitive cell contains at least four molecules. A group-theoretical analysis of the structural mechanisms induced by the irreducible representations of the $P6_3/mmc$ space group of phase I yields structures that fulfill the above conditions [30].

From a classical crystallographic standpoint, the hcp structure of phase I can be viewed as a symmetrical distribution of H_2 molecules containing seven energetically equivalent orientations [54] (Figure 7). If the transformation to phase II is displacive, then a partially ordered structure with symmetry $Cmcm$ and eight molecules in the primitive unit cell results. The increase in the number of molecules per unit cell relative to phase I is consistent with the low-frequency librons and with the second Raman and the sharp IR vibron bands observed in phase II [8, 30]. Accordingly, at the I $\rightarrow$ II transition the molecules adopt one among the seven possible orientations existing in phase I, and may order over two. This may in turn induce a topological frustration giving rise to incommensurability **a**, consistent with superlattice reflections observed by neutron and x-ray diffraction of phase II of D_2 [38]. Extending the above analysis to this phase leads to a structure with space group $Cmc2_1$ with eight molecules per unit cell. The seven different orientations with the molecular axes pointing to the midpoints of the upper and lower tetrahedra of a trigonal bipyramid are all realized. The strong IR vibron absorption of phase III relative to phase II [30] then arises from the crystallographically distinct nature of the hydrogen atoms within a molecule in these structures. However, the quantum disorder in phase II produces symmetry equivalence of the atoms in a molecule, whereas in phase III the "classical" ordering of the molecules breaks the symmetry as a result of the stronger intermolecular interactions (*i.e.*, charge transfer).

Two distinct irreducible representations of the $P6_3/mmc$ space group describe phases II and III. The $Cmcm$ structure would then be stabilized for phase II; the The $Cmc2_1$ structure is isotranslational to that of the lower pressure phase [55]. A Landau model for the transitions between these structures can be developed [54]. For the transition from $P6_3/mmc$ to $Cmc2_1$ the symmetry-breaking mechanisms associated with the order parameters take place simultaneously and full ordering occurs progressively on compression. A theoretical phase diagram can be drawn that differs from experiment unless one or more invariant points in addition to the I-II-III triple point exist along the I-III line [46]. An additional phase I' separated from phase I by a first-order transition line that merges with phases I and III at a second triple point [46] is consistent with this analysis. The different symmetries of phases I and III, exclude passing continuously from phase I to III beyond a proposed critical point. If we assume that phases I' and III are isosymmetric, their structures correspond to different equilibrium values of the order parameters [54]. A partially ordered and possibly incommensurate structure with space group $Cmcm$, and a fully ordered isotranslational ferroelectric structure with lower $Cmc2_1$ symmetry follow naturally if the transitions are displacive.

A number of different structures of phases II and II display have been proposed on the basis of first-principles calculations. Those that display no direct group-subgroup relationship with the structure of phase I include the $Pca2_1$ ($Z = 8$) structure proposed for phase II [56, 57, 58] and for the related $Pca2_1$ ($Z = 16$) structure considered for phase III [42]. Both structures can be

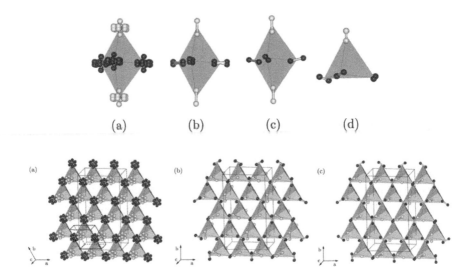

Figure 7. *Predicted crystal structures for* H_2 *assuming non-reconstructive transitions. Top: Local orientational ordering schemes. Bottom: (a) phase I; (b) phase II; (c) phase III [54].*

derived from the $Pa\bar{3}$ $(Z = 8)$ structure of H_2, which has been also considered for phase II formed from p-H_2 [30, 59]. Other structures considered theoretically for phase II include those with space groups $A2/a$ $(Z = 64)$ [47] and $P2_1/c$ $(Z = 8)$ [60], and for phase III $Cmc2_1$ $(Z = 8)$ [61], $Cmca$ $(Z = 12)$ [62], and $C2/c$ $(Z = 24)$ [62]. Transformations to these structures from phase I are reconstructive. Recent x-ray diffraction measurements of d-spacings across the II-III transitions is consistent with a non-reconstructive transition [63].

3.3   Electronic properties

Calculations carried out in the 1970s suggested that metallization of the solid at low temperatures proceeds first by band gap closure within the molecular solid (*i.e.*, prior to dissociation) [64, 65]. When phase III was discovered there was great interest in whether it was metallic because numerous theoretical calculations predicted band gap closure at these pressures. The problem theoretically was obtaining a crystal structure at the relevant pressures that gave an insulating phase (*i.e.*, without band overlap), that is also consistent with spectroscopic data [66, 67].

Measurements of the refractive index and its dispersion provide constraints on the direct band gap of the material through oscillator models. These measurements indicated gap closure occur above 300 GPa at $\leq$ 300 K [8]. Direct measurements of resistivity showed that hydrogen remains insulating at low temperature to at least 230 GPa [68]. Those same methods show a wealth of transitions to metallic and superconducting states in other elements at these pressures. In contrast, there is a drop in resistivity and a conductivity onset at 140 GPa in reverberating shock wave experiments, as discussed below [69].

Measurements of optical conductivity by infrared absorption and reflectivity spectra showed that the material remains in an insulating state at 150 GPa and ≤300 K [44, 45]. An alternative to measuring the IR signature of the metallization is tracking the absorption edge in order to constrain the band gap starting at zero pressure. Visible absorption measurements have been performed to ∼300 GPa. Evidence for a decreasing band gap was reported in optical measurements by Loubeyre et al. [70]. These are consistent with early observations using heterogeneous hydrogen-ruby composite samples at ultrahigh pressures [71], though the pressure could not be measured due to technical reasons (see discussion in Ref. [8]). There is also evidence for resonance Raman effects in these studies, confirming that electronic excitations of the hydrogen are sampled at optical wavelengths [70, 71]. As mentioned above, the most recent measurements, for which the most reliable pressures were determined, show that molecules persist to 320 GPa. A new technique, involving the use of x-ray Raman spectroscopy, directly measures the band gap above the absorption edge of the diamond. Tests of this technique to ∼10 GPa show that the initial threshold is close to that measured in the ambient pressure low-temperature solid by one-photon absorption spectroscopy in the vacuum ultraviolet region of the spectrum [18].

4 High pressures and temperatures

Figure 9 summarizes salient features of the high P-T phase diagram of hydrogen from a combination of static and dynamic compression experiments and theory. Experimental studies to 15 GPa and 526 K have been made using high *P-T* Brillouin scattering and resistive heating to give the EOS for the fluid and the melting curve [72]. The effective potentials developed for low or modest temperatures could be used to reproduce high *P-T* behavior. In particular, high-temperature behavior on shock compression could be predicted. These solid-state data provided a baseline for the high *P-T* fluid EOS by showing that the results could reproduce the pressure-density relations for the Hugoniot of the fluid [73].

Recent theory points to a pressure for band gap closure in the range of 380-450 GPa [70, 74], while transformation to a monatomic solid is predicted at even higher pressures (500 GPa) [74]. At temperatures above the Fermi energy (1 Ry or 13.6 eV), hydrogen becomes a fully ionized mixture of protons and electrons [75]. Information in the intermediate region, called the regime of warm dense matter, is far less certain. Breakthroughs in dynamic compression measurements on hydrogen took place in the 1990s. A conductivity onset was identified at 140 GPa and 2500 K in reverberating shock wave experiments

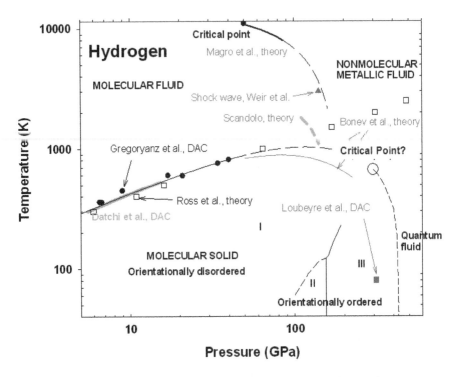

Figure 8. *Observations of darkening of hydrogen at multimegabar pressures Left: Optical micrograph from [71]; (see also [8]). Middle: Photomicrograph at 320 GPa [70]. Right: Measured visible absorption as a function of pressure [70].*

on fluid hydrogen [69]. The electrical conductivity was reported to increase by four orders of magnitude, which has been interpreted as signaling the onset of an insulator-metal transition of the Mott type [76].

Measurements to ∼300 GPa at room and lower temperature indicate that hydrogen still remains an insulating molecular solid, as described above [70, 74]. The molecular dissociation line is predicted to end at a critical point at about 50 GPa and 10,000 K [75, 77]. This has given rise to the current phase diagram (Figure 9). At low temperature and much higher pressures, the existence of a metallic superfluid is predicted [79]. Evidence was reported for an unusual compressibility of the fluid [80, 81]. The high *P-T* dynamic compression results have been complemented by numerous simulations of behavior in this range [82, 83, 84, 85]. These results have led to a new series of questions about high *P-T* behavior, including the EOS (solid and fluid), the melting relations, bonding state, structural changes in the fluid, plasma phase transitions and plasma properties. These questions have driven the continuied

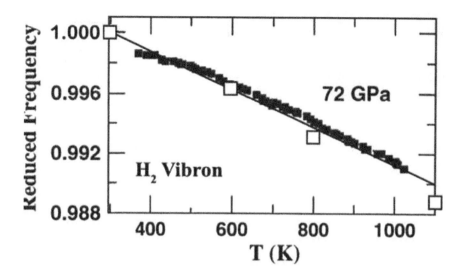

Figure 9. *High P-T phase diagram of hydrogen (adapted from [51]).*

development of high *P-T* techniques.

4.1 Melting curve

The form of the melting curve is of fundamental interest. Melting of hydrogen was measured up to 10 GPa and fit to phenomenological melting laws. A melting maximum was predicted by a Kechin model fit based on subsequent optical and *P-T* scan data to 15 GPa by Datchi et al. [86]. Gregoryanz et al. [87] measured the melting curve to 45 GPa, and the results were consistent with the earlier Kechin law fit. However, fits to Kraut-Kennedy and Simon models, which give no maxima, are in qualitative agreement with RRY potential simulations [87, 88]. First-principles calculations also predict a maximum in the melting line, largely consistent with the Kechin fit extrapolations [83]. The negative melting slope requires the fluid to become denser than the solid, further implying softer intermolecular potentials due to intermolecular charge transfer and possible partial dissociation. Theoretical predictions of the melting maximum [83] have been interpreted in terms of an enhancement of the intermolecular coupling due to the charge transfer in liquid molecular hydrogen. Scandolo [82] attributed the maximum to the intersection of the molecular to nonmolecular liquid-liquid transition line with the melting line. The liquid-liquid transition has been examined in more recent work by Morales et al [89]. The transition has been confirmed in more recent quasi-isentropic compression experiments [90]. Bonev et al [83] explicitly calculated

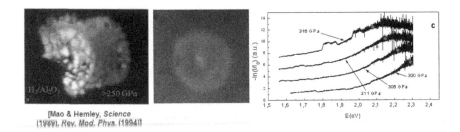

Figure 10. *Temperature dependence of the Raman vibron frequency at high pressure [37, 87].*

the melting line and ascribed the turnover in the melting curve to increased softening of the intermolecular potential in the dense fluid relative to the solid.

The melting line has been examined in subsequent experiments. A sharp maximum at ∼65 GPa and 1050 K was claimed in studies to ∼81 GPa, [91], as a result that departs considerably from the previous curves [86, 87]. Eremets and Trojan [92] reported data to 150 GPa that are in apparent agreement with the Kechin fits to the lower pressure data and first-principles theory. Time resolved Raman and speckle measurements have been carried out that are consistent with the neagive slope of the melting curve [93]. It has been established that hot compressed hydrogen has high mobility and reactivity [86, 87], so partial or complete loss of hydrogen in the sample cavity as a result of diffusion or chemical reactions complicates the interpretation of static high P-T experiments. This loss of hydrogen can also occur at higher pressures without heating, and this effect complicates the claims of static pressurization of hydrogen well above 300 GPa [94], where evidence for actocal containment of the sample was not provided. In fact, recent experiments demonstrate that hydrogen can react to form hydrides with relatively inert metals such as W, Re, and Ir [95, 96].

4.2   High P-T vibrational dynamics

Vibron spectra measured by Raman scattering at high pressures and temperatures have a variety of phenomena [87]. There is evidence for increases in anharmonicity above 100 GPa and the Raman spectrum changes very little upon melting in the higher pressure range (∼50 GPa). Feldman et al. [37] presented a simple physical model to explain these observations and the extent to which they provide information on the dissociation and other transitions at higher P-T conditions. To address questions about the high P-T vibrational dynamics, we return to the vibron Hamiltonian introduced above,

$$H = \sum_i W_i - \sum_{ij} \epsilon'(i,j;T)|\phi_i><\phi_j|, \tag{2}$$

where the hopping coefficient ϵ' now has an explicit temperature dependence, which is given as

$$\epsilon' = \int g(r; T, P)\epsilon'_{static}(r), \tag{3}$$

$$\epsilon'_{static} = A(P)(r_0^{eq}/r)^{6.8}, \tag{4}$$

$$g(r) = \frac{1}{\sigma(2\pi)^{1/2}} exp(-(r - r^{eq})^2/(2\sigma^2)) \tag{5}$$

Here $g(r)$ is the pair correlation function, dependent on temperature and atom pair, and σ is the relative mean square displacement [37]. The function $A(P)$ is determined empirically from the IR-Raman frequency differences to yield

$$\epsilon'_{RT} = 0.5(r_0^{eq}/r^{eq})^{6.8}, \tag{6}$$

where r^{eq} is the equilibrium nearest neighbor distance at a given pressure and room temperature. The value of the exponent is lower than that used in Ref. [97] (6.89), similarly lower exponents have also been successfully employed in other analyses.

The calculations give excellent agreement with experiment for dv/dT over a broad range of pressures [37] (Figure 10). The nearly constant value of dv/dT between 25 and 100 GPa found experimentally [87] is well represented by the theory and is in accord with the lack of pressure dependence documented in laser heating Raman measurements discussed below. However, neither the observed strong upturn above 150 GPa nor the decreasing behavior can be explained if quasiharmonic or higher order anharmonic effects are taken into account. In particular, the question of vibrational coupling versus bond weakening can be addressed using the tight-binding Van Krandendonk model. Thus the results are consistent with the lack of bond weakening to ~100 GPa found in the early IR measurements [97].

Raman scattering of the vibron combined with with laser heating probes the interatomic (or intramolecular) potential. Hot bands constrain the potential and match those of the isolated molecule. With increasing temperature, the Raman vibron feature develops hot bands corresponding $v_s 1 \rightarrow Z$ and $Z \rightarrow 3$ transitions. These highter vibrational states sample the anharmonic region of the interatomic potential [51]. There is no direct evidence for an appreciable loss of bound states (compare Ref. [98]), though the anharmonicity increases with pressure [96]. A key problem is confining hot and reactive H_2 in the sample chamber at high pressure long enough to make meaningful measurements as discussed above. New methodologies have been developed to achieve good local confinement of the hydrogen with *in situ* measurements made to ~130 GPa and 2500 K. These measurements provide evidence for a marked decrease in vibron frequencies in the hot dense fluid above 60 GPa [96].

5 Conclusions

Dense hydrogen is rich in phenomena and continues to drive developments in high pressure research. Knowledge of properties at low pressure is crucial for understanding and predicting high P-T behavior where those properties are dramatically altered. Major experimental and theoretical challenges remain regarding crystal structures and electronic properties up to 300 GPa (or ~14-fold compression at room temperature and below). Further experimental work is required to identify the existence of possible solid phases in this pressure range, in addition to phases I, II, and III. This work is especially important for understanding the behavior of hydrogen at these pressures but at higher temperatures, including the origin of the possible negative slope of the melting line [83, 84, 87, 91]. Particulary important developments are the use of time-resolved high P-T methods (e.g., Ref. [96]) and combined static/dynamic pressure techniques [99]. Meanwhile, theory continues to reveal potentially new physical phenomena [79]. These provide challenges for experiment, specifically including combined magnetic fields and low temperature at ultrahigh pressures. New discoveries are within reach,including exotic new low-temperature physics and high P-T regimes of astrophysical interest.

References

[1] Cavendish, H., Three papers, containing experiments on factitious air, *Phil. Trans*, 56, 141, 1766.

[2] Wigner, E., and Huntington, H.B., On the possibility of a metallic modification of hydrogen, *J. Chem. Phys.* 3, 764, 1935.

[3] Ashcroft, N.W., Metallic hydrogen: A high-temperature superconductor?, *Phys. Rev. Lett.*, 21, 1748, 1968.

[4] Brovman, E.G., Kagan, Yu., and Kholas, A. Structure of metallic hydrogen at zero pressure, *JETP*, 34, 1300, 1972.

[5] Babaev, E., Sudbø, A., and Ashcroft, N.W., A superconductor to superfluid phase transition in liquid metallic hydrogen, *Nature*, 421, 666, 2004.

[6] Hemley, R.J., Crabtree, G.W. and Buchanan, M.V., Energy challenges for materials in extreme environments, *Phys. Today*, 62, 32, 1991.

[7] DeMarcus, W.C., The constitution of Jupiter and Saturn, *Astron. J.*, 63, 2, 1958.

[8] Mao, H.K. and Hemley, R.J., Ultrahigh-pressure transitions in solid hydrogen, *Rev. Mod. Phys.*, 66, 671, 1994.

[9] Ceperley, D.M. and Alder, B., Ground state of solid hydrogen at high pressures, *Phys. Rev. B*, 36, 2092, 1987.

[10] Hylleraas, E.A., Über die Elektronenterme des Wasserstoffmoleküles, *Z. Phys.*, 71, 739, 1931.

[11] Heitler, W. and London, F., Wechselwirkung neutraler Atome und homoopolare Bidung nach der Quantummmechanik, *Z. Physik*, 44, 455, 1927.

[12] Dewar, J., Sur la solidification de l'hydrogene, *Ann. Chim. Phys.*, 18, 145, 1899.

[13] Silvera, I.F., The solid molecular hydrogens in the condensed phase: fundamentals and static properties, *Rev. Mod. Phys.*, 52, 393, 1980.

[14] Kolos, K. and Wolniewicz, L., Improved theoretical ground-state energy of the hydrogen molecules, *J. Chem. Phys.*, 49, 404, 1968.

[15] Van Kranendonk, J. and Karl, G., Theory of the rotational and vibrational excitations in solid parahydrogen, and frequency analysis of the infrared and Raman spectra, *Rev. Mod. Phys.*, 40, 531, 1968.

[16] Van Kranendonk, J., *Solid Hydrogen*, New York, Plenum, 1983.

[17] Sharp, T.E., Potential-energy curves for molecular hydrogen and its ions, *Atom Data Nucl. Data*, 12, 119, 1970.

[18] Inoue, K., Kanzaki, H., and Suga, S., Fundamental absorption spectra of solid hydrogen, *Solid State Comm.*, 30, 627, 1979.

[19] Keesom, W. H., de Smedt, J., and Mooy, H.H., On the crystal structure of para-hydrogen at liquid helium temperature, *Proc. Kon. Akad. V. Wetensch. Amster.*, 33, 813, 1930.

[20] Mao, H.K. and Bell, P.M., Observations of hydrogen at room temperature (25 degrees C) and high pressure (to 500 K), *Science*, 203, 1004, 1979.

[21] Sharma, S.K., Mao, H.K., and Bell, P.M., Raman measurements of hydrogen in the pressure range 0.2-630 kbar at room temperature, *Phys. Rev. Lett.*, 44, 886, 1980.

[22] Mao, H.K. et al., Synchrotron x-ray diffraction measurements of single-crystal hydrogen to 26.5 GPa, *Science*, 239, 1131, 1988.

[23] Glazkov, V.P. et al., Neutron-diffraction study of the equation of state of molecular deuterium at high pressures, *JETP Lett.*, 47, 763 1988.

[24] James, H.M., Walk-counting method, with an application to energy bands and impurity states in close-packed lattices, *Phys. Rev.*, 164, 1153, 1967.

[25] James, H.M. and van Kranendonk, J., Theory of the anomalous intensities in the vibrational Raman spectra of solid hydrogen and deuterium, *Phys. Rev.*, 164, 1159, 1967.

[26] Eggert, J. H., Mao, H.K., and Hemley, R.J., Observation of two-vibron bound-to-unbound transition in solid deuterium at high pressure, *Phys. Rev. Lett.*, 70, 2301, 1993.

[27] Feldman, J.L. et al., Vibron excitations in solid hydrogen: a generalized binary random alloy problem, *Phys. Rev. Lett.*, 74, 1379, 1995.

[28] Silvera, I.F., and Wingaarden, R.J., New low-temperature phase of molecular deuterium at ultrahigh pressure, *Phys. Rev. Lett.*, 47, 39, 1981.

[29] Mazin, I.I. et al., Quantum and classical orientational ordering in solid hydrogen, *Phys. Rev. Lett.*, 78, 1066, 1997.

[30] Goncharov, A.F. et al., New high-pressure excitations in parahydrogen, *Phys. Rev. Lett.*, 80, 101, 1998.

[31] Hemley, R.J. et al., Spectroscopic studies of p-H_2 to above 200 GPa, *J. Low Temp. Phys.*, 110, 75, 1998.

[32] Eggert, J. H. et al., Pressure-enhanced ortho-para conversion in solid hydrogen up to 58 GPa, *Proc. Nat. Acad. Sci.*, 96, 12269, 1999.

[33] Pravica, M.G. and Silvera, I.F., NMR study of ortho-para conversion at high pressure in hydrogen, *Phys. Rev. Lett.*, 81, 4181, 1998.

[34] Strzhemechny, M.A. and Hemley, R.J., New ortho-para conversion mechanism in dense solid hydrogen, *Phys. Rev. Lett.*, 85, 5595, 2000.

[35] Hemley, R.J. and Mao, H.K., Progress in cryocrystals to megabar pressures, *J. Low Temp. Phys.*, 122, 33, 2002.

[36] Hemley, R.J. et al., Equation of state of solid hydrogen and deuterium from single-crystal x-ray diffraction to 26.5 GPa., *Phys. Rev. B*, 42, 6458, 1990.

[37] Feldman, J., Johnson, K., and Hemley, R.J., Vibron hopping and bond anharmonicity in hot dense hydrogen, *J. Chem. Phys.*, 130, 054502, 2009.

[38] Goncharenko, I.N. and Loubeyre, P., Neutron and x-ray diffraction study of the broken symmetry phase transition in solid deuterium, *Nature*, 435, 1206, 2005.

[39] Hemley, R.J. et al., Vibron effective charge in dense hydrogen, *Europhys. Lett.*, 37, 403, 1997.

[40] Edwards, B. and Ashcroft, N.W. Spontaneous polarization in dense hydrogen, *Nature*, 388, 652, 1997.

[41] Hanfland, M., Hemley, R.J., and Mao, H.K., Novel infrared vibron absorption in solid hydrogen at megabar pressures, *Phys. Rev. Lett.*, 70, 3760, 1993.

[42] Kohanoff, J. et al., Dipole-quadrupole interactions and the nature of phase III of compressed hydrogen, *Phys. Rev. Lett.*, 83, 4097, 1999.

[43] Souza, I. and Martin, R.M., Polarization and strong infrared activity in compressed solid hydrogen, *Phys. Rev. Lett.*, 81, 4452, 1998.

[44] Hemley, R.J. et al., Synchrotron infrared spectroscopy to 0.15 eV of H_2 and D_2 at megabar pressures, *Phys. Rev. Lett.*, 76, 1667, 1996.

[45] Chen, N., Sterer, E., and Silvera, I.F., Extended infrared studies of hydrogen at high pressure, *Phys. Rev. Lett.*, 76, 1663, 1996.

[46] Goncharov, A.F. et al., Invariant points and phase transitions in deuterium at megabar pressures, *Phys. Rev. Lett.*, 75, 2514, 1995.

[47] Surh, M.P. et al., Ab initio calculations for solid molecular hydrogen, *Phys. Rev. B*, 55, 11330, 1997.

[48] Baer, B.J., Evans, W.J., and Yoo, C.S., Coherent anti-Stokes Raman spectroscopy of highly conmpressed solid deuterium at 300K: Evidence for a new phase and implications for the band gap,*Phys. Rev. Lett.*, 98, 235503, 2007; Erratum: 102, 209901, 2009.

[49] Hemley, R.J., Mao, H.K., and Hanfland, M., Spectroscopic investigations of the insulator-metal transition in solid hydrogen, in *Molecular Systems Under High Pressure* Pucci, R. and Piccitto, G., Eds. Elsevier Science, North Holland, Amsterdam, 1991, 223.

[50] Goncharov, A.F., Hemley, R.J., and Mao, H.K., to be published.

[51] Goncharov, A.F. and Crowhurst, J.C., Raman spectroscopy of hot compressed hydrogen and nitrogen: Implications for the intramolecular potential, *Phys. Rev. Lett.*, 96, 055504 2006.

[52] Hemley, R.J. and Mao, H.K. Phase transition in solid molecular hydrogen at ultrahigh pressures, *Phys. Rev. Lett.*, 61, 857, 1988.

[53] Lorenzana, H.E., Silvera, I.F., and Goettel, K.A. Evidence for a structural phase transition in solid hydrogen at megabar pressures, *Phys. Rev. Lett.*, 63, 2080, 1989.

[54] Toledano, P. et al, Symmetry breaking in dense hydrogen: mechanisms for the transitions to phase II and phase III *Phys. Rev. Lett.*, 103, 105301, 2009.

[55] Stokes, H.T. and Hatch, D.M., *Isotropy Space Groups of the 230 Crystallographic Space Groups*, World Scientific, Singapore, 1988.

[56] Kohanoff, J. et al., Solid molecular hydrogen: the broken symmetry phase, *Phys. Rev. Lett*, 78, 2783, 1997.

[57] Nagao, K., Takezawa, T., and Nagara, H., Ab initio calculation of optical-mode frequencies in compressed solid hydrogen, *Phys. Rev. B*, 59, 13741, 1999.

[58] Stadele, M. and Martin, R., Metallization of molecular hydrogen: predictions from exact-exchange calculations, *Phys. Rev. Lett.*, 84, 6070, 2000.

[59] Cui, T. et al., Rotational ordering in solid deuterium and hydrogen: A path integral Monte Carlo study, *Phys. Rev. B*, 55, 12253, 1997.

[60] Edwards, B. and Ashcroft, N.W., Order in dense hydrogen at low temperatures, *Proc. Nat. Acad. Sci.*, 101, 4013, 2004.

[61] Kitamura, H. et al., Quantum distribution of proton in solid molecular hydrogen at megabar pressures, *Nature* 403, 259, 2000.

[62] Pickard, C.J. and Needs, R.J., Structure of phase III of solid hydrogen, *Nature Phys.*, 3, 473, 2007.

[63] Akahama, Y. et al., Evidence from x-ray diffraction of orientational ordering in phase III of solid hydrogen at pressures up to 183 GPa, *Phys. Rev. B*, in press.

[64] Ramaker, D.E., Kumar, L., and Harris, F.E., Exact-exchange crystal Hartree-Fock calculations of molecular and metallic hydrogen and their transitions, *Phys. Rev. Lett.* 34, 812, 1975.

[65] Friedli, C, and Ashcroft, N.W., Combined representative method for use in band structure calculations: application to highly compressed hydrogen, *Phys. Rev. B*, 16, 662, 1977.

[66] Ashcroft, N.W., Optical response near a band overlap: Application to dense hydrogen, in *Molecular Systems Under High Pressure* Pucci, R. and Piccitto, G., Eds. Elsevier Science, North Holland, Amsterdam, 1991.

[67] Kaxiras, E., Broughton, J., and Hemley, R.J., Onset of metallization and related transitions in solid hydrogen, *Phys. Rev. Lett.*, 67, 1138, 1991.

[68] Hemley, R.J., Eremets, M.I., and Mao, H.K., Progress in experimental studies of insulator-metal transitions at multimegabar pressures, in *Frontiers of High Pressure Research II*, Hochheimer, H.D. et al., Eds., Kluwer, Amsterdam, 2002, p. 201.

[69] Weir, S.T., Mitchell, A.C., and Nellis, W.J., Metallization of fluid molecular hydrogen at 140 GPa (1.4 Mbar), *Phys. Rev. Lett.*, 76, 1860, 1996.

[70] Loubeyre, P., Ocelli, F., and LeToullec, R., Optical studies of hydrogen to 320 GPa and evidence for black hydrogen, *Nature*, 416, 613, 2002.

[71] Mao, H.K. and Hemley, R.J., Optical observations of hydrogen above 200 gigapascals: evidence for metallization by band overlap, *Science*, 244, 1462, 1989.

[72] Matsuishi, K. et al., Equation of state and intermolecular interactions in fluid hydrogen from Brillouin scattering at high pressures and temperatures, *J. Chem. Phys.*, 118, 10683, 2003.

[73] Duffy, T.S. et al., Sound velocity in dense hydrogen and the interior of Jupiter, *Science*, 263, 1590, 1994.

[74] Goncharov, A.F. et al., Spectroscopic studies of the vibrational and electronic properties of solid hydrogen to 285 GPa, *Proc. Nat. Acad. Sci.*, 98, 14234, 2001.

[75] Saumon, D. and Chabrier, G., Fluid hydrogen at high density: Pressure ionization, *Phys. Rev. A*, 46, 2084, 1992.

[76] Nellis, W.J., Weir, S.T., and Mitchell, A.C., Minimum metallic conductivity of fluid hydrogen at 140 GPa (1.4 Mbar), *Phys. Rev. B*, 59, 3434, 1999.

[77] Magro, W.R. et al., Molecular dissociation in hot, dense hydrogen, *Phys. Rev. Lett.*, 76, 1240. 1243, 1996.

[78] Goncharov, A.F., and Crowhurst, J.C., Proton delocalization under extreme conditions of high pressure and temperatures, *Phase Trans.*, 80, 1051, 2007.

[79] Babaev, E., Sudbø A., and Ashcroft, N.W., Observability of a projected new state of matter: A metallic superfluid,*Phys. Rev. Lett.*, 95, 105301, 2005.

[80] Collins, G.W. et al., Measurements of the equation of state of deuterium at the fluid insulator-metal transition, *Science*, 281, 1178, 1998.

[81] Knudson, M.D. et al., Equation of state measurements in liquid deuterium to 70 GPa, *Phys. Rev. Lett.*, 87, 22501, 2001.

[82] Scandolo, S., Liquid-liquid phase transition in compressed hydrogen from first-principles simulations, *Proc. Nat. Acad. Sci.*, 100, 3051, 2003.

[83] Bonev, S.A. et al., A quantum fluid of metallic hydrogen suggested by first-principles calculations, *Nature*, 431, 669, 2004.

[84] Grinenko, A. et al., Probing the hydrogen melting line at high pressures by dynamic compression, *Phys. Rev. Lett.* 101, 194801, 2008.

[85] Vorberger, J. et al., Hydrogen-helium mixtures in the interiors of giant planets, *Phys. Rev. B*, 75, 024206, 2007.

[86] Datchi, F., Loubeyre, P., and LeToullec, R., Extended and accurate determination of the melting curves of argon, helium, ice (H_2O), and hydrogen (H_2), *Phys Rev. B*, 61, 6535, 2000.

[87] Gregoryanz, E. et al., Raman spectroscopy of hot dense hydrogen, *Phys. Rev. Lett.*, 90, 175701, 2003.

[88] Ross, M., Ree, F.R., and Young, D.A., The equation of state of molecular hydrogen at very high density, *J. Chem. Phys.*, 79, 1487, 1983.

[89] Morales, M.A. et al., Evidence for a first-order liquid-liquid transition in high-pressure hydrogen from ab initio simulations, *Proc. Nat. Acad. Sci.*, doi:10.1073/pnas.1007309107, 2010.

[90] Fortov, V. E. et al., Phase transition in a strongly nonideal deuterium plasma generated by quasi-isentropical compression at megabar pressures, *Phys. Rev. B*, 99, 185001, 2007.

[91] Deemyad, S. and Silvera, I.F., Melting line of hydrogen at high pressures, *Phys. Rev. Lett.*, 100, 155701, 2008.

[92] Eremets, M.I. and Trojan, I.A., Evidence of maximum in the melting curve of hydrogen at megabar pressure, *JETP*, 89, 198, 2009.

[93] Subramanian, N., to be published.

[94] Narayana, C. et al., Solid hydrogen at 342 GPa: no evidence for an alkalai metal, *Nature*, 393, 46, 1998.

[95] Strobel, T., Somayazulu, M., and Hemley, R.J., Novel pressure-induced interactions in silane-hydrogen, *Phys. Rev. Lett.*, 103, 065107, 2009.

[96] Subramanian, N. et al., Raman spectroscopy of hydrogen confined at extreme conditions, *J. Phys. Conf. Series*, 215, 012057, 2010.

[97] Hanfland, M. et al., Synchrotron infrared spectroscopy at megabar pressures: vibrational dynamics of hydrogen to 180 GPa, *Phys. Rev. Lett.*, 69, 1129, 1992.

[98] Ashcroft, N.W., Pairing instabilities in dense hydrogen, *Phys. Rev. B*, 41, 10963, 1990.

[99] Loubeyre, P. et al., Coupling static and dynamic compressions: First measurements in dense hydrogen, *High Press. Res.*, 24, 25, 2004.

[100] Mao, H.K. and Hemley, R.J., Hydrogen at high pressures, *Am. Sci.*, 80, 234, 1992.

[101] Silvera, I.F. and Goldman, V. V., The isotropic intermolecular potential for H_2 and D_2 in the solid and gas phases, *J. Chem. Phys.*, 69, 4209, 1978.

Index